Blackwell's, Loughborough, 22 July 03
(Springer "Yellow" Sale)
RRP = £63

Grundlehren der mathematischen Wissenschaften 313

A Series of Comprehensive Studies in Mathematics

Springer
Berlin
Heidelberg
New York
Barcelona
Budapest
Hong Kong
London
Milan
Paris
Santa Clara
Singapore
Tokyo

Paul Malliavin

Stochastic Analysis

Springer

Cataloging-in-Publication Data applied for

Die Deutsche Bibliothek – CIP-Einheitsaufnahme

Malliavin, Paul:
Stochastic analysis / Paul Malliavin. - Berlin ; Heidelberg ; New York ; Barcelona ; Budapest ; Hong Kong ; London ; Milan ; Paris ; Santa Clara ; Singapore ; Tokyo : Springer, 1997
(A series of comprehensive studies in mathematics ; 313)
ISBN 3-540-57024-1
NE: Grundlehren der mathematischen Wissenschaften

Mathematics Subject Classification (1991):
31C15, 31C25, 31C45, 35H05, 46E35, 60G15, 60G35, 60G44, 60G60, 60H05, 60H07, 60H15, 60H30, 60J45, 60J60

First Reprint 2002

ISSN 0072-7830

ISBN 3-540-57024-1 Springer-Verlag Berlin Heidelberg New York

Springer-Verlag Berlin Heidelberg New York
a member of Bertelsmann Springer Scince+Business Media GmbH

Cover design: MetaDesign plus GmbH, Berlin
Typesetting: Typeset in LATEX by Ingeborg Jebram, Heiligkreuzsteinach, and reformatted by Kurt Mattes, Heidelberg, using the MathTime fonts and a Springer TEX macro-package
SPIN: 10866791 41/3111-5 4 3 2 1 Printed on acid-free paper

Preface

An appropriate coverage of the subjects contained in the five parts of this book would need several monographs. We hope that the global treatment presented here may emphasize some of their deep interactions. As far as possible we present self-contained proofs; we have also tried to produce a book that could be used in a graduate course.

Our thread of Ariadne is the introduction into stochastic analysis of the methodology used in classical analysis and differential geometry. Our geometric point of view has obliged us to pay great attention to the foundations. On the other hand our notation, which follows the usual conventions, will allow an experienced worker to look directly at any section of this book, without spending time on the foundational sections.

Each part is constructed according to the following format: a short introduction, a detailed table of contents at the beginning of each chapter of that part and a short note on the literature at the end of each part.

The style of writing oscillates from one part to the next between that of a rather technical monograph in Part II to a broader survey style in Parts IV and V.

I owe a great debt to K. Itô, J.R. Norris, D.W. Stroock for their careful reading of the first draft and for their far-reaching suggestions.

Paris, January 1997 *Paul Malliavin*

Table of Contents

Contents by Chapter

Chapter I. Gaussian Probability Spaces

Definition of a Gaussian probability space, reducibility – Hermite polynomials on $\mathbb{R}$ – Hermite polynomials on $\mathbb{R}^{\mathbb{N}}$ – Numerical model of a Gaussian probability space – Intrinsic geometry on a Gaussian probability space – The Ornstein-Uhlenbeck semigroup, chaos decomposition – The Cameron-Martin representation – Abstract Wiener space.

Chapter II. Gross-Stroock Sobolev Spaces over a Gaussian Probability Space

Continuity of the Cameron-Martin representation over $L^{\infty-0}$ – The space $\mathbb{D}_1^{\infty}$ of differentiable vectors of the Cameron-Martin representation – Gradient operator – Generalized polynomials – Cauchy operator; Krée-Meyer inequality for the gradient – The spaces $\mathbb{D}_1^p$ – Gradient of Hilbert-valued functionals – Recursive approach to higher derivatives and to Krée-Meyer inequalities of higher order – The space $\mathbb{D}_{\infty}$ of smooth functionals, its approximation by C^{∞} cylindrical functionals – Divergence as the adjoint of the gradient – Divergence of smooth vector fields – Shigekawa acyclicity of the complex of differential forms – Appendix: Proof of the L^p inequality for the Hilbert transform.

Chapter III. Smoothness of Laws

Divergence of differentiable flow – Divergence as the adjoint operator of a derivation – Non-degenerate maps and their covariance matrices – Lifting up vector fields through a non-degenerate map – Pushing down divergences – Hölder regularity under $\mathbb{D}_2^{\infty-0}$ hypothesis – Smoothness under $\mathbb{D}_{\infty}$ hypothesis – Lifting up and pushing down through a non-degenerate $\mathbb{D}_{\infty}$ map – Inverse image of a distribution – Absolute continuity of scalar functionals under $\mathbb{D}_1^1$ hypothesis – Appendix: Computation of derivatives by means of divergence operators – Law of a weakly non degenerated map.

Chapter IV. Foundations of Quasi-Sure Analysis: Hierarchy of Capacities and Precise Gaussian Probability Spaces

Capacities on a numerical model – Tightness of capacities – Quasi-continuous functions – Tchebycheff inequalities – Redefinitions – Positive generalized functionals, their representation by a Borel measure – Equilibrium potentials – Continuity of capacities – Capacitability of Borel sets – Equilibrium measures – Measures of finite energy – Charge and capacity – Slim sets are null sets for all measures of finite energy – Invariance of capacities under a change of numerical model, precise Gaussian probability space – Quasi-sure analysis on an abstract Wiener space.

Chapter V. Differential Geometry on a Precise Gaussian Probability Space

Regular disintegration corresponding to a non-degenerate functional; principle of descent – Partial function associated to a splitting – Finite codimensional projection of a slim set is slim – Implicit function theorem in finite codimension – Differential forms of degree p – Unnormalized conditional expectation of differential forms – Currents; currents along a fibre of a non-degenerate map – Submanifolds; their defining functions – Gauss map of a submanifold, its approximate continuity – Hausdorff area measure of a submanifold – Functional characterization of the area measure – Coarea formula along a non-degenerate map – The operator δ; its commutation with the non-normalized conditional expectation – Oriented submanifold – Stokes' theorem.

Chapter VI. White Noise Stochastic Integrals as Divergences

Gaussian probability space over an L^2 space – Construction of $\operatorname{Seg}(L^2(\Gamma))$ by Wiener-Itô multiple integrals – Fock space, naturality of Wiener – Itô multiple integrals – Stroock-Taylor formula – Nualart-Pardoux construction of the Skorokhod stochastic integral – The Gaveau-Trauber theorem identifying the Skorokhod integral with a divergence – Energy estimate for the Skorokhod integral – Existence of traces and of Stratonovich integrals.

Chapter VII. Itô's Theory of Stochastic Integration

The probability space of Brownian motion and its filtration – Energy identity for stochastic integral of an adapted process – Itô's stochastic integral of an adapted process – Chaos expansion in terms of iterated Itô stochastic integrals – Itô representation of a martingale by a stochastic integral – Clark-Bismut-Ocone representation of a martingale in $\mathbb{D}_1^p$ – Itô calculus on semi-martingales – Covariance under C^∞-maps of the Stratonovich representation of semi-martingales – Change of variables formula – Appendix: Estimates for Brownian martingales.

Chapter VIII. From Ordinary Differential Equations to Stochastic Flow: The Transfer Principle

Stratonovich SDE – Intrinsic stochastic integral – Itô SDE, uniqueness of the Cauchy problem – The Stroock-Varadhan piecewise linear approximation – Algebraic analysis on the group of C^∞-diffeomorphisms: the exponential map, the

Part I

Differential Calculus on Gaussian Probability Spaces

In the elementary theory of $\mathbb{R}^n$-valued random variables, operations on the subclass of random variables having a C^1-density relative to Lebesgue measure are often realized through computations of ordinary differential calculus: for instance, the determination of conditional laws by computing differential forms, the realization of a change of variables by computing Jacobians. Our purpose is to extend this methodology to more general probability spaces.

The Lebesgue measure of $\mathbb{R}^n$ can be characterized by its invariance under the group of translations. Given a probability space Ω, *the quasi-automorphism group* will be a "natural" group of transformations of Ω leaving quasi-invariant the probability measure; this notion is quite general; it will be developed in this book in the context of *a Gaussian probability space*, which means an abstract probability space Ω on which we have a Hilbert space $\mathcal{H}$ of Gaussian random variables. The additive group of $\mathcal{H}$ will define the quasi-automorphism group of Ω. Any unitary isomorphism of $\mathcal{H}$ will then generate an automorphism of the Gaussian probability space structure of Ω. The realization of this unitary invariance as a fact built into the construction of Ω itself is done in Chapter I.

The quasi-automorphism group $\mathcal{H}$ operates on a suitable algebra of random variables. The infinitesimal action of $\mathcal{H}$ will lead to the notion of *$\mathcal{H}$-Sobolev spaces* on Ω. Chapter II will be devoted to the study of the algebra of *smooth random variables* which are the random variables belonging to all those Sobolev spaces.

The *Jacobian* of an $\mathbb{R}^d$-valued smooth random variable is defined in Chapter III; an appropriate lower bound for this Jacobian will imply that the corresponding law has a C^∞-density relative to Lebesgue measure. This theorem will result from an interplay between classical harmonic analysis for Sobolev spaces on $\mathbb{R}^d$ and *elliptic estimates* established in Chapter II for Sobolev spaces on Ω. This interplay will be realized by lifting differential forms by the inverse image and pushing down by conditional expectations.

Chapter I

Gaussian Probability Spaces

Contents: Definition of a Gaussian probability space, reducibility – Hermite polynomials on $\mathbb{R}$ – Hermite polynomials on $\mathbb{R}^{\mathbb{N}}$ – Numerical model of a Gaussian probability space – Intrinsic geometry on a Gaussian probability space – The Ornstein-Uhlenbeck semigroup, chaos decomposition – The Cameron-Martin representation – Abstract Wiener space.

In the 1950's Irving Segal developed for the needs of Quantum Field Theory, an *abstract theory of integration of an abstract Hilbert space*. In the 1960's Leonard Gross built the theory of *Gaussian Borel measures* on an arbitrary Banach space. A very special case of Gross's theory is the classical Wiener space that we shall discuss later. Looking in this chapter for the greatest generality combined with the easiest approach, we shall follow a route close to Segal's approach in a version more concrete than his original paper. This point of view is parallel to recent papers of K. Itô.

One basic object of probability theory is the Boolean algebra B of measurable events. By a theorem of Stone this Boolean algebra can be represented as the family of subsets of a compact space K which are both closed and open. The space K can be thought of as the *Stone spectrum* of B; the Stone spectrum is an *intrinsic* object, but one on which we are unable to work: the compact space K is highly non-separable. A substitute for this approach is Gelfand's theory of Banach algebras. We now take for a basic object the algebra of *bounded random variables*; this is a Banach algebra which is isomorphic to the *algebra of continuous functions* on its *Gelfand spectrum*. Again, the Gelfand spectrum is an intrinsically defined compact space; the Gelfand spectrum is highly non-separable.

In order to keep separability within an intrinsic approach, we will introduce not a single model, but a *family* of separable models. The family will have an intrinsic meaning; the choice of a model in the family will lose the intrinsic character. *Intrinsic properties* will then be defined as properties which are independent of the choice of the model.

Some readers might wish to avoid the axiomatic method developed in this chapter. They should then proceed in the following way: skip the reading of the first seven sections and go directly to the last section where the classical abstract

Wiener space appears as an example; this example in mind, it is possible to follow the main lines of the whole book.

1. Axioms of Gaussian probability spaces

1.1 Definition. A *Gaussian probability space* $(\Omega, \mathcal{A}, \mathcal{P}; H, \mathcal{A}_H^{\perp})$ is given by the following elements
1.1.1. $(\Omega, \mathcal{A}, \mathcal{P})$ a probability space.
1.1.2. A *closed subspace* H of $L^2(\Omega, \mathcal{A}, \mathcal{P})$ such that all the random variables belonging to H have a *centered Gaussian law.*

The σ-field generated by those variables is denoted by $\mathcal{A}_H$.
1.1.3. Another σ-field $\mathcal{A}_H^{\perp}$ independent of $\mathcal{A}_H$ and such that

$$\mathcal{A}_H \otimes \mathcal{A}_H^{\perp} = \mathcal{A}.$$

We shall call $\mathcal{A}_H^{\perp}$ the σ-field of the transverse variables or the transverse σ-field.

1.2. Subspaces

Given a Gaussian probability space $(\Omega, \mathcal{A}, \mathcal{P}; H, \mathcal{A}_H^{\perp})$, a *subspace* will be given by

1.2.1. $H' \subset H$ a closed subspace of H.

1.2.2. $\mathcal{A}_{H'}^{\perp}$ a sub σ-field of $\mathcal{A}$ such that $\mathcal{A}_{H'}$ and $\mathcal{A}_{H'}^{\perp}$ are independent. We denote $\mathcal{A}' = \mathcal{A}_{H'} \otimes \mathcal{A}_{H'}^{\perp}$, and we assume

$$\mathcal{A}' \cap \mathcal{A}_H^{\perp} \subset \mathcal{A}_{H'}^{\perp}.$$

1.2.3. Example. Given a Gaussian probability space $(\Omega, \mathcal{A}, \mathcal{P}, H, \mathcal{A}_H^{\perp})$ and given a closed vector subspace H' of H, then there exists $\mathcal{B}$ such that $(\Omega, \mathcal{A}, \mathcal{P}, H', \mathcal{B})$ is a subspace of $(\Omega, \mathcal{A}, \mathcal{P}, H, \mathcal{A}_H^{\perp})$.

Proof. Denote by V the orthogonal complement of H' in H. Then orthogonality in H implies independence. Therefore $\mathcal{A}_V$ is independent of $\mathcal{A}_{H'}$. We take for $\mathcal{B}$ the σ-field generated by $\mathcal{A}_V$ and $\mathcal{A}_H^{\perp}$. □

1.3. Irreducibility

1.3.1. Definition. A Gaussian probability space is *irreducible* if $\mathcal{A}_H = \mathcal{A}$.

1.3.2. Remark. At first sight it might appear strange to work with non-irreducible spaces. This concept is introduced for the following reasons:

(i) The possibility to work with a subspace (as explained in 1.2.3) which will provide approximations by *finite dimensional* Gaussian spaces.
(ii) The general idea that a probability space can always be extended and that the admissible operations have to be stable under the extension of the space.

1.3.3. Remark. We introduce the Hilbert space $G = L^2(\Omega, \mathcal{A}_H^\perp, \mathcal{P})$. Then

$$L^2(\Omega, \mathcal{A}, \mathcal{P}) = L^2\big((\Omega, \mathcal{A}_H, \mathcal{P});\ G\big),$$

where the r.h.s. denotes an L^2 space of G-valued functions.

1.4. Isomorphism

Given two Gaussian spaces $(\Omega, \mathcal{A}, \mathcal{P}, H)$ and $(\Omega', \mathcal{A}', \mathcal{P}', H')$, an isomorphism will be given by an isometry $u : L^2(\Omega', \mathcal{A}', \mathcal{P}') \to L^2(\Omega, \mathcal{A}, \mathcal{P})$ such that u restricted to L^∞ is an algebraic homomorphism and such that $u(H') = H$, $u\left(L^2\left(\Omega', \mathcal{A}_{H'}^\perp, \mathcal{P}'\right)\right) = L^2\left(\Omega, \mathcal{A}_H^\perp, \mathcal{P}\right)$.

Example 1. A bijective measurable map $j : \Omega \to \Omega'$ preserving the probability measure, the spaces H and H', and $\mathcal{A}_H^\perp$ and $\mathcal{A}_{H'}^\perp$, induces such an isomorphism. If the σ-fields $\mathcal{A}$ and $\mathcal{A}'$ are "sufficiently large" in Ω and Ω', all the isomorphisms are of this nature. (This is the case if Ω, Ω' are separable, complete metric spaces and if $\mathcal{A}$ and $\mathcal{A}'$ contain the Borel σ-fields.)

One of the purposes of this chapter is to show that *the equivalence classes under isomorphism of irreducible Gaussian probability spaces $(\Omega, \mathcal{A}, \mathcal{P}, H)$ are classified by the dimension of H.* To prepare the proof of this result, we shall recall in the next two subsections some notation and classical results on Hermite polynomials.

2. Hermite polynomials on $\mathbb{R}$

We consider on $\mathbb{R}$ the normal law

$$\gamma(d\xi) = \frac{1}{\sqrt{2\pi}} \exp\left(-\frac{\xi^2}{2}\right) d\xi.$$

We consider the real Hilbert space associated to the scalar product

$$\int \psi(\xi)\varphi(\xi)\gamma(d\xi) = (\psi \mid \varphi).$$

We denote by $C_0^1(\mathbb{R})$ the set of C^1-functions with compact support. We denote by ∂ the operator of differentiation:

$$(\partial\varphi)(\xi) = \varphi'(\xi).$$

Remark. In mathematical physics the operator ∂ is called the *annihilation operator.*

2.1. Lemma. *Denote by* ∂^* *the operator defined, for* $\varphi \in C^1(\mathbb{R})$, *by*

$$(\partial^*\varphi)(\xi) = -\varphi'(\xi) + \xi\varphi(\xi).$$

Then if $\partial\varphi$ *and* $\partial^*\psi \in L^2(\gamma)$ *we have*

$$(\partial\varphi \mid \psi) = \left(\varphi \mid \partial^*\psi\right).$$

Remark. In mathematical physics, the operator ∂^* is called the *creation operator.*

Proof. If φ has compact support, this identity is obtained by integration by parts. Set $q(\xi) = (1 - |\xi|)^+$, $q_\varepsilon(\xi) = q(\varepsilon\xi)$.Then

$$(\partial(q_\varepsilon\varphi) \mid \psi) = \left(q_\varepsilon\varphi \mid \partial^*\psi\right)$$

and when $\varepsilon \to 0$ the right hand side converges.

The same property of the left hand side will result from $\lim((\partial q_\varepsilon)\varphi \mid \psi) = 0$, which follows from the estimate

$$|((\partial q_\varepsilon)\varphi \mid \psi)| < \varepsilon\|\varphi\|_{L^2}\|\psi\|_{L^2}.$$

□

2.2. Definition of the Hermite polynomials. We define the sequence

$$\begin{aligned} H_0(\xi) &= 1 \\ H_n &= \partial^* H_{n-1} = (\partial^*)^n 1. \end{aligned}$$

By induction on n we see that H_n is a polynomial of degree n and that its term of highest degree is ξ^n.

2.3. Lemma.

$$\partial\partial^* - \partial^*\partial = 1.$$

Proof. A straightforward computation establishes this commutation relation, which is basic in quantum mechanics. □

2.4. Lemma. *Denote by* μ *a probability measure on* $\mathbb{R}^n$ *such that there exists* $c > 0$ *for which*

$$\int_{\mathbb{R}^n} e^{c|y|}\mu(dy) < +\infty$$

then the polynomials in the coordinate functions are dense in $L^p(\mu)$ *for* $p \in [1, +\infty)$.

Proof. Denote by V the L^p-closure of the polynomials. If $V \neq L^p$ we can find $u \in L^q(\mu)$, $u \neq 0$, such that u is orthogonal to V. Consider the formal integral

$$\hat{u}(t) = \int_{\mathbb{R}^n} e^{it.y} u(y)\mu(dy).$$

Now

$$\left|\hat{u}(\sigma + i\tau)\right| \leq \int e^{\tau.y}|u(y)|\mu(dy) \leq \left\|e^{\tau.y}\right\|_{L^p} \|u\|_{L^q} < +\infty$$

which implies the convergence if $|\tau| < \frac{c}{p}$.

Therefore $\hat{u}$ is the restriction to $\mathbb{R}^n$ of a function holomorphic in the tube $\left\{\sigma + i\tau;\ \sigma \in \mathbb{R}^n,\ |\tau| < \frac{c}{p}\right\}$. Furthermore, by the orthogonality to V

$$\left[\frac{\partial}{\partial t_1^{\alpha_1} \dots \partial t_n^{\alpha_n}} \hat{u}\right](0) = 0.$$

By analyticity this implies $\hat{u} = 0$. By the Fourier inversion formula therefore, $u d\mu = 0$ so $u = 0$ a.e. μ, which is a contradiction. □

2.5. Theorem

2.5.1. $\partial H_n = n\ H_{n-1}$.

2.5.2. $(n!)^{-1/2} H_n$ *is an orthonormal basis of* $L^2(\gamma)$.

2.5.3. *Define* $\mathcal{L} = \partial^*\partial$, *then* $\mathcal{L}H_n = n\ H_n$.

2.5.4. *Let* $f \in L^2(\gamma)$. *Assume that all derivatives of* f *belong to* $L^2(\gamma)$, *then the* L^2*-expansion of* f *can be written*

$$f = \sum_{n=0}^{+\infty} \frac{1}{n!} E\left(\partial^n f\right) H_n$$

where $E(u)$ *denotes* $(u \mid 1)$.

2.5.5.

$$\int_{\mathbb{R}^2} H_n(\xi \cos\theta + \eta \sin\theta) H_p(\xi)\gamma(d\xi)\gamma(d\eta) - n!(\cos\theta)^n \delta_n^p$$

(where $\delta_n^p = 0$ *if* $p \neq n$ *and* $\delta_n^n = 1$).

2.5.6.

$$\exp\left(\lambda x - \frac{\lambda^2}{2}\right) = \sum_{n=0}^{+\infty} \frac{\lambda^n}{n!} H_n(x),$$

this convergence being uniform on compact sets in (λ, x). *This identity is called the generating function identity.*

Proof.

(2.5.1) For $n = 1$, $H_1 = \xi$, $\partial H_1 = 1 = H_0$; we proceed by induction on n. Assume that (i) is true for $n < p$. Then

$$\begin{aligned} \partial H_p &= \partial \partial^* H_{p-1} = \partial^* \partial H_{p-1} + H_{p-1}, \quad \text{now by 1.2.2} \\ &= \partial^* (p-1) H_{p-2} + H_{p-1} = p H_{p-1}. \end{aligned}$$

□

(2.5.2)

$$\left(H_s \mid (\partial^*)^\ell 1 \right) = \left(\partial^\ell H_s \mid 1 \right)$$

If $\ell > s$ as H_s is a polynomial of degree s, we have $\partial^\ell H_s = 0$. If $\ell = s$, then $\partial^s H_s = s!$. This proves that $\left\{ \frac{1}{(s!)^{1/2}} H_s \right\}$ is an orthonormal system. Therefore $\{H_s\}$ are linearly independent. They generate by linear combination the vector space of polynomials which is dense in L^2 by 2.4. □

(2.5.3)

$$\partial^* \partial H_n = \partial^* n H_{n-1} = n \partial^* H_{n-1} = n H_n$$

□

(2.5.4) By (ii) every $f \in L^2(\gamma)$ has the following L^2 expansion:

$$f = \sum c_n H_n \quad \text{with} \quad c_n = \frac{1}{n!} (f \mid H_n).$$

As f and f' belong to $L^2(\gamma)$ we have by 2.1

$$(f \mid H_n) = \left(f \mid \partial^* H_{n-1} \right) = (\partial f \mid H_{n-1})$$

and so by induction

$$\ldots = \left(\partial^{n-1} f \mid H_1 \right) = \left(\partial^{n-1} f \mid \partial^* 1 \right) = \left(\partial^n f \mid 1 \right).$$

□

(2.5.5) We remark that $\gamma \otimes \gamma$ is the normal law $\gamma_{\mathbb{R}^2}$ on $\mathbb{R}^2$, therefore it is rotation invariant. We denote by ∂_θ the derivation operator defined on smooth functions on $\mathbb{R}^2$ by

$$(\partial_\theta f) = \cos\theta \frac{\partial f}{\partial \xi} + \sin\theta \frac{\partial f}{\partial \eta}.$$

We denote by ∂_θ^* the adjoint of ∂_θ in $L^2\left(\gamma_{\mathbb{R}^2}\right)$.

We define $j_\theta : R^2 \to R$ by $j_\theta(\xi, \eta) = \xi \cos\theta + \eta \sin\theta$. Given $u \in C^1(\mathbb{R})$, we denote $u_\theta = u \circ j_\theta$. We then have

$$\partial_0 u_0 = (\partial u) \circ j_0 \quad \text{and} \quad \partial_0^* u_0 = (\partial^* u) \circ j_0.$$

The measure $\gamma_{\mathbb{R}^2}$ being invariant by rotation we deduce

$$\partial_\theta^* u_\theta = (\partial^* u) \circ j_\theta.$$

Therefore, denoting by 1 the constant function on $\mathbb{R}^2$, we have

$$\int H_n(\xi \cos\theta + \eta \sin\theta) H_p(\xi) \gamma(d\xi) \gamma(d\eta) = \left((\partial_\theta^*)^n 1 \mid (\partial_o^*)^p 1 \right)_{L^1(\gamma \otimes \gamma)}.$$

If $h < p$ we transpose and get zero because $\partial_o^p (\partial_\theta^*)^n 1 = 0$. The case $n > p$ is treated in the same way. If $n = p$ we have to compute

$$\partial_\theta^n (\partial_o^*)^n 1 = \partial_\theta^n (\xi^n) = n!(\cos\theta)^n.$$

□

(2.5.6) We apply the formula 2.5.4 to

$$f_\lambda(x) = \exp\left(\lambda x - \frac{1}{2}\lambda^2\right)$$

then

$$\partial_x^n f_\lambda = \lambda^n f_\lambda, \qquad E\left(\partial_x^n f\right) = \lambda^n.$$

□

3. Hermite polynomials on $\mathbb{R}^{\mathbb{N}}$

We denote by $\mathbb{N}$ the natural numbers $1, 2, \ldots$ Our aim is to develop on $\mathbb{R}^{\mathbb{N}}$ a theory parallel to the Hermite theory on $\mathbb{R}$ from the last paragraph. As $\mathbb{R}^{\mathbb{N}}$ contains $\mathbb{R}^n$ for all finite n we shall get as by-product a theory on all finite-dimensional spaces.

We shall take on $R^{\mathbb{N}}$ the σ-field $B_{\mathbb{R}^{\mathbb{N}}} = \bigotimes_{k=1}^{+\infty} B_{\mathbb{R}}$, where $B_{\mathbb{R}}$ is the Borel σ-field and we set $\gamma_{\mathbb{N}} = \bigotimes_{k=1}^{+\infty} \gamma$.

We denote by $e_k^*(x) = x^k$ the k-th coordinate function. We denote by e_k the point of $\mathbb{R}^{\mathbb{N}}$ which has zero coordinates with the exception of the k-th coordinate which is 1. Define

$$(\partial_k \varphi)(x) = \lim_{\varepsilon \to 0} \varepsilon^{-1} \left[\varphi(x + \varepsilon e_k) - \varphi(x)\right]$$

$$\partial_k^* \varphi = -\partial_k \varphi + e_k^* \varphi.$$

3.1. Theorem. *Denote by* $\mathbf{p}$ *a positive integer-valued function defined on* $\mathbb{N}$. *Write*

$$|\mathbf{p}| = \sum_{n \in \mathbb{N}} \mathbf{p}(n).$$

Consider the following simplex of functions

$$\mathcal{E} = \{\mathbf{p};\ |\mathbf{p}| < +\infty\}.$$

Define for $\mathbf{p} \in \mathcal{E}$

$$\begin{aligned} \mathbf{p}! &= \prod_{n \in N} (\mathbf{p}(n))! \\ H_{\mathbf{p}} &= \prod_{n} H_{\mathbf{p}(n)} \circ e_n^*. \end{aligned}$$

Then we have

3.1.1. $$\partial_k H_{\mathbf{p}} = p(k) H_{\mathbf{p}-\varepsilon_k}$$

3.1.2. $$\partial_k^* H_{\mathbf{p}} = H_{\mathbf{p}+\varepsilon_k}$$

where $\varepsilon_k(n) = 0$ *if* $n \neq k$, $\varepsilon_k(k) = 1$.

3.1.3. *Define*

$$\mathcal{L} = \sum_{k \in \mathbb{N}} \partial_k^* \partial_k \quad \textit{(Ornstein-Uhlenbeck generator)},$$

then

$$\mathcal{L} H_{\mathbf{p}} = |\mathbf{p}| H_{\mathbf{p}}.$$

3.1.4. $(\mathbf{p}!)^{-1/2} H_{\mathbf{p}}$ *is an orthonormal basis of* $L^1(\Omega, \mathcal{A}, P)$, $\Omega = \mathbb{R}^{\mathbb{N}}$, $\mathcal{A} = B_{\mathbb{R}^{\mathbb{N}}}$, $P = \gamma_{\mathbb{N}}$.

3.1.5. *Denote*

$$\partial_{\mathbf{p}}^* = \prod_k \left(\partial_k^*\right)^{\mathbf{p}(k)},$$

then

$$H_{\mathbf{p}} = \partial_{\mathbf{p}}^* 1.$$

3.1.6. *If* $f = \tilde{f}\left(e_1^*, \ldots, e_k^*\right)$ *is a function of a finite number of coordinates and if* $\tilde{f}$ *and all its partial derivatives belong to* $L^2(\gamma_k)$, *then the* L^2*-expansion of* f *can be written*

$$f = \sum_{p \in \mathcal{E}} \frac{1}{\mathbf{p}!} E\left(\partial_{\mathbf{p}} f\right) H_{\mathbf{p}}$$

(*where* $\partial_{\mathbf{p}} = \prod_{k=1}^{\infty} (\partial_k)^{\mathbf{p}(k)}$).

3.1.7.

$$\int_{R^{\mathbb{N}}\times\mathbb{R}^{\mathbb{N}}} H_{\mathbf{p}}(x\cos\theta + y\sin\theta)H_{\mathbf{q}}(x)d\gamma(x)\otimes d\gamma(y) = \cos\theta^{|p|}\mathbf{p}!\delta_p^q$$

with $\delta_p^q = 0$ *if* $\mathbf{p} \neq \mathbf{q}$, $= 1$ *if* $\mathbf{p} = \mathbf{q}$.

3.1.8. *The linear span of the* $H_{\mathbf{p}}$ *is dense in* L^p, $p \in [1, +\infty)$.

Proof. We remark that if $k \neq k'$ then ∂_{k^*} commutes with $\partial_{k^*}^*$ and ∂_k with $\partial_{k'}$. Therefore the order of terms does not matter in their products.

(3.1.1) Define

$$A_{k,\mathbf{p}} = \prod_{j\neq k} H_{\mathbf{p}(j)} \circ e_j^*$$

then

$$H_{\mathbf{p}} = A_{k,\mathbf{p}} H_{\mathbf{p}(k)} \circ e_k^*$$

$$\partial_k H_{\mathbf{p}} = A_{k,\mathbf{p}}\left(\partial_k\left(H_{\mathbf{p}(k)} \circ e_k^*\right)\right) = A_{k,\mathbf{p}}\mathbf{p}(k)H_{\mathbf{p}(k)-1} = \mathbf{p}(k)H_{\mathbf{p}-\varepsilon_k}.$$

□

(3.1.2)

$$\partial_k^* H_{\mathbf{p}} = A_{k,\mathbf{p}}\left(\left(\partial_k^* H_{\mathbf{p}(k)}\right) \circ e_k^*\right) = H_{\mathbf{p}+\varepsilon_k}.$$

□

(3.1.3)

$$\mathcal{L}H_{\mathbf{p}} = \sum_{k\in\mathcal{E}_{\mathbf{p}}} \partial_k^*\partial_k H_{\mathbf{p}},$$

where $\mathcal{E}_{\mathbf{p}} = \{k;\ \mathbf{p}(k) > 0\}$ (finite sum)

$$\partial_k^*\partial_k H_{\mathbf{p}} = A_{k,\mathbf{p}}\left(\left(\partial_k^*\partial_k H_{\mathbf{p}(k)}\right) \circ e_k^*\right) = \mathbf{p}(k)H_{\mathbf{p}}$$

$$\mathcal{L}\ H_{\mathbf{p}} = \left(\sum \mathbf{p}(k)\right) H_{\mathbf{p}} = |\mathbf{p}|H_{\mathbf{p}}.$$

□

(3.1.4) Suppose $\mathbf{p} \neq \mathbf{p}'$, so that there exists k such that $\mathbf{p}(k) \neq \mathbf{p}'(k)$. We have then

$$H_{\mathbf{p}} = A_{k,\mathbf{p}}\left(H_{\mathbf{p}(k)} \circ e_k^*\right), \qquad H_{\mathbf{p}'} = A_{k,\mathbf{p}'}\left(H_{\mathbf{p}'(k)} \circ e_k^*\right).$$

Then by the Fubini theorem

$$(H_{\mathbf{p}} \mid H_{\mathbf{p}'}) = \int_{\mathbb{R}^{N-k}} A_{k,\mathbf{p}} A_{k,\mathbf{p}'} \left[\int_R H_{\mathbf{p}(k)}(\xi) H_{\mathbf{p}'(k)}(\xi) \gamma(d\xi) \right]$$

as the integral in the bracket is zero. We get then the orthogonality. The normality can be shown in the same way.

The completeness results from the following argument. Denote by $\mathcal{A}_n$ the σ-field generated by the e_k^*, $k \in [1, n]$; we define in this way a filtration of the probability space. As

$$\lim_n E^{\mathcal{A}_n}(f) = f$$

for all $f \in L^2(\gamma_n)$, it is sufficient to prove that, denoting $\mathcal{E}_n = \{\mathbf{p} \in \mathcal{E}, \mathbf{p}(k) = 0 \text{ if } k > n\}$, the $\{H_{\mathbf{p}}\}$, $\mathbf{p} \in \mathcal{E}_n$, constitute a basis of $L^2(\gamma_n)$.

Then as γ_n satisfies the hypothesis of 2.4, we have that the polynomials are dense in $L^2(\gamma_n)$. By orthogonality the $\{H_{\mathbf{p}}\}$ are linearly independent. The polynomial $H_{\mathbf{p}}$ has a unique monomial of total degree $|\mathbf{p}|$, which is given by $\prod_k \left(e_k^*\right)^{\mathbf{p}(k)}$. Therefore the $\{H_{\mathbf{p}}\}$, $p \in \mathcal{E}_n$ constitute a basis of polynomials in n-variables.

(3.1.5) We shall prove this formula by an induction on $r = |\mathbf{p}|$. For $r = 0$ the formula is true. Assume that the formula is true for $r < r_0$. Then if $|\mathbf{p}| = r_0$

$$H_{\mathbf{p}} = \prod_k H_{\mathbf{p}(k)} \circ e_k^*.$$

Introduce $\mathbf{p}_1 = \mathbf{p} - \varepsilon_{k_0}$ where k_0 is such that $\mathbf{p}(k_0) > 0$. Then $|\mathbf{p}_1| = r - 1$,

$$H_{\mathbf{p}_1} = \partial_{\mathbf{p}_1}^* 1,$$

and by (3.1.2)

$$\partial_{k_0}^* H_{\mathbf{p}_1} = H_{\mathbf{p}}.$$

□

(3.1.6)

$$f = \sum_{\mathbf{p} \in \mathcal{E}} c_{\mathbf{p}} H_{\mathbf{p}}$$

with

$$c_{\mathbf{p}} = \frac{1}{\mathbf{p}!} (f \mid H_{\mathbf{p}}) = \frac{1}{\mathbf{p}!} \left(f \mid \partial_{\mathbf{p}}^* 1\right) = \frac{1}{\mathbf{p}!} (\partial_{\mathbf{p}} f \mid 1).$$

(3.1.8) The polynomials are dense in L^p, $p \in [1, 2]$, because they are dense in L^2. Denote by V the linear space of polynomials in L^p, $p \in]2, +\infty[$. Assume that $\bar{V} \neq L^p$. Then there exists $h \in L^q$ such that $E(hu) = 0$ for every polynomial u. There exists a sequence of polynomials $u_n \xrightarrow{L^q} h$. Then we shall have

$$|E(u_n u)| \le \|h - u_n\|_{L^q} \|u\|_{L^p}.$$

Take $u = \mathrm{sig}(u_n)|u_n|^{p-1}$, we get

$$\|u_n\|_{L^p}^p \le \|h - u_n\|_{L^q} \|u_n\|_{L^p}$$

or

$$\le \|h - u_n\|_{L^q} \|u_n\|_{L^p}.$$

Then $u_n \to 0$ in L^p, hence a fortiori in L^q, therefore $h = 0$.

3.2. The Hermite algebra

The Hermite polynomials in $\mathbb{R}$ constitute a linear basis of the ring of polynomials. Therefore the product of polynomials can be expressed in this basis.

Theorem. *We have*

$$H_s H_\ell = \sum_{r=|s-\ell|}^{s+\ell} c_{s,\ell}(r) H_r$$

where

3.2.1.
$$c_{s,\ell}(r) = \frac{\Gamma(s+1)\Gamma(\ell+1)}{\Gamma(g-\ell+1)\Gamma(g-s+1)\Gamma(g-r+1)}$$

with $g = \frac{1}{2}(\ell+s+r)$ *and when* g *is not an integer* $c_{s,\ell}(r) = 0$.

Proof. Using the generating function identity 2.5.6, we have

$$\begin{aligned} e^{\lambda\lambda'} \sum_{n=0}^{+\infty} \frac{(\lambda+\lambda')^n}{n!} H_n(x) &= \exp\left((\lambda+\lambda')x - \frac{(\lambda+\lambda')^2}{2} + \lambda\lambda'\right) \\ &= \sum_{p,p'=0}^{+\infty} \frac{\lambda^p \lambda'^{p'}}{p!p'!} H_p(x) H_{p'}(x) \end{aligned}$$

We replace $e^{\lambda\lambda'}$ by its Taylor series and then mutiply out the Taylor series on the l.h.s.; on identifying the coefficients of monomials in λ, λ' we obtain the desired identity. □

Extension to $\mathbb{R}^{\mathbb{N}}$. We define the *structural constants*

3.2.2.
$$C_{\mathbf{p},\mathbf{p}'}(\mathbf{p}'') = \prod_k c_{\mathbf{p}(k),\mathbf{p}'(k)}(\mathbf{p}''(k)).$$

3.2.3. Corollary.

$$H_{\mathbf{p}} H_{\mathbf{p}'} = \sum c_{\mathbf{p},\mathbf{p}'}(\mathbf{p}'') H_{\mathbf{p}''}.$$

Remark. The coefficient $c_{\mathbf{p},\mathbf{p}'}(\mathbf{p}'')$ determines an algebraic structure on $\mathbb{R}^{\mathcal{E}}$.

4. Classification of irreducible Gaussian probability spaces

4.1. The fundamental algebra

Given a Gaussian probability space $(\Omega, \mathcal{A}, \mathcal{P};\ H, \mathcal{A}_H^{\perp})$, we shall denote by $\tilde{\mathcal{A}}_H$ the *algebra generated by the random variables* in H. We can describe $\tilde{\mathcal{A}}_H$ as the set of random variables which can be written as $Q(G_1, \dots, G_q)$ where $G_j \in H$ and where Q is a polynomial in q variables. As $\{G_i\}$ are Gaussian variables, $\tilde{\mathcal{A}}_H \subset L^p(\Omega, \mathcal{A}, P)$ for every $p < +\infty$.

This section is based on the following two observations:

(i) An irreducible Gaussian probability space is fully determined by its fundamental algebra $\tilde{\mathcal{A}}_H$.
(ii) Once a basis for H is chosen, $\tilde{\mathcal{A}}_H$ is canonically isomorphic to an Hermite algebra.

4.2. Lemma. *Let $(\Omega, \mathcal{A}, \mathcal{P}, H)$ be an irreducible Gaussian probability space, then, for every $p \in [1, +\infty[$, $\tilde{\mathcal{A}}_H$ is dense in $L^p(\Omega, \mathcal{A}_H, \mathcal{P})$.*

Proof. Consider a finite-dimensional subspace $S \subset H$. Then, by Lemma 2.4, we have that $\tilde{\mathcal{A}}_S$ is dense $L^p(\Omega, \mathcal{A}_S, \mathcal{P})$. As

$$L^p(\Omega, \mathcal{A}_H, \mathcal{P}) = \overline{\cup_{S \subset H} L^p(\Omega, \mathcal{A}_S, \mathcal{P})}$$

the lemma follows. □

4.3. The numerical model

An irreducible Gaussian probability space $(\Omega, \mathcal{A}, \mathcal{P}, H)$ where we have chosen a basis of H, will be called a numerical model.

We say that an isomorphism u between two numerical models is *numerical* if u sends the second basis to the first.

4.4. Theorem. 4.4.1. *Every numerical automorphism of a numerical model is the identity.*
4.4.2. *Two numerical models are isomorphic if their Hilbert spaces of Gaussian variables have the same dimension.*

Proof. **(4.4.1)** We have an automorphism which preserves the basis e_k of H. Therefore, as u is an *algebra automorphism*,

$$u(H_{\mathbf{p}}) = H_{\mathbf{p}}.$$

As $H_{\mathbf{p}}$ are dense in $L^2(\Omega, \mathcal{A}, P)$ it follows that u is the identity. □

(4.4.2) Consider two numerical models $(\Omega, \mathcal{A}, \mathcal{P};\ H)$, $(\Omega', \mathcal{A}', \mathcal{P}';\ H')$, then there exists a unique map λ of $H' \to H$ which sends the basis of H' into the basis of H. We extend this map to a linear map

$$\mathcal{A}_{H'} \xrightarrow{\tilde{\lambda}} \mathcal{A}_H$$

which is defined by

$$\tilde{\lambda}(H'_{\mathbf{p}}) = H_{\mathbf{p}}$$

where

$$H'_{\mathbf{p}} = \prod_k H_{\mathbf{p}(k)} \circ e'^{*}_k$$

$$H_{\mathbf{p}} = \prod_k H_{\mathbf{p}(k)} \circ e^{*}_k.$$

As the algebraic structure of Hermite polynomials is given by the *structural constants* $c_{p,p'}(p'')$ defined in 3.2.2, it follows that $\tilde{\lambda}$ is an *algebra homomorphism.* Furthermore, by irreducibility, the linear span of the $H'_{p'}$ is dense in the corresponding L^2 space, so $\tilde{\lambda}$ extends to an isometry u between the L^2-spaces.

4.5. Corollary. *Two numerical models of the same dimension are canonically isomorphic. We can therefore legitimately say that in a given dimension, the numerical model is a unique object.*

Proof. If we had two numerical isomorphisms, by composition we would get a numerical automorphism which would have to be the identity. □

4.6. Corollary. *The irreducible Gaussian probability spaces* $(\Omega, \mathcal{A}, \mathcal{P}, H)$ *are classified by the dimension of* H.

Proof. When we have an irreducible Gaussian probability space, it becomes a numerical model by the *choice* of a basis in H. We then apply 4.4.2. □

5. Intrinsic geometry on Gaussian probability space

5.1. We address ourselves to the following question: *Given a Gaussian probability space* $(\Omega, \mathcal{A}, \mathcal{P};\ H)$, *what are the objects which are naturally associated to this structure?*

We have already introduced

- $\tilde{\mathcal{A}}_H$ the fundamental algebra of random variables generated by H.
- $\mathcal{A}_H$ the σ-field generated by H.

We can *choose* a basis of H. Then $(\Omega, \mathcal{A}_H; \mathcal{P}, H)$ will become a numerical model for which a full description is provided in Section 4.

By an *intrinsic property* we mean a property which does not depend on the *choice* of a basis of the Hilbert space H.

An elementary analogue of our present situation is following: granted known *Cartesian analytic geometry*, then *Euclidean geometry* could be recaptured as those properties which are invariant by a change of Cartesian axes.

Because $L^2(\Omega, \mathcal{A}_H, \mathcal{P})$ is isomorphic to a numerical model, the search for intrinsic properties is equivalent to the search for properties of a numerical model which are invariant under a change of basis.

Two intrinsic objects will be introduced in the next section, the Ornstein-Uhlenbeck semigroup and the canonical representation of H.

5.2 Unitary equivalence

Given two bases e_k, e'_s of H, there exists an orthogonal matrix with an infinite number of rows and columns a_s^k such that

$$e'_s = \sum_k a_s^k e_k.$$

For fixed s, we consider the following series J_s

$$J_s = \sum_k a_s^k \xi_k.$$

We denote by Ω the probability space constituted by $\left(\mathbb{R}^{\mathbb{N}}, \mathcal{B}_{\mathbb{R}^{\mathbb{N}}}, \gamma_{\mathbb{N}}\right)$; on Ω the ξ_k are independent normal variables; therefore the partial sums of this series constitute an L^2- martingale which converges almost surely to an L^2 function $J_s(\xi)$. The law of the $J_s(*)$ is the law of a sequence of independent normal variables; therefore the map $J : \Omega \to \Omega$ defined by $\xi_* \to J_*(\xi)$ is an isomorphism of Gaussian probability spaces.

5.3. Application: the Segal model

5.3.1. Theorem. *Given a separable Hilbert space $\mathcal{H}$, there exists a canonical Gaussian irreducible probability space $(\Omega, \mathcal{A}, P, \mathcal{H})$ which has as a space of Gaussian variables the space $\mathcal{H}$ itself.*

We denote by $\mathrm{Seg}(\mathcal{H})$ *this Gaussian probability space.*

Furthermore $\mathcal{H} = \mathcal{H}_1 \oplus \mathcal{H}_2$ *implies* $\mathrm{Seg}(\mathcal{H}) = \mathrm{Seg}(\mathcal{H}_1) \times \mathrm{Seg}(\mathcal{H}_2)$.

Proof. We consider the probability space $\Omega = \left(\mathbb{R}^{\mathbb{N}}, \mathcal{B}_{\mathbb{R}^{\mathbb{N}}}\right)$ and the Gaussian probability space $\mathcal{G}$ defined on Ω by the Hermite algebra. Every orthonormal basis $\{e_k\}$ of $\mathcal{H}$ generates an isomorphism of $\mathcal{H}$ onto l^2. The change of basis induces the isomorphism of $\mathcal{G}$ described in 5.2; this isomorphism preserves the intrinsic geometry of $\mathcal{G}$ which is therefore canonically associated to $\mathcal{H}$.

5.3.2. Remark. The set $\mathbb{R}^{\mathbb{N}}$ can be considered as the *spectrum* of the Hermite algebra; this spectrum is separable, in contrast to the Gelfand or the Stone spectra considered in the introduction of this chapter. The price to pay for this separability is that we have to consider as many spectra as we have bases of $\mathcal{H}$; we must therefore work up to the isomorphisms of spectra induced through 5.2 by those changes of basis.

5.3.3. Notation Given $h \in H$ we shall denote by $W(h)$ the random variable on $\mathrm{Seg}(H)$ associated to $h \in H$.

6. The Ornstein-Uhlenbeck semigroup

Theorem. *To a given Gaussian probability space* $(\Omega, \mathcal{A}, \mathcal{P};\ H, \mathcal{A}_H^{\perp})$ *is associated in a canonical way an Hermitian semigroup of contractions*

6.1. $$P_t : L^2(\Omega, \mathcal{A}, P) \to L^2(\Omega, \mathcal{A}, P)$$

which satisfies

6.2. $$P_t(f) > 0 \quad \text{if} \quad f > 0.$$

6.3. $$\begin{cases} P_t\left(L^2(\mathcal{A}_H)\right) \subset L^2(\mathcal{A}_H) \\ P_t(uv) = uP_t(v) \ \text{ if } \ u \in L^2\left(\mathcal{A}_H^{\perp}\right), \quad v \in L^2(\mathcal{A}_H) \end{cases}$$

(which means that P_t *has no action in the transverse variables).*

6.4. *The spectral decomposition of* P_t *is of the form*

6.5. $$L^2(\Omega, \mathcal{A}, P) = \overset{\perp}{\underset{n\in\mathbb{N}}{\oplus}} C_n$$

where $P_t f = e^{-nt} f$ *if* $f \in C_n$. C_n *is called the chaos of order n*

$$\mathcal{C}_0 = \left\{f : E^{\mathcal{A}_H} f = f\right\}.$$

6.6. *If the space is irreducible, by transfer from a numerical model,* $\mathcal{C}_n$ *is the closed subspace of* L^2 *generated by the Hermite polynomials* H_p *with* $|p| = n$.

6.7. *For* $p \in [1, +\infty[$, P_t *operates as a semigroup of contractions on* L^p

6.8. $$\|P_t f - f\|_{L^p} \to 0 \quad \text{when} \quad t \to 0.$$

Proof. Assume that we have constructed P_t on $L^2(\Omega, \mathcal{A}_H, P)$. Then as $\{uv\}$ where $u \in L^2\left(\mathcal{A}_H^{\perp}\right)$, $v \in L^2(\mathcal{A}_H)$ generate linearly $L^2(\mathcal{A})$, we will define $\tilde{P}_t$

(i) $$\tilde{P}_t(uv) = uP_t(v).$$

Therefore we are reduced to proving the theorem for an irreducible Gaussian space.

Construction of P_t in the irreducible case

Let $\left(\mathbb{R}^{\mathbb{N}}, B_{\mathbb{R}^{\mathbb{N}}}, \gamma_{\mathbb{N}}, \ell_2\right)$ be a numerical model. Then we define P_t on L^2 by its bilinear form given by the *Mehler formula*

6.9. $$(P_t f \mid g) = \int_{\mathbb{R}^{\mathbb{N}} \times \mathbb{R}^{\mathbb{N}}} f(x\cos\theta + y\sin\theta)g(x)\gamma_{\mathbb{N}}(dx)\gamma_{\mathbb{N}}(dy)$$

where $\cos\theta = e^{-t}$, $t > 0$, $\theta \in \left[0, \frac{\pi}{2}\right[$.

We have to show that 6.9 is invariant under an orthogonal change of coordinates defined by

$$x'_k = \sum_s \alpha_k^s x_s, \; y'_s = \sum_s \alpha_k^s y_s$$

where α_k^s is an orthogonal matrix. Then x'_k, being a linear combination of the Gaussian variables x_s is also a Gaussian variable. We have

$$\begin{aligned} E\left((X'_k)^2\right) &= \sum_s (\alpha_k^s)^2 = 1 \\ E\left(x'_k x'_\ell\right) &= \sum_s \alpha_k^s \alpha_\ell^s = 0. \end{aligned}$$

Therefore the x'_k are independent, normal variables. This means that the measure γ_{N} is invariant by the orthogonal transformation $\left(\alpha_k^s\right)$.

As $(x \cos\theta + y \sin\theta) \to (x' \cos\theta + y' \sin\theta)\ \gamma_N(dx) \otimes \gamma_N(dy)$ a.e. this implies, with the invariance of γ_{N}, the invariance of 6.9 which therefore has an intrinsic meaning.

Now by Hölder's inequality

$$\begin{aligned} &(P_t f \mid g)_{L^2(\gamma_{\mathrm{N}})} \\ &\quad \le \|g\|_{L^r(\gamma_{\mathrm{N}})} \left(\int_{R^{\mathrm{N}} \times R^{\mathrm{N}}} |f(x\cos\theta + y\sin\theta)|^s d\left(\gamma_{\mathrm{N}} \otimes \gamma_{\mathrm{N}}\right)\right)^{1/s}. \end{aligned}$$

Consider now the map

$$q_\theta : \mathbb{R}^{\mathrm{N}} \times \mathbb{R}^{\mathrm{N}} \to \mathbb{R}^{\mathrm{N}}$$

defined by

$$q_\theta(x, y) = x\cos\theta + y\sin\theta.$$

This map satisfies

$$(q_\theta)_* \left(\gamma_{\mathrm{N}} \otimes \gamma_{\mathrm{N}}\right) = \gamma_{\mathrm{N}}.$$

We have only to do again a variance analysis

$$E\left((x_k\cos\theta + y_k\sin\theta)^2\right) = \cos^2\theta + \sin^2\theta = 1$$

$$E\left((x_k\cos\theta + y_k\sin\theta)\left(x_{k'}\cos\theta + y_{k'}\sin\theta\right)\right) = 0.$$

Then

$$\int |f\left(x\cos\theta + y\sin\theta\right)|^s\, d\gamma_{\mathrm{N}}(x) d\gamma_{\mathrm{N}}(y) = \|f\|^s_{L^s(\gamma_{\mathrm{N}})}.$$

We get finally

(ii) $$|\left(P_t f | g\right) \le \|f\|_{L^s} \|g\|_{L^r} \quad \frac{1}{r} + \frac{1}{s} = 1, \qquad r \in [1, +\infty].$$

Then (ii) implies that the bilinear form is bounded on $L^2 \times L^2$. This bilinear form is symmetric:

(iii) $$(P_t f \mid g) = (f \mid P_t g).$$

In fact we make on $\mathbb{R}^N \times \mathbb{R}^N$ the change of variables

$$x' = x\cos\theta + y\sin\theta \qquad y' = x\sin\theta + y\cos\theta.$$

A study of the covariance shows that the measure $\gamma_N \otimes \gamma_N$ is preserved by this change in variable. We have

$$x = x'\cos\theta - y'\sin\theta,$$

as the measure $\gamma_N(dy)$ is invariant by the change of variable $y \to -y$, we get (iii). The bilinear form $(f, g) \to (P_t f \mid g)$ is therefore symmetric, bounded by 1. It defines without ambiguity an Hermitian operator P_t on $L^2(\Omega, \mathcal{A}, P)$ of norm ≤ 1. It follows from 6.9 that $(P_t f \mid g) > 0$ if $f, g > 0$, which proves 6.2.

Take a numerical model. Now, by 3.1.7, we have

(iv) $$P_t H_p = e^{-nt} H_p.$$

This means that the basis of $L^2(\Omega, \mathcal{A}, P)$ associated to the numerical model diagonalizes P_t. From this diagonalization we see the semigroup property

(v) $$P_t \circ P_{t'} = P_{t+t'}.$$

Denote $\mathcal{E}_n = \{p \in \mathcal{E};\ |p| = n\}$, $C_n =$ the closed linear span of H_p with $p \in \mathcal{E}_n$. We deduce from (iv) that $P_t(h) = e^{-nt}h$ for all $h \in C_n$. As $L^2(\Omega, \mathcal{A}, P) = \overset{\perp}{\bigoplus} C_n$, this means that

$$C_n = \ker\left(I - e^{nt} P_t\right).$$

As the right hand side is intrinsic, this means that C_n is intrinsic. Furthermore for every numerical model $H_p \in C_n$ for $p \in \mathcal{E}_n$ which implies 6.6. Now when $t \to +\infty$, $P_t H_p \to 0$ if $p \neq 0$. This means that $P_t f \to E(f)$ which proves 6.5. Finally the inequality (ii) implies 6.7.

We remark that 6.8 is true for $p = 2$ granted the spectral decomposition.

Given $f \in L^p$ we can find $g \in L^2 \cap L^\infty$ such that

$$\|f - g\|_{L^p} < \varepsilon.$$

Furthermore the inequality $-\|g\|_{L^\infty} \leq g(x) \leq \|g\|_{L^\infty}$ implies that

$$\|P_t g\|_{L^\infty} \leq \|g\|_{L^\infty}.$$

As $\|P_t g - g\|_{L^2} \to 0$, this implies that $P_t g$ converges in probability towards g. As $|P_t g - g|$ is dominated by $2\|g\|_{L^\infty}$, Lebesgue's dominated convergence theorem implies that

$$\|P_t g - g\|_{L^p} \to 0$$

which implies 6.8 by the estimate

$$\|P_t f - f\|_{L^p} \leq 2\|f - g\|_{L^p} + \|P_t g - g\|_{L^p}.$$

□

7. The canonical representation of H

7.1. Definition.

$$L^{\infty-0}(\Omega,\mathcal{A},\mathcal{P})=\bigcap_{p<+\infty}L^p(\Omega,\mathcal{A},\mathcal{P}),\quad L^{1+0}(\Omega,\mathcal{A},\mathcal{P})=\bigcup_{p>1}L^p(\Omega,\mathcal{A},\mathcal{P}).$$

By Hölder's inequality $L^{\infty-0}(\Omega,\mathcal{A};\mathcal{P})$ is an algebra. It is the largest algebra which is *contained in* $L^1(\Omega,\mathcal{A},P)$.

Theorem. *To any Gaussian probability space* $\left(\Omega,\mathcal{A},\mathcal{P},H,\mathcal{A}_H^{\perp}\right)$, *there is associated a canonical map*

$$\rho: H\to \operatorname{End}\left(L^{\infty-0}(\Omega,\mathcal{A},\mathcal{P})\right). \tag{7.2.}$$

ρ is a representation of the additive group of H which means:

$$\rho(h+h')=\rho(h)\rho(h') \tag{7.3.}$$

$$\|\rho(h)f\|_{L^p}\le \exp\left(2p^2\|h\|_H^2\right)\|f\|_{L^{2p}} \tag{7.4}$$

$$\begin{cases}\rho(h)\left(L^{\infty-0}(\mathcal{A}_H)\right)\subset L^{\infty-0}(\mathcal{A}_H)\\ \rho(h)(uv)=u\big(\rho(h)v\big)\ \text{ if }\ u\in L^{\infty-0}\left(\mathcal{A}_H^{\perp}\right),\quad v\in L^{\infty-0}(\mathcal{A}_H)\end{cases} \tag{7.5}$$

(which means the triviality of the action of ρ in the transverse variables)

$$\lim_{\varepsilon\to0}\frac{\rho(\varepsilon h)-I}{\varepsilon}=M_h \tag{7.6}$$

where M_h is the multiplication operator on $L^{\infty-0}(\Omega,\mathcal{A},\mathcal{P})$ by the random variable $W(h)$.

Remark. We use the same notational convention as in 5.3.3; $W(h)$ denotes the random variable on Ω associated to $h\in H$.

Proof. Assume that ρ is defined on $L^{\infty-0}(\Omega,\mathcal{A}_H,\mathcal{P})$, it will then be extended to $\tilde\rho$ defined on $L^{\infty-0}(\Omega,\mathcal{A},\mathcal{P})$ by the formula

$$\tilde\rho(h)(uv)=u\big(\rho(h)v\big)\qquad u\in L^{\infty-0}\left(\mathcal{A}_H^{\perp}\right),\quad v\in L^{\infty-0}(\mathcal{A}_H).$$

Therefore we may restrict ourselves to the case of an irreducible Gaussian space. □

We shall need the following classical result on the numerical model.

7.7. Cameron-Martin Theorem. *Given* $f \in L^{1+0}\left(\mathbb{R}^{\mathbb{N}}, B_{\mathbb{R}^{\mathbb{N}}}\right)$ *and given* $\alpha \in \mathbb{R}^{\mathbb{N}}$ *we denote*

$$(\tau_\alpha f)(x) = f(x+\alpha).$$

Suppose $\alpha \in \ell^2$, *then there exists* $k_\alpha \in L^{\infty-0}(\gamma_N)$ *such that*

$$\int_{\mathbb{R}^{\mathbb{N}}} (\tau_\alpha f)\, d\gamma_{\mathbb{N}} = \int_{\mathbb{R}^{\mathbb{N}}} f k_\alpha d\gamma_{\mathbb{N}}.$$

Furthermore,

$$k_\alpha(x) = \exp\left(\sum_{k\in\mathbb{N}} \alpha_k x_k - \frac{1}{2}\|\alpha\|_{\ell^2}^2\right)$$

$$\|k_\alpha\|_{L^p} \le \exp\left(\frac{p^2}{2}\|\alpha\|_{\ell^2}^2\right)$$

$$\lim_{\varepsilon\to 0} \varepsilon^{-1}\left(k_{\varepsilon\alpha}(x) - 1\right) = \sum_{k\in\mathbb{N}} \alpha_k x^k.$$

Proof. Assume that f depends only upon the first r coordinates. Define

$$M_s(x) = \exp\left(\sum_{k=1}^{s} \alpha_k x_k - \frac{1}{2}\sum_{k=1}^{s} \alpha_k^2\right).$$

We prove, using elementary change of variable on $\mathbb{R}^s$, that

$$E(\tau_\alpha f) = E(M_s f) \quad \text{if} \quad s \ge r.$$

Furthermore M_s is a martingale and $E(M_s) = 1$. Replacing $a_k \to \lambda\alpha_k$ we get

$$\begin{aligned} E\left(\exp\left(\lambda\sum_{k=1}^{s} \alpha_k x_k - \frac{\lambda^2}{2}\sum_{k=1}^{s}\alpha_k^2\right)\right) &= 1 \\ E\left(\exp\left(p\sum_{k=1}^{s}\alpha_k x_k\right)\right) &\le \exp\left(\frac{p^2}{2}\|\alpha\|_{\ell^2}^2\right). \end{aligned}$$

Therefore $\{M_s\}$ is an L^p-martingale which converges to an element k_α such that

$$\|k_\alpha\|_{L^p} \le \exp\left(\frac{p^2}{2}\|\alpha\|_{\ell^2}^2\right).$$

This inequality proves the Cameron-Martin theorem. □

7.8. Construction of $\rho(h)$

We choose an orthonormal basis of H, which defines a map $j : H \to \ell^2$ and represents $(\Omega, \mathcal{A}, P)$ on the numerical model $\left(\mathbb{R}^{\mathbb{N}}, B_{\mathbb{R}^{\mathbb{N}}}, \gamma_{\mathbb{N}}\right)$. We denote by q the induced map $L^{\infty-0}\left(\mathbb{R}^N;\ B_{\mathbb{R}^n}, \gamma_{\mathbb{N}}\right) \mapsto L^{\infty-0}(\Omega, \mathcal{A}, P)$. We define $\rho(h)$ by the formula

$$\rho(h) = q \circ \tau_{j(h)} \circ q^{-1}.$$

Then the properties 7.3, 7.4, 7.6 will result by transfer from the Cameron-Martin theorem. We have to prove only 7.2. In other words, that the representation ρ *does not depend* of the choice of the basis of H that we have made. A change of basis on H is reflected on the numerical model by an orthogonal transformation c_k^i, then $\alpha'_k = c_k^i \alpha_i$, $x'_k = c_k^i x_i$ and the transformation $x \mapsto x + \alpha$ mapped to the translation $x' \mapsto x' + \alpha'$. Going back to $(\Omega, \mathcal{A}, \mathcal{P};\ H)$ we find $\rho(h) = \rho'(h)$.

7.9. Absence of an intrinsic group structure

The reader might be amazed by the cautious way we have defined $\rho(h)$. Why do not say simply *that* $\mathbb{R}^{\mathbb{N}}$ *is a vector space*! Of course $\mathbb{R}^{\mathbb{N}}$ is an obvious vector space, the change of numerical model is expressed by a linear change of coordinates described in 5.2, but the corresponding series converges only *almost surely*; this means that the honest expression $\omega_1 + \omega_2$ in a numerical model can be transformed by a change of coordinate into $\omega_1 +$ "∞" $\overset{?}{=}$ "∞". Ω has no canonical group structure; H does not operate canonically as a group of transformations on Ω.

8. Example: Abstract Wiener spaces

In this paragraph we want to replace the space $\mathbb{R}^{\mathbb{N}}$ used in the construction of the numerical models by some Banach space obtained by a completion of $\mathcal{H}$.

8.1. Cylindrical functions

We denote by Δ the set of finite-dimensional subspaces of $\mathcal{H}$; to each $\delta \in \Delta$, we associate the orthogonal projection $P_\delta : \mathcal{H} \to \delta$; the cylindrical functions $f : \mathcal{H} \to \mathbb{R}$ are the functions of the form $f = F \circ P_\delta$ for some $\delta \in \Delta$ and where F is a Borel function defined on δ. We denote by $\mathcal{S}$ the vector space of cylindrical functions. We define a *mean* by

$$E(f) = \int_\delta F \, d\gamma_\delta,$$

where γ_δ denotes the standard Gaussian variable on the Euclidean space δ and where the integral of the r.h.s. is assumed to be absolutely convergent.

Lemma. *The mean is not σ-additive which means that there is no (σ-additive) measure γ on $\mathcal{H}$ such that $E(f) = \int_{\mathcal{H}} f \, d\gamma$.*

Proof. We construct a strictly increasing sequence of subspaces $\delta_n \in \Delta$; we choose a sequence $A_n \subset \delta_n$ of Borel sets satisfying

i) the compatibility condition $P_{\delta_n}(A_{n+1}) = A_n$
ii) $\lim_n \gamma_n(A_n) > 0$.

Denote by F_n the indicator function of A_n and by $f_n = F_n \circ P_{\delta_n}$ the corresponding cylindrical function. Then $\{f_n\}$ is a decreasing sequence of functions satisfying $\lim_n E(f_n) > 0$. If the mean was σ-additive then, by the Lebesgue theorem, $\lim f_n \in \mathcal{S}$. Since $\lim f_n \notin \mathcal{S}$, this proves the lemma. □

This example shows the necessity to enlarge the space of test functions; we shall obtain this enlargement by enlarging $\mathcal{H}$.

8.2. Radonifying norms: the Hilbert space case

This paragraph is a pedagogical preparation for 8.3; it is possible to skip it and to go directly to 8.3.

Given an Hermitian operator u, of the Hilbert-Schmidt class (which means that $\mathrm{trace}(u^*u) < \infty$), such that $u(h) = 0$ implies $h = 0$, we define a new norm on $\mathcal{H}$ by the formula

$$|h| = ||u(h)|| \qquad \text{where} \quad ||h|| = \sqrt{(h|h)}.$$

By the spectral theory of Hilbert-Schmidt operators the eigenfunctions of u form an orthonormal basis e_k of $\mathcal{H}$ such that

$$u(e_k) = \lambda_k e_k \quad \text{where the } \lambda_k \text{ are real and satisfy} \quad \sum_k \lambda_k^2 < \infty.$$

We denote by B the Hilbert space obtained by completion of $\mathcal{H}$ for the norm $|*|$.

8.2.1. Theorem. *Denote by $\{\xi_k\}$ a sequence of independent normal variables, then almost surely the series $\Phi(\xi) = \sum_k \xi_k e_k$ converges in B.*

Proof. We introduce the following processes:

$$M_n = \left| \sum_{k=0}^{n} \xi_k e_k \right|^2 ,$$

then M_n is a submartingale; from $E(M_n) < \sum \lambda_k^2 < \infty$, we deduce that $\lim M_n < \infty$ almost surely. We have

$$E\left(\left|\sum_{k=q}^{q+r}\xi_k e_k\right|^2\right) = E\left(M_{q+r} - M_q\right)$$

then by the Cauchy criterion the convergence of the sequence $\{M_n\}$ implies the convergence of the series Φ. □

8.2.2. Corollary. *Denote by Ω the probability space $(\mathbb{R}^{\mathrm{N}}, \mathcal{B}_{\mathbb{R}^{\mathrm{N}}}, \gamma_{\mathbb{R}^{\mathrm{N}}})$, then $\Phi : \Omega \to B$ is a measurable map; therefore $\Phi_*(\gamma_{\mathbb{R}^{\mathrm{N}}})$ is a Borel measure, which we denote by μ.*

8.3. Gross' radonifying norms

Definition. A *radonifying norm* $|*|$ on $\mathcal{H}$ will be a norm such that there exists an enumerable family of finite-dimensional, mutually orthogonal, subspaces $\delta_k \in \Delta$, generating a dense linear subspace of $\mathcal{H}$ and with the following property. Consider on each space δ_n the standard Gaussian variable Γ_n associated to the Euclidean metric induced by the Hilbert space structure of H. Choose all the Γ_n independent; we require that

(8.3.1.) $$\sum_n P(\{|\Gamma_n| > n^{-2}\}) < \infty.$$

Theorem. *Almost surely the series $\Phi = \sum \Gamma_n$ is norm-convergent in B, the completion of $\mathcal{H}$ for the radonifying norm.*

Proof. By the Borel-Cantelli lemma, almost surely $|\Gamma_n| \le n^{-2}$ for n large enough.
The conclusions of Corollary 8.2.2 remain valid for the Banach space B. □

8.4. Cylindrical integration through radonification

The natural injection $j : \mathcal{H} \to B$ has dense image; its adjoint $j' : B' \to \mathcal{H}$ is therefore injective with dense image.

We denote $\Delta'' = \{\delta \in \Delta;\ \delta \subset B'\}$. Choosing an orthornormal basis δ_j of $\delta \in \Delta$, we have $P_\delta(h) = \sum(h|\delta_j)\delta_j$. For $\delta \in \Delta''$ the projection P_δ extends to B through the formula $\Pi_\delta(b) = \sum < b, \delta_j > \delta_j$. We denote by $\mathcal{S}'$ the vector space of cylindrical functions of the form $f = F \circ P_\delta$ where $\delta \in \Delta''$ and set $\tilde{f} = F \circ \Pi_\delta$, then

$$E(f) = \int_B \tilde{f}\, d\mu.$$

The norm $\sqrt{E([f]^2)}$ defines on $\mathcal{S}$ an inner product space structure. This structure $\mathcal{S}'$ is dense in $\mathcal{S}$; therefore the map $f \to \tilde{f}$ has an extension to an isometry of $\mathcal{S}$ into $L^2_\mu(B)$. This isometry preserves L^2 scalar products; in particular, it preserves the scalar product by 1: means are transformed into integrals.

8.5. Cameron-Martin theorem and Mehler formula

The fact that B is a vector space containing H makes it possible to avoid the discussion of 7.9. Then the Cameron-Martin Theorem 7.7 can be written

$$(8.5.1.)\quad \int_B g(b+h)\ \mu(db) \quad = \quad \int_B g(b)k_h(b)\ \mu(db)$$
$$\text{where}\quad k_h(b) = \exp\left(\tilde{h}(b) - \frac{1}{2}||h||^2\right).$$

Applying the Fubini theorem to 6.9, we get the Mehler formula:

$$(8.5.2.)\quad [P_t f](b_0) = \int_B f\left(e^{-t}b_0 + \sqrt{1-e^{-2t}}b\right)\mu(db) \qquad \forall f \in L^1_\mu.$$

8.6. Comparison with the construction of Seg($\mathcal{H}$) given in 5.3

We construct a preferred basis $\{e_k\}$ of $\mathcal{H}$: in the Hilbert space case this basis will be constituted by the eigenfunctions of u; in the Banach case by putting together orthonormal bases chosen arbitrarily in each δ_n. This basis determines a numerical model of Seg($\mathcal{H}$); the series Φ of 8.2 is a measurable map from this numerical model to B. The methodology of the radonifying norms is essentially equivalent to working with a fixed numerical model; it ignores fully the equivalence relation induced on the numerical models by changes of basis of $\mathcal{H}$; in particular we have no invariance under unitary isomorphisms of $\mathcal{H}$.

As $\mathbb{R}^{\mathbb{N}}$ has a natural vector space structure, it is possible in a fixed numerical model to write the formulae of Section 8.5. In conclusion, the methodology of numerical models has no operational disadvantages; on the other hand, it gives more insight into the fact that we work on objects defined up to some equivalence. The unitary equivalence proved in 5.2 in the context of an irreducible Gaussian probability space has no proper counterpart in the context of an abstract Wiener space.

8.7. Example of Brownian motion

Wiener's original construction of Brownian motion emphasizes the fact that

8.7.1. *Increments of Brownian motion on disjoint intervals are independent Gaussian variables.*

Using 5.3.1 this independence will correspond to orthogonality in the associated Hilbert space. We introduce the Hilbert space

$$(8.7.2)\quad H^1([0,1];\ \mathbb{R}^d) \quad = \quad \left\{\varphi \in C([0,1];\ \mathbb{R}^d); \varphi(\tau) = \int_0^\tau \dot{\varphi}(\lambda)\ d\lambda\right.$$
$$\left.\text{where}\quad ||\varphi||^2_{H^1} = \int_0^1 |\dot{\varphi}(\tau)|^2_{\mathbb{R}^d}\ d\tau < \infty\right\}.$$

8.7.3. Proposition. *The evaluation map $e_\tau : \phi \to \phi(\tau)$ is a continuous map from $H^1 \to \mathbb{R}^d$.*

Identifying linear forms on H^1 with elements of H^1, we have the orthogonality relation

$$\left(e_{\tau_2} - e_{\tau_1} \mid e_{\tau_4} - e_{\tau_3}\right) = 0 \quad \text{for} \quad \tau_1 \leq \tau_2 \leq \tau_3 \leq \tau_4.$$

Proof. By Schwarz' inequality $|e_\tau(\phi)| \leq \sqrt{\tau}||\phi||_{H^1}$.

Denote $\hat{e}_\tau(\lambda) = \tau_\wedge\lambda$, then $e_\tau(\phi) = (\phi|\hat{e}_\tau)_{H^1}$. The identity

$$\frac{d}{d\lambda}\left[e_{\tau_2} - e_{\tau_1}\right] = 1_{[\tau_1,\tau_2]}$$

proves the claimed orthogonality. □

8.7.4. Radonifying norms

We consider on H^1 the norm $|\phi| = \sup|\phi(\tau)|$. We denote by H^1_n the subspace of H^1 constituted by the functions whose derivatives are constant on all the dyadic intervals of the form $\left(k2^{-n}, (k+1)2^{-n}\right)$. We denote by δ_n the orthogonal complement of H^1_{n-1} in H^1_n. Then $\delta_n = \left\{\phi \in H^1_n; \phi(k2^{-n+1}) = 0 \text{ for all } k \in \mathbb{N}\right\}$.

Theorem. *The condition 8.3.1 is satisfied and almost surely the series Φ is norm-convergent in the Banach space of continuous functions.*

Proof. As $\phi \in \delta_n$ implies that ϕ vanishes on the dyadic rationals of order $(n-1)$ we have

$$|\phi| \leq \sup_k \int_{k2^{-n}}^{(k+2)2^{-n}} |\dot{\phi}|_{\mathbb{R}^d}\, d\lambda = 2^{-n/2} \sup_k |\Lambda|_{\mathbb{R}^d},$$

where Λ is the standard Gaussian variable on $\mathbb{R}^d$. If all the components of Λ are smaller in absolute value than ρ/d, then $|\Lambda|_{\mathbb{R}^d}$ is smaller than ρ; this remark reduces our majoration to the case of scalar valued Gaussian variables; then 8.3.1 results from the following inequality:

$$P\left(|\Gamma_n| > n^{-2}\right) < 2^n P\left(|\Lambda|_{\mathbb{R}^d} > 2^{n/4}\right) < 2d2^n \int_{d^{-1}2^{n/4}}^{+\infty} \exp\left(-\frac{t^2}{2}\right) dt.$$

□

8.7.5. Definition. Wiener measure is the Borel measure μ on $C\left([0,1]; \mathbb{R}^d\right)$ which is the image under the function Φ of $\gamma_{\mathbb{R}^N}$.

8.7.6. Remark. Introduce for $\alpha \in]0, 1]$ the following Hölder norm:

$$|g|_\alpha = |g(0)| + \sup \frac{|g(\tau) - g(\tau')|}{|\tau - \tau'|^\alpha}$$

The continuous functions ψ which satisfy $|\psi|_\alpha < \infty$ constitute the Banach space Y_α of Hölder functions of order α.

Then the same computation will show that, for $\alpha < \frac{1}{2}$ the series Φ converges in Y_α; therefore *almost surely the paths of Brownian motion are Hölderian of order α, for $\alpha < \frac{1}{2}$.*

We can realize Wiener measure as a Borelian measure μ_α on the Banach space Y_α. All these realizations have H^1 as Cameron-Martin space. There is no reason to prefer one realization to another: from the point of view of the abstract Wiener space theory, an intrinsic definition of Brownian motion does not exist.

By contrast, the Segal model associated to the Hilbert space H^1 gives an intrinsic construction of Brownian motion.

9. Convention of notations

In the remainder of the book we shall limit ourselves essentially to irreducible Gaussian probability spaces. The irreducible Gaussian probability space $(\Omega, \mathcal{A}, P, H)$ will be often abbreviated as X. The incorrect formulae 8.5.1 and 8.5.2 will be often used instead of their formal elaboration in paragraph 7.

Chapter II

Gross-Stroock Sobolev Spaces over a Gaussian Probability Space

Contents: Continuity of the Cameron-Martin representation over $L^{\infty-0}$ – The space $\mathbb{D}_1^\infty$ of differentiable vectors of the Cameron-Martin representation – Gradient operator – Generalized polynomials – Cauchy operator; Krée-Meyer inequality for the gradient – The spaces $\mathbb{D}_1^p$ – Gradient of Hilbert-valued functionals – Recursive approach to higher derivatives and to Krée-Meyer inequalities of higher order – The space $\mathbb{D}_\infty$ of smooth functionals, its approximation by C^∞ cylindrical functionals – Divergence as the adjoint of the gradient – Divergence of smooth vector fields – Shigekawa acyclicity of the complex of differential forms – Appendix: Proof of the L^p inequality for the Hilbert transform.

In the theory of harmonic analysis on $\mathbb{R}^n$ a basic fact is the invariance of the Lebesgue measure under translation. Therefore translations act on $L^2(\mathbb{R}^n)$ as unitary operators; the Sobolev space W_1^2 can then be defined as the domain of the infinitesimal generator of this action.

After a preliminary paragraph on the topology of $L^{\infty-0}$, we shall proceed in the same way over a Gaussian probability space. We have defined in I.7 a canonical action of the Cameron-Martin space H on $L^{\infty-0}$. Then the domain of the infinitesimal generator of this action will lead us to the basic Sobolev space $\mathbb{D}_1^\infty$ on which the differential D will be defined. An important fact is that the differential Df of a *scalar* function will be an *H-valued* function.

The infinitesimal generator $\mathcal{L}$ of the Ornstein-Uhlenbeck semigroup leads to the operator $\wedge = (-\mathcal{L}+1)^{-1/2}$. This operator, which is defined on L^2 by the spectral decomposition of $\mathcal{L}$, can in fact be extended to an endomorphism of L^p. We shall prove in Section 3 the following basic inequality of Krée-Meyer

$$\|Df\|_{L^p} \le c\|\Lambda^{-1} f\|_{L^p}.$$

The left-hand side is a *vector valued* function. Its norm can nevertheless be controlled by the norm of the *scalar* valued function on the right-hand side.

In Section 4 we shall investigate differentials of higher order; using functional calculus on the Ornstein-Uhlenbeck semigroup we again have the estimate

$$\|D^n f\|_{L^p} \le c\|\Lambda^{-n} f\|_{L^p}, \qquad c = c(n, p).$$

A hierarchy of Sobolev spaces $\mathbb{D}_r^p$ will be defined $\big(p \in (1, +\infty),\ r \geq 0$ an integer$\big)$. In Section 5 we shall construct an approximation procedure for the space $\mathbb{D}_r^p$ by Sobolev spaces over finite-dimensional spaces; this approximation procedure will be used as a mechanism to extend the usual rules of differential calculus on $\mathbb{R}^d$.

In Section 6 we define smooth vector fields as the elements of $\mathbb{D}_1^p(X, H)$. The notion of *divergence* of a vector field is a natural generalization of the Cameron-Martin theorem. We have the following basic fact: *every smooth vector field has a divergence.*

Finally, we shall prove for a *differential form* ω the identity of Shigekawa

$$\|d\omega\|_{L^2}^2 + \|d^*\omega\|_{L^2}^2 = \|D\omega\|_{L^2} + \|\omega\|_{L^2}^2$$

which will imply in part III the energy identity for Itô's stochastic integral.

1. Topology on $L^{\infty-0}$

We recall that

$$L^{\infty-0}(\Omega, \mathcal{A}, P) = \bigcap_p L^p(\Omega, \mathcal{A}, P).$$

It is sufficient to take the intersection on a sequence $p_j \to +\infty$. We shall choose $p_j = 2j$ the sequence of even integers. Then $L^{\infty-0}$ appears as the intersection of a countable decreasing family of Banach spaces. We define a *distance* on $L^{\infty-0}$ by the formula

$$d(f_1, f_2) = \sum_{j=2}^{+\infty} 2^{-j} \left(\|f_1 - f_2\|_{L^{2j}} \wedge 1\right)$$

(where $a \wedge b$ denotes $\min(a, b)$).

More generally we shall work with a countable family of Banach spaces $\{B_\alpha\}$ where $\alpha \in S$. For every $\alpha, \beta \in S$, we write $\beta > \alpha$ when $B_\beta \subset B_\alpha$. We will *assume* card$\{\beta \in S, \beta > \alpha$ is not true$\}$ is finite.

In the example of $L^{\infty-0}$, S is the set $\mathbb{N}$ of natural integers and the order relation is the usual relation on $\mathbb{N}$. We introduce the complication to deal with an unspecified countable set of indices. Later on, this will allow us to treat $\mathbb{D}_\infty^\infty$ as well as $L^{\infty-0}$. We define

$$B_\infty = \bigcap_{\alpha \in S} B_\alpha$$

and we define a metric on B_∞ by

1.1. $$d(f, f') = \sum_{\alpha \in S} \varepsilon_\alpha \left(\|f - f'\|_\alpha \wedge 1\right)$$

where $\varepsilon_\alpha > 0$ for all α, $\sum \varepsilon_\alpha < +\infty$.

We will assume that the norms on $\{B_\alpha\}$ are such that $\|f\|_\beta \geq \|f\|_\alpha$ if $\beta > \alpha$.

1.2. Proposition. *$d(f, f') \to 0$ if $\|f - f'\|_\alpha \to 0$ for all α, and B_∞ is a complete metric space.*

Proof. The first assertion will result from the following remarks (i) and (ii).

(i) Given ε and α, there exists η such that

$$d(f_1, f_2) < \eta \quad \text{implies} \quad \|f_1 - f_2\|_\alpha < \varepsilon.$$

(ii) Given ε' there exists η' and a finite subset $S_1 \subset S$ such that

$$\|f_1 - f_2\|_{L^\alpha} < \eta' \quad \text{for all} \quad \alpha \in S_1 \quad \text{implies} \quad d(f_1, f_2) < \varepsilon'.$$

To prove (i), we remark that

$$\varepsilon_\alpha^{-1} d(f_1, f_2) \geq \|f_1 - f_2\|_\alpha$$

if the l.h.s. is smaller than 1.

To prove (ii) we remark that for any finite subset S_1

$$d(f_1, f_2) \leq c_s \sum_{\alpha \in S_1} \|f_1 - f_2\|_\alpha + \sum_{\beta \notin S_1} \varepsilon_\beta.$$

A Cauchy sequence for d is a Cauchy sequence for all the norms $\| \ \|_\alpha$, and therefore converges in B_∞. □

1.3. Proposition. *$L^{\infty-0}$ is a topological algebra which means that the product is a continuous operation.*

Proof.

$$\begin{aligned}\|f_n g_n - f_\infty g_\infty\|_{L^j} &\leq \|(f_n - f_\infty) g_n\|_{L^j} + \|(g_\infty - g_n) f_\infty\|_{L^j} \\ &\leq \|f_n - f_\infty\|_{L^{2j}} \|g_n\|_{L^{2j}} + \|g_\infty - g_n\|_{L^{2j}} \|f_\infty\|_{L^{2j}}.\end{aligned}$$

□

1.4. Tame approximation

We shall call a *tame system* any increasing family of finite dimensional subspaces of H: $V_1 \subset V_2 \subset \ldots \subset V_n \ldots$ whose union is dense in H. Given a subspace $V \subset H$ we denote by $\mathcal{A}_V$ the σ-field of Ω generated by the elements of V. We have

Proposition. *Fix $f \in L^{\infty-0}$, then $E^{\mathcal{A}_{V_n}}(f)$ converges in $L^{\infty-0}$ to $E^{\mathcal{A}_H}(f)$.*

Proof. $E^{\mathcal{A}_{V_n}}(f)$ is an L^p-martingale and $E^{\mathcal{A}_H}(f)$ is its final value. □

Notation. We shall sometimes abbreviate $E^{\mathcal{A}_V}$ by E^V.

1.5. Proposition. *The canonical representation (defined in I.7) is a continuous map from $H \times L^{\infty-0}$ to $L^{\infty-0}$.*

Proof. As the action of ρ on the transverse variable is trivial, we shall deal with an irreducible Gaussian space. We have by I.7.4

$$\|\rho(h)(f-g)\|_{L^p} \le \exp\left(2p^2\|h\|^2\right)\|f-g\|_{L^{2p}}$$

which proves that for a fixed h

$$f \to \rho(h)f$$

is continuous, uniformly in h in a fixed ball of H. □

Fix $f \in L^{\infty-0}$, let ε, ρ be given; then we can find (by 1.4) a subspace of finite dimension V_n such that

$$\|f' - f\|_{L^p} < \varepsilon, \quad \text{where} \quad f' = E^{V_n}(f).$$

Decompose $h = v + w$ with $v \in V_n$, $w \in V_n^{\perp}$

$$\tau_h(f') = \tau_v(f').$$

By a classical result of harmonic analysis on the *finite-dimensional* Euclidean space V_n we have that there exists $\eta > 0$ such that

(i) $$\|\tau_v(f') - f'\|_{L^p} < \varepsilon \quad \text{for} \quad \|v\| < \eta.$$

Now if $\|h\| < \eta$, we have

$$\|\tau_h(f) - f\| < \|\tau_h(f - f')\| + \|\tau_h f' - f'\| \le 2\varepsilon + \varepsilon \exp\left(2p^2\eta^2\right) < 4\varepsilon$$

if we have taken η small enough.

1.6. Extension to Hilbert-valued functions

We denote by G a *separable* Hilbert space. Then the space $L^p(X; G)$ is defined as the space of measurable maps from X to G such that $\|f\| \in L^p(X)$. The same definition is made for $L^{\infty-0}(X; G)$. We shall develop a reduction mechanism to the scalar case which will be useful later.

We denote $X = \left(\Omega, \mathcal{A}, H, \mathcal{A}_H^{\perp}\right)$ our given Gaussian probability space. We introduce the Segal model of G; denote it by $(\tilde{\Omega}, \tilde{\mathcal{A}}', G)$. We define a new Gaussian probability space

$$Y = \left(\Omega \times \tilde{\Omega}', \quad \mathcal{A} \otimes \tilde{\mathcal{A}}, \quad H, \mathcal{A}_H^{\perp} \otimes \{\tilde{\Omega}\}\right)$$

(where $\{\tilde{\Omega}'\}$ denotes the trivial σ-field on $\tilde{\Omega}$, having only two elements).

Lemma. *$L^p(X; G)$ can be identified by an isometric embedding to a closed subspace of $L^p(Y)$.*

Proof. It is known that the elementary functions of the form

$$u = \sum_k g_k 1_{A_k}$$

are dense in $L^p(X; G)$. We denote by $W(g_k)$ the Gaussian variables associated to g_k (see I.5.3.3). If we define

$$u' = \sum_k W(g_k) 1_{A_k}$$

then the map $u \to u'$ realizes the isometric embedding. □

1.7. Generalized polynomials

1.7.1. Definition. We have defined in I.6 the *chaos decomposition* of $L^2 = \overset{\perp}{\otimes} \mathcal{C}_n$. We shall consider the vector space $\mathcal{P}$ generated *algebraically* by $\{\mathcal{C}_n\}$. Now an element $f \in \mathcal{P}$ is of the form

(i) $$f = \sum_{k \leq r} u_k \qquad \text{with} \quad u_k \in \mathcal{C}_k, \quad u_r \neq 0.$$

We shall call r the degree of f. We denote by $\mathcal{P}_r$ the vector space of *generalized polynomials* of degree $\leq r$.

Choosing a basis of H, we shall denote by $\mathcal{H}_r$ the vector space of finite linear combinations of Hermite polynomials H_p with $|p| \leq r$. Obviously $\mathcal{H}_r$ depends on the choice of the basis; nevertheless we have the following statement

(ii) $\mathcal{P}_r =$ closure in L^2 of $\mathcal{H}_r$; furthermore $\mathcal{P}_r$ is intrinsic.

1.7.2. Proposition. *If $f \in \mathcal{P}_r$ and $g \in \mathcal{P}_s$ then $fg \in \mathcal{P}_{r+s}$. Furthermore there exists a constant c_k such that*

(i) $$\|fg\|_{L^2} \leq c_k \|f\|_{L^2} \|g\|_{L^2}, \qquad f, g \in \mathcal{P}_k.$$

Proof. We shall prove first (i) where $f, g \in \mathcal{H}_k$. We have by I.3.2.2

$$H_{p'} H_{p''} = \sum c_{p',p''}(p) H_p.$$

It follows by inspection of I.3.2.2 that

$$c_{p',p''}(p) \leq k!$$

given p', p'' such that $|p'|$, $|p''| \leq k$. Furthermore

$$\operatorname{card}\left\{p, c_{p',p''}(p) \neq 0\right\} \leq 2^k.$$

Then

$$\begin{aligned} f &= \sum \alpha\left(f, p'\right) H_{p'} \\ g &= \sum \alpha\left(g, p''\right) H_{p''} \end{aligned}$$

Therefore

$$\alpha(fg, p) = \sum \alpha\left(f, p'\right) \alpha\left(g, p''\right) c_{p',p''}(p).$$

By Schwarz' inequality

$$|\alpha(fg;\ p)|^2 \leq \left(2^k k!\right)^2 \sum_{p',p''} \left[\alpha(f, p')\alpha(g, p'')\right]^2 \eta(p, p', p'')$$

where $\eta(p, p', p'') = 1$ if $c_{p',p''}(p) > 0$ and 0 otherwise.

Now take the sum over p

$$\|fg\|_{L^2}^2 \leq \left(2^k k!\right)^2 \sum_{p',p''} \left(\alpha(f, p')\alpha(g, p'')\right)^2 \left(\sum_p \eta(p, p', p'')\right).$$

The last sum is still smaller than 2^k. We therefore obtain

$$\|fg\|_{L^2} \leq 2^{3k/2} k! \|f\|_{L^2} \|g\|_{L^2}.$$

If $u, v \in \mathcal{P}_k$, there exists f_n, $g_n \in \mathcal{H}_k$ such that $u_n \xrightarrow[L^2]{} u$, $v_n \xrightarrow[L^2]{} v$. Then $u_n v_n \xrightarrow[L^2]{} uv$. As $u_n v_n \in \mathcal{H}_{2k}$ we have $uv \in \mathcal{P}_{2k}$. □

1.7.3. Proposition. *All the L^p norms $p \in]1, +\infty[$ are equivalent on $\mathcal{P}_r$.*

Proof. We shall first prove the equivalence of the L^2 norm with the L^{2^q} norm. Consider the following proposition (π_q): there exists a function $c_q(r)$ such that

$$\|f\|_{L^{2^q}} \leq c_q(r)\|f\|_{L^2} \qquad \text{for} \quad f \in \mathcal{P}_r.$$

Obviously (π_0) is true. Assume π_q is true for $q < q_0$. Then

$$\|f\|_{L^{2^{q_0}}}^2 = \|f^2\|_{L^{2^{q_0-1}}} < c_{q_0-1}(2r)\|f^2\|_{L^2}$$

and by 1.7.2

$$\leq c'\|f\|_{L^2}^2.$$

Then the L^2 norms for $p > 2$ are equivalent to the L^2 norm. By duality the same result is proved for $p \in]1, 2[$. □

1.7.4. Corollary. $\mathcal{P} \subset L^{\infty - 0}(X)$.

1.7.5. Corollary. *The orthogonal projection P_{C_k} on the chaos of order k defines a bounded operator in L^p for $p \in]1, +\infty[$.*

Proof. Consider first the case $p \in [2, +\infty[$. Then we have

$$\|P_{C_k} f\|_{L^2} \le \|f\|_{L^2} < \|f\|_{L^p}$$

and

$$\|P_{C_k} f\|_{L^p} \le c_k \|P_{C_k} f\|_{L^2} .$$

Now if $p \in]1, 2[$ there exists $g \in L^q$ with $\|g\|_{L^q} = 1$ such that

$$\|P_{C_k} f\|_{L^p} = \left(f \mid P_{C_k} g\right) \le c_k \|f\|_{L^p} .$$

□

1.7.6. Remark. The map $h \to W(h)$ defines an isometric bijection from H to C_1 with the L^2-norm.

Proof. The Hermite polynomials of degree 1 on $R^{\mathbb{N}}$ are $\{e_k^*\}$, $k \in \mathbb{N}$. □

2. Definition of $\mathbb{D}_1^\infty(\Omega)$ and its elementary properties

2.1. Definition. We shall say that a function $f \in L^{\infty-0}(X)$ belongs to $\mathbb{D}_1^\infty(X)$ if

(i) there exists $Df \in L^{\infty-0}(X; H)$ such that
(ii) given $h \in H$ then $\lim_{\varepsilon \to 0} 1/\varepsilon\left[f(x + \varepsilon h) - f(x)\right]$ exists in $L^{\infty-0}(X)$ and is equal to $(Df \mid h)$.

We shall call Df the differential of f.

2.2. Metric on $\mathbb{D}_1^\infty$

We define

$$\|f\|_{\mathbb{D}_1^p} = \|f\|_{L^p(\Omega)} + \|Df\|_{L^p(\Omega; H)} .$$

The distance on $\mathcal{D}_1^\infty$ is defined by

$$d(f, f') = \sum_{j \in \mathbb{N}} 2^{-j} \left(\|f - f\|_{D_1^j} \wedge 1\right) .$$

2.3. Proposition. *$\mathbb{D}_1^\infty(\Omega)$ is an algebra.*

Proof. Given $f_1, f_2 \in \mathbb{D}_1^\infty$, obviously $f_1 + f_2 \in \mathbb{D}_1^\infty$. On the other hand

$$\varepsilon^{-1}\left[f_1(x + \varepsilon h) f_2(x + \varepsilon h) - f_1(x) f_2(x)\right] = \left[\varepsilon^{-1}\left(f_1(x + \varepsilon h) - f_1(x)\right)\right] f_2(x + \varepsilon h) + \left[\varepsilon^{-1}\left(f_2(x + \varepsilon h) - f_2(x)\right)\right] f_1(x).$$

Now as $L^{\infty-0}$ is a topological algebra we obtain convergence of the right hand side and the identity

$$D(f_1 f_2) = f_1 Df_2 + f_2 Df_1 .$$

□

2.4. Corollary. *Given a basis $\{e_k\}$ of H we denote by $\mathcal{H}$ the space of linear combinations of a finite number of Hermite polynomials. Then $\mathcal{H} \subset \mathbb{D}_1^\infty$.*

Proof. Given $h_0 \in H$ we denote by $h_0(x)$ the random variable $W(h_0)$ associated to h_0. Then

$$h_0(x + \varepsilon h) - h_0(x) = \varepsilon(h \mid h_0)$$

which proves that $h_0 \in \mathbb{D}_1^\infty$ and that

$$Dg_0 = q_0$$

where q_0 is the constant map $\Omega \to H$ defined by $x \to h_0$.

In particular, if $\{e_k\}$ denotes a chosen basis, we have $e_k \in \mathbb{D}_1^\infty$. Therefore by 2.2.3 the algebra $\mathcal{H}$ generated by the e_k is contained in $\mathbb{D}_1^\infty$. □

2.5. Theorem (of integration by parts). *For $f \in \mathbb{D}_1^\infty$*

$$E(D_h f) = E\big(W(h) f\big).$$

Proof. We choose a numerical model; we apply the Cameron-Martin Theorem I.7

$$E\left(\tau_{\varepsilon h}(f)\right) = E\left(f k_\varepsilon\right)$$

with

$$k_\varepsilon(x) = \exp\left(\varepsilon h(x) - \frac{\varepsilon^2}{2}\|h\|_H^2\right).$$

Now

$$\begin{aligned} E(D_h f) &= \lim_{\varepsilon \to 0} E\left(\varepsilon^{-1}\left[\tau_{\varepsilon h}(f) - f\right]\right) \\ &= \lim_{\varepsilon \to 0} E\left(\varepsilon^{-1}(k_\varepsilon - 1) f\right), \end{aligned}$$

then $\varepsilon^{-1}(k_\varepsilon - 1) \to W(h)$ in $L^{\infty-0}$. □

2.6. Notation. Denote by $C^1_{\text{pol}}(\mathbb{R}^d)$ the C^1-functions having polynomial growth together with its first derivatives.

Proposition (functional calculus). *Given $f_1 \ldots f_d \in \mathbb{D}_1^\infty(X)$ and $F \in C^1_{\text{pol}}(\mathbb{R}^d)$, define $g = F(f_1 \ldots f_d)$. Then $g \in \mathbb{D}_1^\infty(X)$.*

Proof.

$$Dg = \sum_{i=1}^{d} \frac{\partial F}{\partial \xi_i}(f) D f_i.$$

□

3. Krée-Meyer inequalities

3.1. Symbolic calculus on $\mathcal{L}$

By I.6.4 the operator $e^{t\mathcal{L}}$ has its spectral decomposition in $L^2(X)$ given by

$$e^{t\mathcal{L}} = \sum_{n=0}^{+\infty} e^{-nt} P_{C_n}.$$

Therefore, for any function φ bounded on the negative integers the operator

3.1.1. $$\varphi(\mathcal{L}) = \sum \varphi(-n) P_{C_n}$$

is a bounded operator on $L^2(X)$.

3.1.2. Definition (of the inverse of the Cauchy operator). The operator Λ (inverse of the Cauchy operator) is defined by

$$\Lambda = \sum (n+1)^{-1/2} P_{C_n}.$$

Proposition.

$$\Lambda = \pi^{-1/2} \int_0^{+\infty} t^{-1/2} e^{-t} P_t \, dt$$

Proof. By projection on each chaos this formula reduces to the identity

$$(n+1)^{-1/2} = \pi^{-1/2} \int_0^{+\infty} e^{-(n+1)t} t^{-1/2} dt.$$

□

3.1.3. Proposition. *For each $p \in [1, +\infty]$, Λ can be extended to an endomorphism of L^p of norm* 1.

Proof.

$$\|\Lambda\|_{\mathrm{End}(L^p)} \le \frac{1}{\pi^{1/2}} \int_0^{+\infty} e^{-t} \|P_t\|_{\mathrm{End}(L^p)} t^{-1/2} dt.$$

As $\|P_t\|_{\mathrm{End}(L^p)} \le 1$ we have the domination of the integral by 1. The norm is equal to 1 because $\Lambda \mathbf{1} = \mathbf{1}$. □

3.2. Theorem. *Given $p \in]1, +\infty[$ there exists a constant c_p such that*

$$\|\Lambda f\|_{\mathbb{D}_1^p} \le c_p \|f\|_{L^p}$$

for all $f \in \mathcal{H}$.

Proof. As Λ and D do not operate on the transverse variables we will only consider the case of an irreducible Gaussian space. We can write the Mehler formula in the form

$$\left((\cos\theta)^{-\mathcal{L}}f\right)(x) = \int f(x\cos\theta + y\sin\theta)d\mu(y).$$

Making the change of variable $t = |\log\ \cos\theta|$ in 3.1.2, we get

$$\Lambda = (\pi)^{-1/2}\int_0^{\pi/2}(\cos\theta)^{-\mathcal{L}}\,|\log\ \cos\theta|^{-1/2}\sin\theta\ d\theta.$$

As $P_t f \in \mathcal{H}$ for $f \in \mathcal{H}$ we have that $(\cos\theta)^{-\mathcal{L}}f$ belongs to $\mathcal{P}$ and therefore to $\mathbb{D}_1^\infty$. Therefore we can compute

$$\begin{aligned} I_{\theta,h}(x) = D_h^{(x)}\left[(\cos\theta)^{-\mathcal{L}}f\right](x) &= \cos(\theta)\mathbf{E}^y\left[(D_h f)(x\cos\theta + y\sin\theta)\right] \\ &= \frac{\cos\theta}{\sin\theta}\mathbf{E}^y\left[D_h^{(y)}\left(f(x\cos\theta + y\sin\theta)\right)\right] \end{aligned}$$

where $D^{(x)}$ denotes the differential in x, $D^{(y)}$ the differential in y. By integration by parts in y (see 2.5) we get

$$I_{\theta,h}(x) = \cot(\theta)\mathbf{E}^y\left(W(h)(y)f(x\cos\theta + y\sin\theta)\right).$$

Then *formally*

$$(D_h\Lambda f)(x) =$$
$$\int_0^{\pi/2}\mathbf{E}^y\left(W(h)(y)f(x\cos\theta + y\sin\theta)\right)|\log\ \cos\theta|^{-1/2}\cos\theta\frac{d\theta}{\pi^{1/2}}$$

but this integral diverges at $\theta = 0$. In order to avoid this singularity we introduce

$$\Lambda_\varepsilon f = (\pi)^{-1/2}\int_\varepsilon^{\pi/2}(\cos\theta)^{-\mathcal{L}}\,|\log\ \cos\theta|^{-1/2}\cos\theta\ d\theta.$$

Denote

$$\varphi(\theta) = \pi^{-1/2}\,|\log\ \cos\theta|^{-1/2}\cos\theta$$

$$(D_h\Lambda_\varepsilon f)(x) = \mathbf{E}^y\left(W(h)(y)\int_\varepsilon^{\pi/2} f(x\cos\theta + y\sin\theta)\varphi(\theta)d\theta\right).$$

If we change $y \to -y$ the probability $\mu(dy)$ is preserved. Therefore

$$(D_h\Lambda_\varepsilon f)(x) = \mathbf{E}^y\left[W(h)(-y)\left[\int_\varepsilon^{\pi/2} f(x\cos\theta - y\sin\theta)\varphi(\theta)d\theta\right]\right].$$

Define

$$\psi_\varepsilon(\theta) = \frac{1}{2}\varphi(\theta)\mathbf{1}(|\theta| > \varepsilon)\ \text{signum}\ (\theta).$$

We get

$$(D_h \Lambda_\varepsilon f)(x) = \mathbf{E}^y \left[W(h)(y) \left[\int_{-\pi/2}^{+\pi/2} f(x\cos\theta + y\sin\theta)\psi_\varepsilon(\theta)d\theta \right]\right].$$

Denote

$$J_\varepsilon(x,y) = \int_{-\pi/2}^{+\pi/2} f(x\cos\theta + y\sin\theta)\psi_\varepsilon(\theta)d\theta.$$

We get

$$(D_h \Lambda_\varepsilon f)(x) = (W(h) \mid J_\varepsilon(x,\cdot))_{L^2(Y)}\,.$$

Therefore

$$(D\Lambda_\varepsilon f)(x) = P_{C_y^1}(J_\varepsilon(x,y))$$

where $P_{C_y^1}$ denotes the projection onto the chaos of degree 1 in the variable y. By 1.7.5

$$\begin{aligned} \|D\Lambda_\varepsilon f(x)\|_H^p &\leq c\mathbf{E}^y\left(|J_\varepsilon(x,y)|^p\right) \\ \|D\Lambda_\varepsilon f\|_{L^p(X,H)}^p &\leq c\,\|J_\varepsilon\|_{L^p(X\times Y)}^p\,. \end{aligned}$$

Denote by R_θ the rotation operator on $X \times Y$ defined by

$$R_\theta \begin{pmatrix} x \\ y \end{pmatrix} = \begin{pmatrix} x\cos\theta & + & y\sin\theta \\ -x\sin\theta & + & y\cos\theta \end{pmatrix};$$

R_θ is a measure-preserving transformation. Define

$$u_{x,y}(\theta) = f\left(R_\theta(x,y)\right),$$

then by invariance of the measure

$$\int \left|u_{x,y}(\theta)\right|^p d\mu(x) \otimes d\mu(y)$$

is constant. Therefore

$$\int_{\mathbb{T}} \left|u_{x,y}(\theta)\right|^p d\theta < +\infty \quad \text{a.e. in } (x,y).$$

Remark now that

$$\psi_\varepsilon(\theta) = \left(\frac{1}{(\pi)^{1/2}}\cot\left(\frac{\theta}{2}\right) + r(\theta)\right)\mathbf{1}_{|\theta|>\varepsilon}$$

where $r(\theta) \in L^\infty(\mathbb{T})$. By the theory of the Hilbert transform (see the appendix at the end of this chapter) we get that for all $\varepsilon > 0$

3.2.1. $$\left\|u_{x,y} * \psi_\varepsilon\right\|_{L^p(\mathbb{T})} < c_p \|u_{x,y}\|_{L^p(\mathbb{T})} = c_p\|f\|_{L^p}$$

and $u_{x,y} * \psi_\varepsilon$ converges in L^p when $\varepsilon \to 0$. We have

$$\left(u_{x,y} * \psi_\varepsilon\right)(\theta) = \int_{\mathbb{T}} u_{R_\theta(x,y)}(\theta')\psi_\varepsilon(\theta')d\theta'.$$

Then by integrating over (x, y) in 3.2.1 we get

$$\int_{\mathbb{T}\times X\times Y} d\theta \otimes d\mu(x) \otimes d\mu(y) \left[\int_{\mathbb{T}} u_{R_\theta(x,y)}(\theta')\psi_\varepsilon(\theta')d\theta'\right]^p$$
$$\leq c \left\|u_{x,y}(\theta)\right\|^p_{L^p(X\times Y\times\mathbb{T})}.$$

By the Fubini theorem we can write the left hand side as

$$\int_{\mathbb{T}} d\theta \left[\int_{X\times Y} \left|\int_{\mathbb{T}} u_{R_\theta(x,y)}(\theta')\psi_\varepsilon(\theta')d\theta'\right|^p d\mu(x) \otimes d\mu(y)\right].$$

But by invariance of the measure $\mu(dx) \otimes \mu(dy)$ the expression in brackets is constant in θ. We can therefore *drop* the integration on θ and get again the *same majoration*

$$\int_{X\times Y} \left|\int_{\mathbb{T}} u_{x,y}(\theta')\psi_\varepsilon(\theta')d\theta'\right|^p d\mu(x) \otimes d\mu(y) \leq c_p \|f\|^p_{L^p}$$

or

$$\|D\Lambda_\varepsilon f\|^p \leq c_p \|f\|^p_{L^p}.$$

Furthermore as $f \in \mathcal{P}$, when

$$\|\Lambda_\varepsilon f - \Lambda f\|_{L^p} \to 0, \qquad \|D\Lambda_\varepsilon f - D\Lambda f\|_{L^p} \to 0$$

for all p. Therefore

$$\|D\Lambda f\|_{L^p} \leq c_p \|f\|_{L^p}$$

which is the difficult majoration. Finally by 3.1.3

$$\|\Lambda f\|_{L^p} \leq c_p \|f\|_{L^p}.$$

This last inequality establishes 3.2. □

Remark. In the theory of singular integrals on $\mathbb{R}^d$, Calderon-Zygmund have treated odd convolution kernels of homogeneity degree $-d$ as continuous integrals of the Hilbert-transform taken on all straight lines fibering $\mathbb{R}^d$.

The preceding proof (which is due to G. Pisier) consists in fibering the product of a Gaussian space with itself through a family of circles; again the Hilbert transform on each of those circles appears as the corner-stone of the desired estimates.

We shall now reverse the inequality 3.2.

3.3. Theorem. *Given* $p \in]1, +\infty[$

$$\|f\|_{L^p} \leq c_q \|\Lambda f\|_{D^p_1}$$

for all $f \in \mathcal{P}$, *where* q *denotes the conjugate exponent of* p *and where* c_q *has been defined in 2.3.2.*

Proof. The proof will be based on a duality argument which will be prepared by the next two propositions. □

3.3.1. Proposition. *If $F, G \in \mathcal{P}$ then*

$$(-\mathcal{L}F \mid G)_{L^2} = \int \big((DF)(x) \mid (DG)(x)\big)_H d\mu(x) = (DF \mid DG)_{L^2(X;H)}.$$

Proof. We choose a basis $\{e_k\}$ of H and the associated numerical model. Then we take for F and G two Hermite polynomials H_p and H_q

$$\left(-\mathcal{L}\, H_p \mid H_q\right) = |p| \left(H_p \mid H_q\right) = |p| \delta_q^p \|H_p\|_{L^2}^2$$

$$\begin{aligned} DH_p &= \sum_k p(k) H_{p-\varepsilon_k} e_k \\ DH_q &= \sum q(k') H_{q-\varepsilon_{k'}} e_{k'} \end{aligned}$$

$$\left(DH_p \mid DH_q\right)(x) = \left(\sum_k p(k) q(k) H_{p-\varepsilon_k} H_{q-\varepsilon_k}\right)(x)$$

$$\left(DH_p \mid DH_p\right)_{L^2(X;H)} = \sum \left(p(k)\right)^2 \left\|H_{p-\varepsilon_k}\right\|_{L^2}^2$$

$$\left\|H_{p-\varepsilon_k}\right\|_{L^2}^2 = (p - \varepsilon_k)! = p! \frac{1}{p(k)},$$

then

$$\left(DH_p \mid DH_p\right)_{L^2(X;H)} = \left(\sum_k p(k)\right) \left\|H_p\right\|_{L^2}^2 = |p| \left\|H_p\right\|_{L^2}^2 .$$

By linearity the theorem is proved for *finite linear combinations* of Hermite polynomials. These combinations are dense in C_n. As $\mathcal{L}$ is a bounded operator in $L^2(C_n)$ we get the lemma. □

3.3.2. Corollary. *The norm $\mathbb{D}_1^2$ defines on $\mathcal{P}$ the structure of an inner product space.*

Proof. The bilinear form associated to the $\mathbb{D}_1^2$ norm is

$$\langle f, g\rangle = (Df \mid Dg)_{L^2(X,H)} + (f \mid g)_{L^2(X)}.$$

□

3.3.4. *Proof of Theorem 3.3.*

Let $f, g \in \mathcal{P}$, then $f, g \in L^2$. Define $F = \Lambda g$, $G = \Lambda g$. Then $F, G \in \mathcal{P}$. Using the spectral decomposition in L^2, Λ^{-1} is well defined on $\mathcal{P}$. We have

$$(f \mid g)_{L^2} = \left(\Lambda^{-1} F \mid \Lambda^{-1} G\right) = \left(\Lambda^{-2} F \mid G\right).$$

We remark that $\Lambda^{-2} = -\mathcal{L}+1$; as this can be seen from the chaos decomposition. Therefore by 3.3.1

$$\begin{aligned}(f \mid g) &= (-\mathcal{L}F \mid G) + (F \mid G) = (DF \mid DG) + (F \mid G)\\ (f \mid g) &= (D\Lambda f \mid D\Lambda g) + (\Lambda f \mid \Lambda g).\end{aligned}$$

Now

$$\|f\|_{L^p} = \sup_{\|g\|_{L^q}\le 1} (f \mid g).$$

Theorem 3.2 gives us

$$\|D\Lambda g\|_{L^q} + \|\Lambda g\|_{L^q} \le c_q \qquad \text{if} \quad \|g\|_{L^q} \le 1.$$

Therefore

$$\|f\|_{L^p} \le c_q\,(\|D\Lambda f\|_{L^p} + \|\Lambda f\|_{L^p}) \le c_q \|\Lambda f\|_{L^p}.$$

□

3.4. Definition. $\mathbb{D}_1^p(X)$ will be the completion of $\mathbb{D}_1^\infty(X)$ for the norm $\mathbb{D}_1^p$.

3.5. Main theorem. 3.5.1. *The operator Λ realizes a Banach space isomorphism between $\mathbb{D}_1^p$ and L^p.*
3.5.2. *$\mathcal{P}$ is dense in $\mathbb{D}_1^p$.*

Proof. We denote by $\tilde{\mathbb{D}}_1^p$ the completion of $\mathcal{P}$ for the norm $\| \quad \|_{\mathbb{D}_1^p}$. Then $\tilde{\mathbb{D}}_1^p$ is a Banach space. Just as the operator Λ realizes a norm preserving bijection of $\mathcal{P} \subset \tilde{\mathbb{D}}_1^p$ with $\Lambda\mathcal{P} \subset L^p$, Λ realizes a Banach isomorphism of $\tilde{\mathbb{D}}_1^p$ with the closure of $\Lambda\mathcal{P}$ in L^p. But $\Lambda\mathcal{P} = \mathcal{P}$ and $\mathcal{P}$ is dense in L^p. Then Λ realizes a Banach space isomorphism between $\tilde{\mathbb{D}}_1^p$ and L^p.

We will prove 3.5.2 which will imply $\mathbb{D}_1^p = \tilde{\mathbb{D}}_1^p$ and will establish 3.5.1. On $\mathbb{D}_1^\infty$ we consider the $\mathbb{D}_1^2$-bilinear form

$$\rho(f, g) = E(DF \mid Dg) + E(fg).$$

Then this bilinear form extends to $\mathbb{D}_1^p \times \mathbb{D}_1^q$ where $p^{-1} + q^{-1} = 1$.

3.5.3. Lemma. *If $f \in \mathbb{D}_1^\infty$, $g \in \mathcal{P}$, then*

$$\rho(f, g) = E\big((-\mathcal{L}g + g)f\big)$$

where $\mathcal{L}$ is the Ornstein-Uhlenbeck operator.

Proof. As $f, g \in D_1^\infty$, then $fg \in D_1^\infty$. Furthermore $Dg \in \mathbb{D}_1^\infty(X; H)$. Then we choose $\{e_k\}$ to be an orthonormal basis of H

$$E\left(D_{e_k}\left(f D_{e_k} g\right)\right) = E\left(f W(e_k) D_{e_k} g\right) = E\left(D_{e_k} f \; D_{e_k} g\right) + E\left(f D_{e_k}^2 g\right).$$

(The second derivative of a polynomial is computed according to the formula I.3.1.1). Then we sum over k. For only a finite number of indices is $D_{e_k} g$ non-zero. Therefore the summation involves only a finite number of indices and can be performed. □

3.5.4. Corollary. *Given $f \in \mathbb{D}_1^p$, assume that $\rho(f,g) = 0$ for all $g \in \mathcal{P}$, then $f = 0$.*

Proof. $\rho(f,g) = 0 = E\big(f(-\mathcal{L}g + h)\big)$.

Let $g = \sum c_p H_p$, then $-\mathcal{L}g + g = \sum c_p(|p|+1)H_p$. Therefore $(-\mathcal{L}+1)\mathcal{P} = \mathcal{P}$. Then f orthogonal to all polynomials, $f \in L^p$ implies $f = 0$. □

3.5.5. *Proof of 3.5.2.* We define a map $j : f \to \ell_f$, $j : \mathbb{D}_1^p \to \left(\tilde{\mathbb{D}}_1^q\right)^*$ by

$$\ell_f(k) = \rho(f,k) \qquad k \in \tilde{\mathbb{D}}_1^q.$$

By 3.5.4, j is *injective*. Write Λ_p for the map $L^p \to \tilde{\mathbb{D}}_1^p$ defined by $f \to \Lambda f$. Then by duality $\Lambda_q^* \left(\tilde{\mathbb{D}}_1^q\right)^* \to L^p$. Then $\Lambda_p \Lambda_q^*$ gives an isomorphism of $\left(\tilde{\mathbb{D}}_1^q\right)^*$ onto $\tilde{\mathbb{D}}_1^p$. Consider $j_1 = \Lambda_p \circ \Lambda_q^* \circ j$. Then j_1 gives an *injective* map of $\mathbb{D}_1^p \to \tilde{\mathbb{D}}_1^p$. Finally denote by s the natural injection of $\tilde{\mathbb{D}}_1^p \to \mathbb{D}_1^p$; it remains to show that $j_1 \circ s =$ identity. As $j_1 \circ s$ is continuous it is sufficient to show that for every polynomial u, $j_1 u = u$.

We associate to u the linear form $\ell_u = j(u)$; we have

$$\ell_u(f) = \rho(u,f).$$

Denote

$$\Lambda_q^* \ell_u = z$$

where z is defined by

$$E(zk) = f(u, \Lambda k).$$

We want to prove that $\Lambda z = u$ which is equivalent to

$$E\big((\Lambda u)k\big) = \rho(u, \Lambda k)$$

or

$$\rho(u, \Lambda k) = \big((-\mathcal{L}+1)u \mid \Lambda k\big).$$

As $\Lambda = (-\mathcal{L}+1)^{-1/2}$ is an Hermitian operator, we get by transposition

$$\rho(u, \Lambda k) = \big((-\mathcal{L}+1)^{1/2} u \mid k\big) = E\big((\Lambda u)k\big).$$

□

3.5.6. Corollary. *The space D_1^p is a space of functions which can be identified with a dense subspace V of L^p. Furthermore the operator D is closable in L^p and its domain is V.*

Proof. A Cauchy sequence u_n of polynomials for the D_1^p norms can be written as $u_n = \Lambda f_n$, where f_n is a Cauchy sequence for the L^p norm, which therefore converges in L^p to f_∞. As Λ is a bounded operator in L^p, this implies that u_n is a Cauchy sequence in L^p and converges to u_∞; in this way we have constructed a map $D_1^p \to L^p$. This map is injective: by duality the orthogonal complement of its kernel will be the image of the operator Λ operating on L^q; as Λ is injective on polynomials, this image contains all the polynomials; therefore the image of L^q by Λ is dense in L^q, this fact establishes the injectivity on L^p.

The closability of D means that if the Cauchy sequence in D_1^p of polynomials u_n satisfies $u_\infty = 0$, then $Du_n \to 0$: in fact by the injectivity of Λ we have $f_\infty = 0$, so $\|f_n\|_{L^p} \to 0$, and since Λ realizes an isomorphism $\|u_n\|_{D_1^p} \to 0$.

The closability of D means that D is a well-defined operator on V. □

3.6. Extension to Hilbert-valued functions

3.6.1. Tensor product of two Hilbert spaces

(i) *Case of L^2 spaces*

Consider two probability spaces X_1, X_2. Denote $X = X_1 \times X_2$.

Definition. Given two closed subspaces $V_1 \subset L^2(X_1)$, $V_2 \subset L^2(X_2)$, then

$V_1 \otimes V_2 =$ closure in $L^2(X)$ of the vector space generated by $\{v_1(x_1)v_2(x_2)\}$ $v_i \in V_i$.
$\| \ \|_{V_1 \otimes V_2} =$ norm induced by $L^2(X)$.

Example. Take $V_1 = L^2(X_1)$, $V_2 = L^2(X_2)$, then

$$L^2(X_1) \otimes L^2(X_2) = L^2(X).$$

(ii) *Associated endomorphism*

If $K \in L^2(X)$, then K defines a homomorphism of $L^2(X_1)$ into $L^2(X_2)$ by the formula

$$(A_K f_1)(x_2) = \int_{X_1} K(x_1, x_2) f(x_1) P(dx_1).$$

Then we have the following classical *Hilbert-Schmidt* result.

Proposition.

$$\text{Trace}\left(A_K^* A_K\right) = \int [K(x_1, x_2)]^2 \, P(dx_1) P(dx_2)$$

and

$$L^2(X_1) \otimes L^2(X_2) \simeq \left\{A : L^2(X_1) \to L^2(X_2); \quad \text{trace } A^* A < +\infty\right\}.$$

(iii) *Case of an abstract Hilbert space*

Given G_1, G_2 two abstract Hilbert spaces, we construct their *Segal* models $\mathrm{Seg}(G_1) = X_1$, $\mathrm{Seg}(G_2) = X_2$. Then $G_1 \to \mathbf{G_1} \subset L^2(X_1)$, $G_2 \to \mathbf{G_2} \subset L^2(X_2)$. We define $G_1 \otimes G_2$ as $\mathbf{G}_1 \otimes \mathbf{G}_2$ where this last product has been defined in (i).

We have again

Proposition.

$$G_1 \otimes G_2 \cong \left\{A : G_1 \to G_2; \quad \text{trace } (A^*A) < +\infty\right\}.$$

Notation.

$$\|A\|_{H.S}^2 := \quad \text{trace} \quad (A^*A).$$

(iv) Proposition. *Given $A : G_1 \to G_2$, denote by $\{e_k\}$ an orthonormal basis of G_1, then*

$$\|A\|_{H.S}^2 = \|A\|_{G_1 \otimes G_2}^2 = \sum_k \|A(e_k)\|_{G_2}^2 .$$

Proof. Take $f_1 \ldots f_q \ldots$ to be an orthonormal basis of G_2, then

$$\sum_k \|A(e_k)\|_{G_2}^2 = \sum_{k,s} (Ae_k \mid f_s)^2$$

$$a_s^k = (Ae_k \mid f_s) .$$

Then

$$a_k^s = \left(A^* e_k \mid f_s\right)$$

and

$$\text{trace } A^*A = \sum_{k,s} a_k^s \, a_k^s.$$

□

3.6.2. Gradient of a Hilbert valued function

Given a separable Hilbert space G, denote $Y = \mathrm{Seg}(G)$ its Segal model. We have defined in 1.6 a map

$$j : L^p(X; G) \to L^p(X \times Y).$$

This map j realizes the following identification

$$L^p(X; G) \simeq \left\{u \in L^p(X \times Y); \quad u(x, \cdot) \in C_1^y \ \text{ for almost all } \ x\right\}$$

where we denote by C_1^y the chaos of degree 1 in the variable y.

(i) Definition.

$$\mathbb{D}_1^\infty(X;G) = \{u \in D_1^\infty(X \times Y); \quad u(x,\cdot) \in \mathcal{C}_1^y\}$$

(ii) Given $u \in D_1^\infty(X \times Y)$ and $h \in H$ then the derivative $D_h u \in L^{\infty-0}(X \times Y)$. Choose an orthonormal basis e_k of H, denote $\partial_k = D_{e_k}$, then

$$\|Du\|_H^p = \left[\sum_k (\partial_k u)^2\right]^{p/2}.$$

Therefore

$$\|Du\|_{L^p(X\times Y)}^p = \int_X P(dx) \int_Y \left(\sum (\partial_k u)^2(x,y)\right)^{p/2} P(dy).$$

But

$$(\partial_k u)(x,\cdot) \in \mathcal{C}_1^y.$$

As the L^2-norm and the L^p-norm are equivalent on $\mathcal{C}_1^y$, we have

$$\|Du\|_{L^p(X\times Y)}^p \approx \int_X P(dx) \left[\int_Y \sum [(\partial_k u)(x,y)]^2 \, P(dy)\right]^{p/2}.$$

We have therefore proved

(iii) Lemma. *Let $f \in \mathbb{D}_1^\infty(X;G)$, then*

$$\|Df\|_{L^p(X\times Y)} \approx \int_X \|Df\|_{H\otimes G}^p(x) P(dx).$$

(iv) Theorem. *$f \in \mathbb{D}_1^\infty(X,G)$ if and only if $f \in L^{\infty-0}(X;G)$ and there exists $Df \in L^{\infty-0}(X; H\otimes G)$ such that $\lim_{\varepsilon\to 0} \varepsilon^{-1}\big(f(x+\varepsilon h)-f(x)\big)$ exists in $L^{\infty-0}(X;G)$ and is equal to $\big(Df(x)\big)(h)$.*

Remark. For simplicity we shall denote $\big(Df(x)\big)(h)$ by $(Df \mid h)(x)$.

Remark. We could have taken (iv) as a definition of $\mathbb{D}_1^\infty(X;G)$. We have prefered to take the definition (i) in order to show the *naturality* of the Hilbert-Schmidt norm $H \otimes G$.

(v) Definition of the norm $\mathbb{D}_1^p(X;G)$.

$$\|f\|_{\mathbb{D}_1^p(X;G)}^p = \|f\|_{L^p(X;G)}^p + \|Df\|_{L^p(X;H\otimes G)}^p.$$

3.6.3. Theorem. *For $p \in]1,+\infty[$*

$$\|\Lambda f\|_{\mathbb{D}_1^p(X;G)} \approx \|f\|_{L^p(X;G)}.$$

Proof. Consider $f \to \mathbf{f} \in L^p(X \times Y)$ and apply Theorem 3.5 to $\mathbf{f}$. □

4. Differentials of higher order

4.1. Definition. We shall proceed by induction on the order of r of differentiation.
Order 2:

$$\mathbb{D}_2^p(X;G) \doteq \{f \in \mathbb{D}_1^p(X;G); \quad Df \in \mathbb{D}_1^p(X; H \otimes G)\}$$

$$D^2 f = D(Df), \qquad \text{and} \quad D^2 f(x) \in H \otimes H \otimes G.$$

Order r:

$$\mathbb{D}_r^p(X;G) = \{f \in \mathbb{D}_{r-1}^p(X;G); \quad D^{r-1} f \in \mathbb{D}_1^p\left(X; H^{\otimes(r-1)} \otimes G\right)\}$$

$$D^r f = D\left(D^{r-1} f\right), \qquad \text{and} \quad D^r f(X) \in H^{\otimes r} \otimes G.$$

4.2. Definition of a norm

$$\|f\|_{\mathbb{D}_r^p} = \|f\|_{D_{r-1}^p} + \|D^{r-1} f\|_{\mathbb{D}_1^p(X; H^{\otimes(r-1)} \otimes G)}.$$

We get by a straightforward computation involving the binomial coefficients

$$\|f\|_{\mathbb{D}_r^p} = \sum_{s=0}^{r} \binom{r}{s} \|D^s f\|_{L^p}.$$

4.3. Theorem (Krée-Meyer inequalities). *For* $p \in]1, +\infty[$ *there exists* $c = c(p,r)$ *such that*

$$c^{-1}\|f\|_{L^p} \le \|\Lambda^r f\|_{\mathbb{D}_r^p} \le c\|f\|_{L^p}.$$

Proof. In order to realize an easy induction on r, we shall introduce the following operator:

$$\Gamma : \mathbb{D}_1^p(X;G) \to L^p\big(X; G \otimes (H \otimes G)\big)$$

defined by

$$\Gamma f = \begin{pmatrix} f \\ Df \end{pmatrix}.$$

4.3.1. Lemma.

$$\frac{1}{c}\|f\|_{\mathbb{D}_r^p} \le \|\Gamma^r f\|_{L^p} \le c\|f\|_{\mathbb{D}_r^p}.$$

Proof. Consider the case $r = 1$

$$\|\Gamma f\|_{L^p} = \left\| \left(\|f\|_G^2 + \|Df\|_{H\otimes G}^2\right)^{1/2} \right\|_{L^p} \leq 2^{1/2} \left(\|f\|_{L^p} + \|Df\|_{L^p}\right)$$
$$\|\Gamma f\|_{L^p}^p \leq \frac{1}{2} \left(\|f\|_{L^p}^p + \|Df\|_{L^p}^p\right).$$

Proceed by induction on r. Assume the lemma to be true for $s < r$. We have

$$\|\Gamma^r f\|_{L^p} = \|\Gamma u\|$$

with $u = \Gamma^{r-1} f$. Then

$$\|\Gamma u\|_{L^p} \sim \|u\|_{\mathbb{D}_1^p}$$
$$\|u\|_{\mathbb{D}_1^p} = \|u\|_{L^p} + \|Du\|_{L^p}.$$

By the inductive hypothesis

$$\|u\|_{L^p} = \|\Gamma^{r-1} f\|_{L^p} \sim \|f\|_{\mathbb{D}_{r-1}^p}.$$

On the other hand

$$\|Du\|_{L^p} = \|D\Gamma^{r-1} f\|_{L^p}.$$

We have $D\Gamma^{r-1} = \Gamma^{r-1}D$, therefore

$$\|Du\|_{L^p} = \|\Gamma^{r-1} Df\|_{L^p} \simeq \|Df\|_{\mathbb{D}_{r-1}^p},$$

then

$$\|\Gamma^r f\|_{L^p} \approx \|f\|_{D_{r-1}^p} + \|Df\|_{\mathbb{D}_{r-1}^p} = \|f\|_{\mathbb{D}_r^p}.$$

□

4.3.2. Multiplier lemma. *Consider the following operator M on L^2:*

$$M = \left(\frac{-\mathcal{L}+1}{-\mathcal{L}+2}\right)^{1/2} = \sum_{n=0}^{+\infty} \left(\frac{n+1}{n+2}\right)^{1/2} P_{C_n}.$$

Then M defines a bounded invertible operator on L^p for $p \in]1, +\infty[$.

Proof.

$$M^{-1} = \left(1 + \frac{1}{-\mathcal{L}+1}\right)^{1/2}.$$

By the binomial series we have

$$\Phi(\zeta) = (1+\zeta)^{1/2} = \sum c_n \zeta^n \quad \text{for} \quad |\zeta| < 1.$$

An easy application of Stirling's formula shows that

$$\sum |c_n| < +\infty.$$

On the other hand

$$\left(\frac{1}{-\mathcal{L}+1}\right)^n = \frac{1}{(n-1)!}\int_0^{+\infty} P_t t^{n-1} e^{-t} dt$$

$$\|M^{-1} f\|_{L^p} \le \sum |c_n| \frac{1}{(n-1)!}\int_0^{+\infty} \|P_t f\|_{L^p} t^{n-1} e^{-t} dt = \left(\sum |c_n|\right) \|f\|_{L^p}$$

$$M^{-1} = \left(\frac{-\mathcal{L}+1}{-\mathcal{L}+2}\right)^{-1/2} = \left(1 - \frac{1}{-\mathcal{L}+2}\right)^{-1/2} = \Phi\left(-\frac{1}{\mathcal{L}+2}\right).$$

We have as before

$$\|Mf\|_{L^p} \le$$
$$\left(\sum_n |c_n| \frac{1}{(n-1!)}\int_0^{+\infty} e^{-2t} t^n dt\right) \|f\|_{L^p} = \|f\|_{L^p} \sum_n |c_n| 2^{-n}.$$

□

Proof of Theorem 4.3.

Given a separable Hilbert space G_1 we denote

$$\mathcal{L}^p(X; G_1) \doteq L^p(X, G_1) \oplus L^p(X; H \otimes G_1).$$

We want to construct $A_k \in \operatorname{End}(\mathcal{L}^p(X, G_1))$ such that

(i) $$\Lambda^{-k} \Gamma \Lambda^k = A_k \Gamma$$

We first construct A_k on the polynomials. Using the chaos decomposition we see that

$$A_k = \begin{pmatrix} \text{Identity} & 0 \\ 0 & M^k \end{pmatrix}$$

satisfies identity (i). Now define

$$\begin{aligned} B_n &= \Gamma^n \Lambda^n = \Gamma^{n-1} \Gamma\ \Lambda^{n-1} \Gamma \\ &= \Gamma^{n-1} \Lambda^{n-1} A_{n-1} B_1 = B_{n-1} A_{n-1} B_1. \end{aligned}$$

We get

$$B_n = B_1 A_1 B_1 A_2 B_1 A_3 B_1 \ldots B_1 A_{n-1} B_1.$$

By Theorem 3.5, B_1 is a bounded operator for the L^p norm. It is the same by 4.3.2 for A_k, therefore

$$\|B_n f\|_{L^p} \le c \|f\|_{L^p}.$$

Now we want to show

(ii) $$\|B_n f\|_{L^p} > c \|f\|_{L^p}.$$

We shall prove it by induction on n. Granted Theorem 3.5, the first step $n = 1$ is true. Assume that (ii) is true for $n < r$. We have

$$B_r = B_{r-1} A_{r-1} B_1 .$$

Then

$$\|B_r f\|_{L^p} = \|B_{r-1} A_{r-1} B_1 f\|_{L^p} \geq c \, \|A_{r-1} B_1 f\|_{L^p} .$$

By 4.3.2, A_{r-1} is invertible, therefore

$$\|B_r f\|_{L^p} \geq c \|B_1 f\|_{L^p} \geq c \|f\|_{L^p} .$$

□

4.4. Corollary. *The spaces $\mathbb{D}_r^p$ are Banach spaces* ($p \in 1, +\infty[, \quad r \in \mathbb{N}$).

Proof. We have to prove that $\mathbb{D}_r^p$ is complete for the corresponding norm. But Λ^r realizes a Banach isomorphism of L^p with D_r^p; the result follows from the completeness of L^p. □

4.5. Smooth functions

We shall denote

$$\mathbb{D}_\infty(X) = \bigcap_{p,r} \mathbb{D}_r^p(X).$$

In this intersection $(p, r) \in \mathbb{N}^2$. We have on $\mathbb{N}^2$ the order relation $(p, r) < (p', r')$ if $p < p'$ and $r < r'$. Then $(p, r) < (p', r')$ implies $D_r^p \supset D_{r'}^{p'}$, and the domination of the first norm by the second. Then the construction developed in Section 1 gives rise to a metric on $\mathbb{D}_\infty(X)$. Then $\mathbb{D}_\infty(X)$ *is a complete metric space.*

5. Cylindrical approximations

5.1. Theorem. *Given an irreducible Gaussian probability space $X = (\Omega, H)$ and an increasing sequence V_n of closed subspaces of H, such that $\bigcup_n V_n = H$, denote by $\mathcal{A}_{V_n}$ the σ-field generated by V_n. Given $g \in \mathbb{D}_r^p(X)$, define $g_n = E^{\mathcal{A}_{V_n}}(g)$, then*

$$g_n \in \mathbb{D}_r^p(X)$$

$$\|g_n - g\|_{\mathbb{D}_r^p} \to 0.$$

5.2. Lemma. *Given a subspace V of H, denote by P_v the orthogonal projection of H onto V. Then we have*

$$D \, E^{\mathcal{A}_V} = P_V E^{\mathcal{A}_V} D$$

and the expression on the left exists whenever that on the right does.

Proof. We have to prove that, given $f \in \mathbb{D}_1^p(X)$, then $\left(E^{\mathcal{A}_V} f\right) \in \mathbb{D}_1^p$ and that we have

$$D\left(E^{\mathcal{A}_V} f\right) = P_V\left(E^{\mathcal{A}_V}(Df)\right).$$

The orthogonal decomposition $H = V \oplus V^{\perp}$ implies that

$$\operatorname{Seg}(H) = \operatorname{Seg}(V) \times \operatorname{Seg}(V^{\perp}).$$

By irreducibility, $X = \operatorname{Seg}(H)$, and $x = (y, z)$ with $y \in \operatorname{Seg}(V)$, $z \in \operatorname{Seg}(V^{\perp})$. Then

$$\left(E^{\mathcal{A}_V} f\right)(y) = \int f(y, z) P(dz)$$

$$h = v + w, \quad v \in V, \quad w \in V^{\perp}.$$

The Cameron-Martin representation ρ restricted to $V^{\perp}$ acts as the identity on the variable y. Therefore

$$\rho(w) E^{\mathcal{A}_V} f = E^{\mathcal{A}_V} f$$

which implies

$$D_w\left(E^{\mathcal{A}_V} f\right) = 0.$$

Now

$$\rho(v)\left(E^{\mathcal{A}_V} f\right)(g) = \int \rho(v) f(y, z) P(dz)$$

or

$$D_v\left(E^{\mathcal{A}_V} f\right)(y) = \int (D_v f)(y, z) P(dz).$$

5.3. *Proof of Theorem 5.1.*

We shall prove by induction on r. For $r = 1$ the existence of the following limits

$$E^{\mathcal{A}_n}(g) \to g, \quad E^{\mathcal{A}_n}(Dg) \to Dg, \quad P_{V_n}\left(E^{\mathcal{A}_n}(Dg)\right) \to Dg$$

combined with 5.2 proves the theorem.

Assume the theorem to be true for $s < r$. Given $f \in \mathbb{D}_r^p$, we denote $u = D^{r-1} f$. Then, as $u \in \mathbb{D}_1^p$, we have

$$\left\| E^{\mathcal{A}_n} u - u \right\|_{\mathbb{D}_1^p} \to 0.$$

Furthermore by 5.2

$$P_{V_n}^{\otimes r-1}\left(E^{\mathcal{A}_n}\left(D^{(r-1)} f\right)\right) = D^{(r-1)}\left(E^{\mathcal{A}_n}(f)\right).$$

As $P_{V_n}^{\otimes r}$ tends to the identity, we deduce that

$$\left\| D^r\left(E^{\mathcal{A}_n} f\right) - f \right\|_{L^p} \to 0.$$

The other derivatives of order $< r$ converge by the inductive hypothesis. □

5.4. Proposition (smoothness of $\mathcal{A}_V$). *Given $X = (\Omega; H)$ an irreducible Gaussian probability space and given V a finite-dimensional subspace of H, then there exists a map*

$$j : \Omega \mapsto V, \qquad j \in D_r^{\infty-0}(X; V)$$

such that if $f \in D_r^p(X; V)$ satisfies

$$E^{\mathcal{A}_V} f = f$$

then there exists $u \in \mathbb{D}_r^p(V)$ such that

$$f = u \circ j.$$

Proof. As in 5.2, the splitting $H = V \oplus V^{\perp}$ induces a product decomposition:

$$X = \mathrm{Seg}(V) \times \mathrm{Seg}(V^{\perp}) = Y \times Z.$$

As $\dim V < +\infty$, we have

$$\mathrm{Seg}(V) = \left(V, \mathcal{B}_V, \exp\left(-\frac{|v|^2}{2}\right) \frac{dv_1 \ldots dv_d}{(2\pi)^{d/2}} \right).$$

Then we use Proposition 5.2. □

5.5. Proposition. *If* $\dim V < +\infty$, $Y = \mathrm{Seg}(V)$, *then*

$$\mathbb{D}_r^p(V) \subset W_r^{p,\mathrm{loc}}(V)$$

where $W_r^p(V)$ denotes the usual Sobolev space of V.

As a consequence, if $f \in \mathbb{D}_\infty(V)$ then f is a C^∞ function, all the derivatives $D^1, D^2 \ldots D^k$, are expressible by the partial derivatives of f.

Proof. Obvious.

5.6. Proposition. *Let $g \in D_2^p(X)$, then the second derivative is symmetric:*

$$D_{h_1}\left(D_{h_2} g\right) = D_{h_2}\left(D_{h_1} g\right).$$

Proof. Consider V_n an increasing sequence of finite-dimensional subspaces of H, such that $\overline{\bigcup V_n} = H$. Define

$$g_n = E^{\mathcal{A}_{V_n}}(g).$$

Then

$$\lim_{n\to\infty} D_{h_1}\left(D_{h_2} g\right) = D_{h_1}\left(D_{h_2} g\right).$$

As

$$g_n = j_n \circ u_n \qquad\qquad u_n \in W_r^{p,\mathrm{loc}}(V_n).$$

We have the finite-dimensional calculus

$$D_{v_1^n} D_{v_2^n} u_n = D_{v_2^n} D_{v_1^n} u_n$$

when $v_1^n = P_{V_n}(h_1)$, $v_2^n = P_{V_n}(h_2)$, which implies

$$D_{h_1}\left(D_{h_2} g_n\right) = D_{h_2} D_{h_1}(g_n).$$

□

5.7. Proposition (Symbolic calculus). *Given $\psi \in C^\infty(\mathbb{R}^d)$, ψ of polynomial growth in all its derivatives, then for every system $u^k \in \mathbb{D}_\infty(X)$, $k \in [1,d]$, the function*

$$v(x) = \psi\left(u^1(x), \dots, u^k(x)\right)$$

satisfies

$$v \in \mathbb{D}_\infty(X).$$

Proof. Denote by V_n an increasing family of finite-dimensional subspaces of H and write $u_n^k = E^{\mathcal{A}_{V_n}}(u^k)$, $v_n = E^{\mathcal{A}_{V_n}}(v)$. Then

$$v_n = \psi\left(u_n^1, \dots, u_n^k\right).$$

Therefore by the finite-dimensional calculus $v_n \in C^\infty(V_n)$. By the polynomial growth of the derivative of ψ we have that for given p, r, there exists p' and c such that

$$\|v_n - v_{n'}\|_{\mathbb{D}_r^p} < c \sup_k \left\|u_n^k - u_{n'}^k\right\|_{\mathbb{D}_r^{p'}}.$$

Therefore v_n converges in $\mathbb{D}_\infty$. □

5.8. Corollary. *The space D_∞ is an algebra.*

Remark. The fact that the C^∞ functions defined on $\mathbb{R}^d$ constitute an algebra plays a basic role in calculus; in order to recapture the analogous fact 5.8 we have to work with the hierarchy of all the L^p spaces; in the framework of a theory limited to L^2 we could not obtain this algebraic structure. From a more naïve point of view, 5.8 means that the product of two *smooth random variables* is a smooth random variable.

6. The divergence operator

6.1. Definition. Given $(X; H)$ a Gaussian probability space, given G a separable Hilbert space, given $u \in L^p(X; H \otimes G)$ we shall say that D^*u exists if there exists $v \in L^p(X; G)$ such that

$$E\left((Df \mid u)_{H\otimes G}\right) = E\left((f \mid v)_G\right) \quad \text{for all} \quad f \in D_\infty(X; G).$$

We shall denote v by D^*u.

Remark. The splitting of the range of u as a tensor product containing H as factor is essential in this definition.

6.2. Theorem. *Assume that $u \in \mathbb{D}_1^p(X; H \otimes G)$, then D^*u exists, $(D^*u) \in L^p(X; G)$ and $\|D^*u\|_{L^p} < c_p\|u\|_{\mathbb{D}_1^p}$.*

Proof. We shall need some results on the algebra generated by our basic operators.

6.3. Proposition (Commutation relation).

$$\begin{aligned} DP_t &= e^{-t}P_tD \\ D\Lambda^{-1} &= M\Lambda^{-1}D \end{aligned}$$

where M is the multiplier defined in 4.3.2.

Proof. By Mehler's formula

$$\begin{aligned} (P_tf)(x) &= E^y\left(f\left(xe^{-t}+\left(1-e^{-2t}\right)^{1/2}y\right)\right) \\ (DP_tf)(x) &= e^{-t}E^y\left((Df)\left(xe^{-t}+\left(1-e^{-2t}\right)y\right)\right)=e^{-t}P_tDf. \end{aligned}$$

□

If we multiply each side of the second identity by Λ on the right, we get

$$D=M\Lambda^{-1}D\Lambda.$$

We shall check this formula on the Hermite polynomial associated to a numerical model H_p with $|p|=n$, i.e. $H_p\in\mathcal{C}_n$. Then

$$\Lambda H_p=(n+1)^{-1/2}H_p$$

$$D\Lambda h_p\in\mathcal{C}_{n-1}$$

implies

$$\Lambda^{-1}(D\Lambda H_p)=n^{1/2}.$$

Therefore

$$\tilde{M}\mid_{\mathcal{C}_{n-1}}=\left(\frac{n+1}{n}\right)^{-1/2}$$

which means

$$\tilde{M}\mid_{\mathcal{C}_n}=\left(\frac{n+2}{n+1}\right)^{-1/2}.$$

□

6.4. Theorem (Riesz transform). *There exists an operator*

$$\mathcal{R}:L^p(X)\to L^p(X;H)$$

such that

$$\|\mathcal{R}f\|_{L^p(X,H)}<c\|f\|_{L^p(X)}$$

for which we have

$$D=\mathcal{R}\Lambda^{-1}.$$

Furthermore

$$D=\Lambda^{-1}M\mathcal{R}.$$

Proof. Define

$$\text{(i)} \qquad \mathcal{R} = D\Lambda.$$

Then by the Krée-Meyer inequality

$$\|\mathcal{R}f\|_{L^p} \le c\|f\|_{L^p}.$$

Multiplying (i) by Λ^{-1} on the right we get

$$D = \mathcal{R}\Lambda^{-1}.$$

Now by 6.3

$$D\Lambda^{-1} = M\Lambda^{-1}D$$

or

$$D = M\Lambda^{-1}\mathcal{R}.$$

As M and Λ are diagonalized by the chaos decomposition and therefore commute, we get

$$D = \Lambda^{-1}M\mathcal{R}.$$

□

6.5. *Proof of Theorem 6.2.*

We write the l.h.s. of the duality relation 6.1 using 6.4, we get

$$(Df \mid u) = \left(\Lambda^{-1}M\mathcal{R}f \mid u\right).$$

Now $u \in \mathbb{D}_1^p(X, H)$ implies that $\Lambda^{-1}u$ exists. As Λ^{-1} is a symmetric operator

$$\left(\Lambda^{-1}M\mathcal{R}f \mid u\right) = \left(M\mathcal{R}f \mid \Lambda^{-1}u\right).$$

We have $M^* = M$ for the duality (L^p, L^q); therefore

$$(Df \mid u) = \left(\mathcal{R}f \mid M\Lambda^{-1}u\right) = \left(f \mid \mathcal{R}^*M\Lambda^{-1}u\right),$$

we finally remark that

$$\mathcal{R}^*M\Lambda^{-1}u \in L^p(X; G).$$

□

6.5.1. Corollary. *If $u \in \mathbb{D}_1^p(X, H \otimes G)$, then*

$$D^*u = R^*M\Lambda^{-1}(u).$$

Remark. It would be *false* to say that

$$D^* = \mathcal{R}^*M\Lambda^{-1}.$$

In fact the domain of D^* is *much larger* than $\mathbb{D}_1^p(X; H \otimes G)$ as will be seen later in Chapter VII, using the theory of Itô's stochastic integrals.

6.6. Divergence of a vector field

6.6.1. Definition of a vector field. We shall call an L^p-vector field on X a function

$$A \in L^p(X; H).$$

This definition is natural. In fact, on the Gaussian probability space all the derivatives have been taken along the Cameron-Martin space which plays a role analogous to the *tangent space* in differential geometry.

We shall call $D^*(A)$ the *divergence* of A and we shall denote it by δA.

6.6.2. Watanabe's Theorem. *Given $A \in \mathbb{D}_1^p(X; H)$ there exists $\delta A \in L^p(X)$ such that*

$$E\big((Df \mid A)\big) = E(f\delta A) \qquad \text{for all } \ f \in \mathbb{D}_1^q(X) \ \text{ and}$$

$$\|\delta A\|_{L^p} \leq c_p \|A\|_{\mathbb{D}_1^p(X;H)}.$$

Proof. By specialization of Theorem 6.2.

6.6.3. Computation of δA in the finite-dimensional case

Assume that $\dim H = d < +\infty$; then by a choice of orthonormal coordinates we get $H = \mathbb{R}^d$; $A \in \mathbb{D}_1^p(X, H)$ is given by its components $A^1, \ldots, A^d$. We have then by the definition

$$\int_{\mathbb{R}^d} \sum_{k=1}^{d} \frac{\partial f}{\partial \xi_k} A^k \ \gamma_d(d\xi) = \int_{\mathbb{R}^d} f\delta A \ \gamma_d(\xi)$$

where

$$\gamma_d(d\xi) = \exp\left(-\frac{1}{2}\sum_{k=1}^{d} \xi_k^2\right) \frac{d\xi_1 \ldots d\xi_d}{(2\pi)^{d/2}}$$

is the normal distribution on $\mathbb{R}^d$.

Then by integration by parts, assuming f to be of compact support, we get

(i) $$(\delta A)(\xi) = \sum_{k=1}^{d}\left(-\frac{\partial A^k}{\partial \xi_k} + \xi_k \ A^k\right).$$

In dimension d the r.h.s. of (i) is a sum of d terms. Therefore the statement that *smooth* vector fields have a divergence given by (i) is *tautological.*

In infinite dimension, taking a numerical model, we find a *formal* analogue of (i) where the r.h.s. is now a series. It results from 5.1 and 6.6.2 that this series *converges*. A direct approach to the convergence of this series seems very hard. In a word, the existence of divergence of smooth vector fields is a deep fact; it will be the cornerstone of Chapter III.

6.7. The Shigekawa complex

6.7.1. Definition of $H^* \wedge H^*$. We shall denote by H^* the space of linear forms on H. Of course H^* can be identified to H using the scalar product.

Then $H^* \otimes H^*$ can be identified with the space of bilinear forms B on $H \otimes H$ which are associated to a Hilbert-Schmidt kernel. More precisely, we introduce the Segal model $X = \text{Seg}(H)$, then

(i) $$B(h_1, h_2) = \int_{X^2} K(x_1, x_2) h_1(x_1)(h_2)(x_2) P(dx_1) \otimes P(dx_2)$$

$$\|B\|^2_{H^* \otimes H^*} = \int \|K\|_{L^2(X^2)},$$

the infinimum being taken on the set of all the kernels K satisfying (i). We denote by $H^* \wedge H^*$ the space of bilinear forms which are *antisymmetric*.

6.7.2. Differential forms of degree 1 and 2

H^*, $H^* \wedge H^*$ are two Hilbert spaces. Therefore the following definitions are quite natural.

1. $\mathbb{D}_r^p(X; H^*)$ space of differential forms of degree 1.
2. $\mathbb{D}_r^p(X; H^* \wedge H^*)$ space of differential forms of degree 2.

There is no difficulty in defining differential forms of degree p. This will be done in V.4.

6.7.3. Coboundary operator of a function

Given a function $f \in \mathbb{D}_1^p(X)$, we consider the map

$$h \in H \xrightarrow{q(x)} (D_h f)(x)$$

we have

$$|(D_h f)(x)| \le \|h\| \quad \|Df(x)\|.$$

We can choose a basis $\{e_k\}$ of H. Then we can find a set S such that $P(S) = 0$ and representatives of the class $D_{e_k}(f)$, $\|Df\|$ which are all defined and finite outside S. Then $\langle q(x), e_k \rangle$ is defined if $x \notin S$, and

$$\left| \sum_{k=1}^{m} \alpha_k \langle q(x), e_k \rangle \right| < c(x) \left\| \sum_{k=1}^{m} \alpha_k \; e_k \right\|_H .$$

Then $q(x)$ extends as a continuous linear form on H. Therefore $x \to q(x)$ is a map $X \to H^*$ which belongs to $L^p(X; H^*)$.

We shall denote this map by df and we shall call df the *coboundary* of f.

6.7.4. Coboundary of a form of degree 1

Given $\rho \in \mathbb{D}_1^p(X; H^*)$ we want to define $(d\rho) \in L^p(X, H^* \wedge H^*)$. Given $h \in H$, then $\langle \rho(x), h\rangle$ is a function. We can define its coboundary $d\big(\langle \rho(x), h\rangle\big)$. We get a bilinear form

$$B(h_1, h_2) = \langle d(\langle \rho(x), h_1\rangle), h_2\rangle \,.$$

This bilinear form is not antisymmetric. We will antisymmetrize and define $d\rho$ by

$$\langle d\rho, h_1 \wedge h_2\rangle = \frac{1}{2}\left[\langle d(\langle \rho(x), h_2\rangle), h_1\rangle - \langle d(\langle \rho(x), h_1\rangle, h_2\rangle\right].$$

As we have seen for df it can be proved that

$$d\rho \in L^p\left(X, H^* \wedge H^*\right).$$

6.7.5. Definition of d^*. The spaces $L^2(X; H^*)$, $L^2(X; H^* \wedge H^*)$ are Hilbert spaces. The Hilbert inner product enables us to define d^* formally by the identity

$$E\big((\sigma \mid df)\big) = E\big((d^*\sigma) f\big) \quad \text{where } \sigma \in D_1^2(X; H^*) \text{ and } f \in \mathbb{D}_1^2(X).$$

For $\theta \in \mathbb{D}_1^2(X; H^* \wedge H^*)$ we proceed in the same way

$$E\big((\theta \mid d\rho)\big) = E\big((d^*\theta \mid \rho)\big) \quad \text{for} \quad \rho \in \mathbb{D}_1^2(X; H^*).$$

The existence of d^* for *smooth differential forms* will be settled by the next theorem which also provides an alternative approach to 6.6.2 for the case $p = 2$.

We shall call a *numerical differential form* associated to a numerical model a form ρ such that

$$\langle \rho(x), e_k\rangle = 0 \quad \text{except for a finite number of indices}$$
$$\langle \rho(x), e_k\rangle \quad \text{are polynomials.}$$

6.7.6. Theorem. *We have for every numerical differential form ρ of degree 1 the following identity*

$$\left\|d^*\rho\right\|^2_{L^2(X)} + \|d\rho\|^2_{L^2(X; H^* \wedge H^*)} = \|D\rho\|^2_{L^2(X, H \otimes H^*)} + \|\rho\|^2_{L^2(X; H^*)} \,.$$

Remark. This identity is parallel to the Weitzenböck-Bochner identity in differential geometry. In the Weitzenböck-Bochner notation the preceding theorem would be stated as follows: *The Ricci curvature tensor is equal to* 1.

Proof. As ρ is numerical, only a finite number of coordinates are involved. Therefore it is sufficient to prove the theorem for $\dim H = \ell < +\infty$. Then we denote by d_0^* *the Euclidean adjoint* defined by

$$\int_{\mathbb{R}^d} (d_0^*\rho)\, f\, d\xi = \int_{\mathbb{R}^d} (\rho \mid df) d\xi.$$

As the adjoint d^* is defined by

$$\int_{\mathbb{R}^d} (d^*\rho) f\ \exp\left(\frac{-|\xi|^2}{2}\right) \frac{d\xi}{(2\pi)^{\ell/2}} = \int (\rho \mid df) \exp\left(\frac{-|\xi|^2}{2}\right) \frac{d\xi}{(2\pi)^{\ell/2}},$$

we have the relation

$$d^\rho = \exp\left(\frac{|\xi|^2}{2}\right) d_0^* \left[\exp\left(\frac{-|\xi|^2}{2}\right) \rho\right]$$

ρ being of degree 1 or 2.

6.7.7. Main lemma. *Given a C^2 function φ on $\mathbb{R}^\ell$, denote*

$$\begin{aligned} d_\varphi^* &= e^\varphi (d_0)^* e^{-\varphi} \\ \Box_0 &= d_0^* d + d d_0^* \\ \Box_\varphi &= d_\varphi^* d + d d_\varphi^*. \end{aligned}$$

Then we have for differential forms of degree 1

$$\Box_\varphi = \Box_0 + \mathrm{Hess}(\varphi).$$

More precisely, given a smooth differential form of degree 1

$$\rho = \sum_{k=1}^{\ell} f_i d\xi^i$$

we have

$$(\Box_\varphi \rho)_i = (\Box_0 \rho) + \sum_{j=1}^{\ell} \frac{\partial^2 \varphi}{\partial \xi^i \partial \xi^j}\, \rho_j.$$

Remark. This lemma is parallel to the identities which produce in complex analysis the *vanishing theorem of Kodaira* or *Hörmander's L^2-estimates for the $\bar\partial$ operator.*

Proof. We denote

$$\partial_k = \frac{\partial}{\partial \xi_k} \qquad \partial_{k,j} = \frac{\partial^2}{\partial \xi_k \partial \xi_j}$$

$$d_\varphi^* \rho = d_0^* \rho + \sum_k \rho_k \partial_k \varphi$$

$$d\left(d_\varphi^*\rho\right)_i = \left(dd_0^*\rho\right)_i + \sum_k \rho_k\left(\partial_{i,k}\varphi\right) + \sum_k \partial_k\varphi\ \partial_i\rho_k$$

$$d\rho = \frac{1}{2}\sum_{k,\ell}(\partial_\ell\rho_k - \partial_k\rho_\ell)\,d\xi_\ell \wedge d\xi_k$$

$$d_\varphi^*(d\rho) = d_0^* d\rho + \frac{1}{2}\sum_{k,\ell}(\partial_\ell\rho_k - \partial_k\rho_\ell)\left[(\partial_\ell\varphi)\,d\xi_k - (\partial_k\varphi)\,d\xi_\ell\right].$$

In order to compute the i-th component we have to take either $d\xi_k = d\xi_i$ or $d\xi_\ell = d\xi_i$; we get

$$\left[d_\varphi^*(d\rho)\right]_i = \frac{1}{2}\sum_\ell(\partial_\ell\rho_k - \partial_i\rho_\ell)\,\partial_\ell\varphi - \frac{1}{2}\sum_k(\partial_i\rho_k - \partial_k\rho_i)\,\partial_k\varphi.$$

This i-th component is a linear form in the $\partial_r\varphi$. In order to compute the coefficient of $\partial_r\varphi$ we have to take $\ell = r$ and $k = r$. We get

$$\frac{1}{2}(\partial_r\rho_i - \partial_i\rho_r) - \frac{1}{2}(\partial_i\rho_r - \partial_r\rho_i) = -\partial_i\rho_r$$

which cancels with the last term of $d_\varphi^* d\varphi$. □

6.7.8. *Proof of Theorem 6.7.6.*

By duality we have

$$\left(d^*\rho \mid d^*\rho\right) + (d\rho \mid d\rho) = (\Box\rho \mid \rho).$$

By a well known result on $\mathbb{R}^\ell$, $\Box_0\rho = -\Delta\rho$ where Δ is the Laplacian on R^n. We have

$$(-\Delta\rho \mid \rho) = \|D\rho\|_{L^2}^2.$$

Finally

$$-\text{Hessian}\left(\log\exp\left(\frac{-|\xi|^2}{2}\right)\right) = \text{Identity}.$$

□

6.7.9. Theorem. *Given $\rho \in \mathbb{D}_1^2(X; H^*)$ then*

1. *$d^*\rho$ exists and $(d^*\rho) \in L^2(X)$,*
2. *$d\rho$ exists and $(d\rho) \in L^2(X)$.*

Proof. Approximation by numerical differential forms, and use of the inequality

$$\|d^*\rho\|_{L^2}^2 \le \|D\rho\|_{L^2}^2 + \|\rho\|_{L^2}^2$$

or

$$\|d\rho\|_{L^2}^2 \le \|D\rho\|_{L^2}^2 + \|\rho\|_{L^2}^2.$$

□

Appendix: Proof of the L^p inequality for the Hilbert transform

This inequality is a well known result in harmonic analysis; for sake of completeness we give below a proof.

Lemma 1. *Given a Hölder continuous real-valued function $U(\theta)$, defined on the circle, then*

$$\int_{-\pi}^{\pi} U(\theta-\theta') 1_{|\theta'|>\varepsilon} \operatorname{cotg}\left(\frac{\theta'}{2}\right) \frac{d\theta'}{2\pi} = V_\varepsilon(\theta)$$

converges uniformly as $\varepsilon \to 0$ towards a continuous function V; denote $F = U + iV$, then there exists an holomorphic function f defined in the unit disk such that $f\left(e^{i\theta}\right) = F(\theta)$.

Proof. We remark that $1_{|\theta|>\varepsilon} \operatorname{cotg}(\theta/2)$ is an odd function; its L^2-scalar product with a constant is zero; therefore

$$V(\theta) = \int_{-\pi}^{\pi} \left[U(\theta-\theta') - U(\theta)\right] \operatorname{cotg}\left(\frac{\theta'}{2}\right) \frac{d\theta'}{2\pi}.$$

The Hölder hypothesis implies the absolute convergence of the integral on the r.h.s., and hence the desired uniform convergence.

Define

$$f(z) = \int_{-\pi}^{\pi} \frac{1+ze^{-i\theta'}}{1-ze^{-i\theta'}} U(\theta') \frac{d\theta'}{2\pi}.$$

then by differentiating the r.h.s. we obtain that f is holomorphic in the disk $|z| < 1$; using the polar coordinates $z = \rho e^{i\varphi}$, we have

$$\Re \frac{1+ze^{-i\theta}}{1-ze^{-i\theta}} = \frac{1-\rho^2}{1-2\rho\cos(\theta-\varphi)+\rho^2},$$
$$\Im \frac{1+ze^{-i\theta}}{1-ze^{-i\theta}} = \frac{2\rho\sin(\theta-\varphi)}{(1-\rho)^2+4\rho\sin^2\left(\frac{\theta-\varphi}{2}\right)}.$$

We recognize in the first kernel the Poisson kernel: when $\rho \to 1$, we have $\lim \Re f(\rho e^{i\varphi}) \to U(\varphi)$. The second kernel converges towards cotg $(\theta-\varphi)/2$.

Lemma 2. *Given $f(z) = u(z) + iv(z)$ holomorphic in the unit disk, assume that $f(0) = 0$, then for every even integer p, we have*

$$\|U\|_{L^p} \leq \left(2^{1/p}-1\right)^{-1} \|V\|_{L^p}.$$

Proof. The function f^p is holomorphic and $f^p(0) = 0$. We write the Cauchy formula by integrating f^p on the circle:

$$0 = i^p \int_0^{2\pi} \left[V^p + \sum_{k>0} i^k C_p^k U^k V^{p-k} \right] \frac{d\theta}{2\pi}.$$

We remark that, under the hypothesis that p is even, the first integral is equal to $\|V\|_{L^p}^p$; we dominate the other terms using Hölder's inequality and we obtain the inequality

$$\xi^p \leq (\xi + \eta)^p - \xi^p \quad \text{where} \quad \xi = \|V\|_{L^p}, \quad \eta = \|U\|_{L^p}.$$

We remark that the combination of Lemmas 1 and 2 gives that the Hilbert transform is a bounded operator in L^p when p is an even integer. The general case could be deduced by interpolation for $p \geq 2$ and then by duality for $1 < p < 2$.

Chapter III

Smoothness of Laws

Contents: Divergence of differentiable flow – Divergence as the adjoint operator of a derivation – Non-degenerate maps and their covariance matrices – Lifting up vector fields through a non-degenerate map – Pushing down divergences – Hölder regularity under $\mathbb{D}_2^{\infty-0}$ hypothesis – Smoothness under $\mathbb{D}_\infty$ hypothesis – Lifting up and pushing down through a non-degenerate $\mathbb{D}_\infty$ map – Inverse image of a distribution – Absolute continuity of scalar functionals under $\mathbb{D}_1^1$ hypothesis – Appendix: Computation of derivatives by means of divergence operators – Law of a weakly non degenerated map.

The smoothness of the law of $\mathbb{R}^d$-valued random variables defined on a Gaussian probability space is the subject of this chapter.

The methodology will result from the mixing of the two following ingredients. Firstly, given a smooth functional $g : X \mapsto \mathbb{R}^n$, elementary *differential geometry* methods provide a *lift up* of smooth functions and of differential forms from $\mathbb{R}^d$ to the corresponding objects on X; then problems raised on $\mathbb{R}^d$ can be transformed into problems on X which are hopefully solvable. Once this solution is obtained we have to go back to $\mathbb{R}^d$; the machinery of differential geometry is then inefficient. Nevertheless the *conditional expectation* will now provide a *push down*. Secondly the main idea used to solve the problem on X will be to proceed to *integrate by parts*; this integration by parts can be performed either by a *divergence* or by the *self-adjointness of the Ornstein-Uhlenbeck generator* as will be done in Chapter X.

The first section of this chapter will be devoted to the behaviour of the *divergence operator* under the operations of lifting up and pushing down.

It is obvious that the law of a *constant map* has the worst regularity. The study of smoothness of law needs some hypothesis of *functional independence*. This hypothesis will be expressed in Section 2 by an assumption on the infinitesimal *covariance matrix* and will lead to the notion of a *non-degenerate map*.

The strict positivity of the infinitesimal covariance matrix will appear as a probabilistic counterpart of the notion of pointwise ellipticity in the theory of partial differential equations; the notion of non-degeneracy corresponds to some kind of *ellipticity* in the mean.

The computation of the divergence for the lifting of coordinate vector fields will be done in Section 3. Then, under $\mathbb{D}_1^\infty$ hypotheses and non-degeneracy it will be shown in Section 4 that the law has a Hölder density. In Section 5 the smoothness of the law will be obtained under $\mathbb{D}_\infty(X)$ hypothesis.

In Section 5 the machinery of lifting up and pushing down through a smooth non-degenerate map will be developed: lifting up distributions on $\mathbb{R}^d$, pushing down the Riemannian structure of X.

In Section 6 absolute continuity for the law of random variables in $\mathbb{D}_1^p$ will be considered.

The computation of divergence by *non-anticipative* stochastic integral will lead in Chapter VI to another approach to the regularity of laws. Alternative approaches will appear in Chapter VII and in Chapter X. This repetition shows that the regularity of laws is a central theme of the whole book.

1. General theory of divergences

We have seen in Chapter II, Section 5, the notion of divergence in the framework of a Gaussian probability space. We shall now study the divergence in a more general context, with an emphasis on its functorial properties.

In order to avoid the development of a cumbersome machinery, the statements of the first two paragraphs will be presented in a schematic form which could be applied to a vector space of finite or infinite dimension or to a Gaussian probability space.

1.1. Integrated divergences of a differential flow

We denote by Q a complete metric, separable vector space and by B a vector field on Q. Granted the vector space structure, a vector field is a map $B : Q \to Q$. To this vector field we associate the ordinary differential equation:

1.1.1. $$\frac{dq}{dt} = B_{q(t)}$$

We denote by $U_t^B(q_0)$ "the solution" of (i) such that $q(0) = q_0$. The *existence* and *uniqueness* of this solution have to be *ensured*; a hypothesis of Lipschitz regularity of B is sufficient for the existence and uniqueness for all $t \in \mathbb{R}$ of the solution.

The fact that the vector field B is time-independent implies the group property:

$$U_t^B \circ U_{t'}^B = U_{t+t'}^B, \qquad t, t' \in \mathbb{R}$$

we shall call U^B the *flow* associated to the vector field B. Conversely, given Φ_t a one-parameter group of smooth transformations of Q, then $\Phi_.(q_0)$ is a smooth curve; its tangent at $t = 0$ defines a vector $\tilde{B}_{q_0}$; the vector field $q_0 \mapsto \tilde{B}_{q_0}$ defines the flow $U_t^{\tilde{B}}$; finally $U_t^{\tilde{B}} = \Phi_t$.

Given a probability measure μ on Q, then the flow defines an image measure $(U_t^B)_*\mu$.

The *integrated divergence* of B relative to μ, which is denoted $\mathrm{div}_\mu(B)$, is said to exist when for every $f \in C(Q)$, the space of continuous bounded functions, we have

1.1.2. $$\lim_{\varepsilon \to 0} \varepsilon^{-1} \langle f, (U_{-\varepsilon}^B)_* \mu - \mu \rangle = \int f \; \mathrm{div}_\mu(B) d\mu.$$

We could forget the vector field as shown below.

1.2.1. Definition (of a measurable flow). Given a probability space Γ we shall call a flow a one-parameter group of measurable transformations Φ_t preserving the class of P-zero sets. We suppose furthermore that Φ_t is *weakly continuous* which means that

1.2.2. $$t \mapsto E\left((\Phi_t^* \varphi)\, \psi\right) \quad \text{and} \quad t \to E\left((\Phi_t^* \varphi)^2\right)$$

is continuous for all $\varphi, \psi \in L^\infty$, where

$$(\Phi_t^* \varphi)(\omega) \doteq \varphi(\Phi_t(\omega)).$$

1.2.3. Divergence of a measurable flow

The divergence $\mathrm{div}_P(\varphi)$ will exist if

$$\left\{ \frac{d}{dt} E(\Phi_t^* f) \right\}_{t=0} = E(\mathrm{div}_P(\Phi) f)$$

for all $f \in L^\infty$ where $\mathrm{div}_P(\Phi) \in L^1$.

In the case of a Gaussian probability space, we have constructed the canonical representation ρ on the Cameron-Martin space. Then, $h \in H$ being fixed, we define a flow by $\Phi_t(\omega) = \rho(th)\omega$.

1.2.4. Cruzeiro's Lemma. *Given a flow Φ_t on a probability space Ω, denote*

$$K_t = \frac{d(\Phi_t)_* P}{dP}.$$

Assume that

$$k \doteq \mathrm{div}_\Gamma(\Phi)$$

exists and that $k \in L^\infty$, then

$$K_t(\omega) = \exp\left(\int_0^t k\left(\Phi_{-\xi}(\omega)\right) d\xi \right)$$

and we have

$$\|K_t\|_{L^p}^p \leq E\big(\exp(pq|tk|)\big)$$

where q in the conjugate exponent of p.

Proof. Define for $h \in L^\infty(\Omega)$

$$\Phi_t^* h = h \circ \Phi_t.$$

Then

$$\Phi_{t+\varepsilon}^* = \Phi_\varepsilon^* \circ \Phi_t^*.$$

We have

$$E(\Phi_t^* f) = E(f K_\tau).$$

Then

$$E(f \dot{K}_\tau) = \lim_{\varepsilon \to 0} E\left(\frac{\Phi_\varepsilon^* - 1}{\varepsilon}\, u\right)$$

where

$$u = \Phi_\tau^* f.$$

Then

$$E(f \dot{K}_\tau) = E(u \operatorname{div}_P \varphi) = E(uk)$$
$$E(uk) = E\left((\Phi_\tau^* f)k\right).$$

We have

$$(\Phi_\tau^* f)k = \Phi_\tau^*\left(f \Phi_{-\tau}^* k\right).$$

Finally

$$E(f \dot{K}_\tau) = E\left(\Phi_\tau^*(f \Phi_{-\tau}^* k)\right) = E\left(f(\Phi_{-\tau}^* k) K_\tau\right).$$

We deduce

$$\dot{K}_\tau(\omega) = \left((\Phi_{-\tau}^* k) K_\tau\right)(\omega)$$

and we have proved the first formula.

Denote $\varphi(t,p) = \sup_{|t'|<t} \|K_{t'}\|_{L^p}$. Choose t_0 such that

$$\varphi(t_0, p) = \left\|K_{t_0}\right\|_{L^p} \qquad \text{or} \qquad \varphi(t_0, p) = \left\|K_{-t_0}\right\|_{L^p},$$

then

$$\begin{aligned} [\varphi(t_0,p)]^p = E\left(K_{t_0}^p\right) \quad &\leq \quad E\left(\exp\left(p t_0 \int_0^{t_0} |k| \left(\Phi_{-\xi}(\omega)\right) \frac{d\xi}{t_0}\right)\right) \\ &\leq \int_0^{t_0} E\left(\exp\left(p t_0 |k| \left(\Phi_{-\xi}(\omega)\right)\right)\right) \frac{d\xi}{t_0}. \end{aligned}$$

If we use the change of variable formula to replace $\Phi_{-\xi}(\omega)$ by ω, and then use the Hölder inequality, we get

$$\leq \int_0^{t_0} [[E(\exp(pq \mid t_0|k|))]]^{1/q} (\varphi(\xi, p))^{1/p} \frac{d\xi}{t_0}$$
$$\leq (\varphi(t_0, p))(E(\exp(pqt_0|k|))) \qquad \text{and :}$$
$$(\varphi(t_0, p))^{p-1} \leq [E(\exp(pqt_0|k|))]^{1/q}.$$

This proves the lemma. □

1.2.5. Remark. In Cruzeiro [1], by an appropriate truncation argument, it is shown to be possible to drop the hypothesis $k \in K^\infty$ and to replace it with

$$E(\exp(\lambda|k|)) < +\infty \quad \text{for all} \quad \lambda$$

and then to deduce that Φ is a stochastic flow and that

$$\|K_t\|_{L^p}^p \leq E(\exp(pq|tk|)).$$

1.2.6. Remark. The integrated divergence K_t needs the existence of a flow. In the next paragraph we want to avoid this necessity by using an infinitesimal version of this integrated divergence.

1.3. Adjoint of the differential

We resume the hypothesis of paragraph 1.1. We assume that on Q there is given an algebra $\mathcal{G}$ of bounded "smooth functions". We assume that for every function $f \in \mathcal{G}$, for every vector field B, the limit

$$\lim \varepsilon^{-1} (f(q + \varepsilon B_q) - f(q))$$

exists for the weak topology on $L^1(\mu)$. We denote this limit by $(D_B f)(q)$.

We shall say that the *differential divergence* of B relative to μ exists and we shall denote it by $\delta_\mu(B)$ if we have

1.3.1. $$\int_Q D_B f d\mu = \int_Q f \delta_\mu(B) d\mu$$

for every $f \in \mathcal{G}$, and where $\delta_\mu(B) \in L^1$.

We will assume that $\mathcal{G}$ is dense for the weak topology in $C_b(Q)$.

1.3.2. Proposition. *Assume that* $\operatorname{div}_\mu(B)$ *exists, then* $\delta_\mu(B)$ *exists, and these two expressions are equal.*

Proof.

$$\int f(q) (U_{-\varepsilon}^B \mu)_* (dq) = \int f(U_\varepsilon^B(q)) d\mu(q).$$

Therefore

$$\lim_{\varepsilon\to 0}\int \varepsilon^{-1}\langle f, (U^B_{-\varepsilon}\mu - \mu)\rangle = \int f(q)\mathrm{div}_\mu(B)d\mu(q)$$
$$= \left\{\frac{d}{d\varepsilon}\int f\left(U^B_\varepsilon(q)\right)d\mu(q)\right\}_{\varepsilon=0}$$
$$= \int D_B f\, d\mu.$$

□

1.3.3. Proposition. *If $\delta_\mu(B)$ exists, then*

$$\delta_\mu(uB) = u\delta_\mu(B) - D_B u \quad \text{for every } u \in \mathcal{G}.$$
$$\delta_{k\mu}(B) = \delta_\mu(B) - D_B(\log k) \quad \text{if } k > 0 \text{ and } \log k \in \mathcal{G}.$$

Proof.

$$\int D_{uB} f\, d\mu = \int u D_B f\, d\mu = \int D_B(uf)d\mu - \int f D_B u\, d\mu$$
$$= \int f\left(u\delta_\mu(B) - D_B u\right)d\mu.$$

Denote $v = \log k$ then

$$\int (D_B f)e^v\, d\mu = \int D_B(fe^v)d\mu - \int d(D_B v)e^v\, d\mu$$
$$= \int f\left(\delta_\mu(B) - D_B v\right)e^v\, d\mu.$$

□

2. Non-degenerate maps

We shall now work with a very specific situation. We shall denote by X a Gaussian probability space, by H its Cameron-Martin space.

A map $g \in \mathbb{D}^p_r(X; \mathbb{R}^d)$ will be given by a system of scalar functions $g_1, \ldots, g_d \in \mathbb{D}^p_r(X)$.

2.1. The Malliavin covariance matrix

2.1.1. Definition. We consider the positive symmetric *matrix*

$$(Dg_i(x) \mid Dg_i(x))_H = \sigma_{i,j}(x).$$

The differentials are defined almost surely. Therefore $\sigma_{i,j}$ are well defined random variables, integrable only if $p \geq 2$. We shall call σ the *infinitesimal covariance matrix.*

We denote by $g(x)$ the map $g'(x) : H \to \mathbb{R}^d$ defined by

$$g'(x) \circ h = \left(D_h g^i(x)\right)_{i \in [1,d]}.$$

As H and $\mathbb{R}^d$ are two Hilbert spaces, $\left(g'(x)\right)^* : \mathbb{R}^d \to H$ is well defined. Then

2.1.2. Proposition. *We have* $g'(x)\left(g'(x)\right)^* : \mathbb{R}^d \to \mathbb{R}^d$; *furthermore*

$$g'(x)\left(g'(x)\right)^* = \left(\sigma_{ij}(x)\right)_{i,j \in [1,d]}.$$

Proof.

$$\begin{aligned}
\left(g'(x)\right)^* \xi &= \sum \xi^i Dg^i \qquad \xi \in \mathbb{R}^d \\
g'(x) \cdot h &= \left((Dg^k \mid h)\right)_{k \in [1,d]} \\
g'(x)\left(g'(x)\right)^* &= \sum \xi^i \left(Dg^k \mid Dg^i\right) = \sum_i \sigma_{i,k} \xi^i.
\end{aligned}$$

□

We define

$$\det\left(g'(x)\right) = \left(\det\left(\sigma_{i,j}(x)\right)\right)^{1/2}.$$

The condition $\det g'(x) > 0$ is equivalent to the linear independence of $(Dg_1)(x)$ $i \in [1, d]$.

2.1.3. Definition. The map $g \in \mathbb{D}_1^1(X; \mathbb{R}^d)$ is *non-degenerate* if and only if

$$\left[\det\left(g'(x)\right)\right]^{-1} \in L^{\infty-0}.$$

The map g is *weakly non-degenerate* if

$$\det\left(g'(x)\right) > 0 \qquad \text{almost surely.}$$

2.2. Lifting up functions

Denote by $C_b^1(\mathbb{R}^d)$ the space of continuously differentiable functions which are bounded and with bounded derivatives.

Given $\varphi \in C_b^1(\mathbb{R}^d)$ we shall consider $\tilde{\varphi}(x) = \varphi\left(g(x)\right)$. We shall call $\tilde{\varphi}(x)$ the lift of φ. If $g \in \mathbb{D}_1^1(X; \mathbb{R}^d)$ then $\tilde{\varphi} \in \mathbb{D}_1^1(X)$.

2.3. Lifting up vector fields through a non-degenerate map

Given a vector field z on $\mathbb{R}^d$, we shall call a *lift* of z to X, a vector field $\tilde{Z}$ on X such that

$$g'(x)\tilde{Z}_x = z_{g(x)},$$

then we will have for every $\varphi \in C_b^1(\mathbb{R}^d)$

$$\langle \tilde{Z}, D\tilde{\varphi} \rangle_x = \langle z, d\varphi \rangle_{g(x)}.$$

Introducing ∂_z, $D_{\tilde{Z}}$ the differential operator corresponding to z and $\tilde{Z}$, we have

$$(\partial_z \varphi)\tilde{} = D_{\tilde{Z}} \tilde{\varphi}.$$

Proposition. *Assume that g' is weakly non-degenerate, then there exists a canonical lift Z of z which satisfies*

$$\|Z_x\| \leq \|\tilde{Z}_x\|$$

for all lifts $\tilde{Z}$ of z.

Proof. Define

$$\begin{aligned} \gamma(x) &= \text{inverse matrix of } \sigma(x) \\ Z_x &= (g'(x))^* \gamma(x) z_{g(x)}, \end{aligned}$$

then

$$g'(x)Z_x = g'(x)\left(g'(x)\right)^* \gamma(x) z_{g(x)} = \sigma(x)\gamma(x) z_{g(x)} = z_{g(x)}.$$

Now, given $\varphi \in C_b^1$, we lift φ to $\tilde{\varphi}$ and get

$$(D\tilde{\varphi} \mid Z)_x = \varphi'\big(g(x)\big) \cdot \big(g'(x) \cdot Z_x\big) = \varphi'\big(g(x)\big) \cdot z_{g(x)}.$$

Denote

$$u(\xi) = \langle \varphi'(\xi), t_\xi \rangle,$$

then we have the identity

$$(D\tilde{\varphi} \mid Z) = \tilde{u}.$$

If $\tilde{Z}_x$ is another lift we will have $\left(\tilde{Z}_x \mid Dg_k(x)\right) = (Z_x \mid Dg_k(x))$. As Z_x belongs to the linear space of $Dg_k(x)$, this implies that

$$\tilde{Z}_x = Z_x + n_x$$

where n_x is orthogonal to Z_x. □

2.4. Pushing down divergence and functoriality of divergences

Theorem. *Denote $\rho = g_*(\mu)$. Keeping the notation of 2.3, we assume furthermore that*

$$\begin{aligned} &\delta_\mu(Z) \quad \textit{exists, then} \\ &\delta_\rho(z) \quad \textit{exists and we have} \\ &\delta_\rho(z) = E^g(\delta_\mu Z) \end{aligned}$$

where E^g denotes the conditional expectation under g.

Proof.

$$I = \int_X (D\tilde{\varphi} \mid Z) d\mu = \int_X \tilde{\varphi}\ \delta_\mu Z dP = \int_{\mathbb{R}^d} E^g(\delta_\mu Z)\varphi\ d\rho.$$

Denote $u = \langle d\varphi, z\rangle$, then we get

$$I = \int_X \tilde{u} d\mu = \int_{\mathbb{R}^d} u d\rho.$$

We get

$$\int_{\mathbb{R}^d} E^g(\delta_\mu Z)\varphi\ d\rho = \int \langle d\varphi, z\rangle d\rho.$$

□

3. Computation of divergence

3.1. Theorem. *Let $g \in \mathbb{D}_2^\infty(X; \mathbb{R}^d)$, denote by σ its covariance matrix. Assume $\left(\det(\sigma)\right)^{-1} \in L^{\infty-0}$. Denote $\gamma = \sigma^{-1}$, define*

$$Z_k = \gamma^* \left(\frac{\partial}{\partial \xi_k}\right) = \sum \gamma_{k,j} D g_j,$$

then

$$\delta Z_k = \sum_{s,j} \left(-\gamma_{k,s}\mathcal{L} g_s + \gamma_{k,s}\left(D^2 g_s\right)\left(D g_j, Z_j\right) + \left(D^2 g_j\right)\left(Z_k, Z_j\right)\right)$$

where $\mathcal{L}$ denotes the Ornstein-Uhlenbeck generator. Furthermore

$$\delta Z_k \in L^{\infty-0}.$$

Proof. We denote $z_j = D g_j$, we apply 1.3.3 using the usual summation convention

$$\delta Z_k = \gamma_{k,j}\delta z_j - D_{z_j}(\gamma_{k,j})$$

$$D_{z_j}(\sigma^{-1}) = -\gamma D_{z_j}(\sigma)\gamma$$

$$\delta Z_k = \gamma_{k,j}\delta z_j + \gamma_{k,r}\left(D_{z_j}\sigma_{r,s}\right)\gamma_{s,j}$$

$$D_{z_j}(\sigma_{r,s}) = D_{z_j}\left(D g_r \mid D g_s\right) = \left(D^2 g_r\right)\left(z_j, z_s\right) + \left(D^2 g_s\right)\left(z_j, z_r\right)$$

$$\gamma_{k,r}\left[D^2 g_r\left(z_j, z_s\right)\right]\gamma_{s,j} = \gamma_{k,r}\left(D^2 g_r\right)\left(z_j, Z_i\right)$$

$$\gamma_{k,r}\left[D^2 g_s\left(z_j, z_r\right)\right]\gamma_{s,j} = \left(D^2 g_s\right)\left(Z_k, Z_s\right).$$

We remark that by II.3.3.1

$$\delta z_j = \delta D g_j = -\mathcal{L} g_j.$$

□

Now $\sigma_{i,j} \in \mathbb{D}_1^\infty(X)$,

$$\gamma = \frac{1}{\det \sigma} P(\sigma)$$

where $P(\sigma)$ is a polynomial in $\sigma_{i,j}$. Therefore $P(\sigma) \in \mathbb{D}_1^\infty$ and as $\left(\det(\sigma)\right)^{-1} \in \mathbb{D}_1^\infty$, we have that $\gamma \in \mathbb{D}_1^\infty$. We deduce then $Z_j \in \mathbb{D}_1^\infty$ and each of the terms appearing in the formula for δZ_k are in $L^{\infty-0}$. □

Remark. Under the hypothesis $g \in \mathbb{D}_2^p$, $\left(\det(\sigma)\right)^{-1} \in L^p$ and $p > 4d$, the same computation shows that $\delta Z_k \in L^r$ with $r = \frac{p}{4d}$.

Corollary. *Write $\rho = (g)_*\mu$, then under the assumptions of 3.1 there exists $q_k \in L^{\infty-0}(\rho)$ such that*

$$\int_{\mathbb{R}^d} \frac{\partial \varphi}{\partial \xi_k}\, d\rho = \int_{\mathbb{R}^d} q_k \varphi\, d\rho$$

for all $\varphi \in C_b^1(\mathbb{R}^d)$ and $k \in [1, d]$.

Proof. We use 3.1 and 2.4. □

4. Regularity under $\mathbb{D}_2^{\infty-0}$ hypothesis

4.1. Theorem. *Let $g \in \mathbb{D}_2^\infty(X; \mathbb{R}^d)$. Assume that $[\det g'(x)]^{-1} \in L^{\infty-0}(X)$, then $(g)_*\mu$ has a density $k(\xi)$ relative to the Lebesgue measure, k is continuous and for every $\beta < 1$ and every $\gamma > 0$*

$$k^\gamma \in \mathcal{H}^\beta(\mathbb{R}^d)$$

where

$$\mathcal{H}^\beta(\mathbb{R}^d) = \left\{ u \in C(\mathbb{R}^d);\ \sup \frac{|u(\xi) - u(\xi')|}{\|\xi - \xi'\|} < +\infty \right\}.$$

Proof. Assume that $d\rho$ has a density:

4.2. $$d\rho = k d\xi.$$

Then denoting $\|\nabla k\| = \left(\sum \left(\frac{\partial k}{\partial \xi_k}\right)^2\right)^{1/2}$ we have

4.3. Lemma.

$$\int_{\mathbb{R}^d} \frac{\|\nabla k\|^p}{k^{p-1}}\, d\xi \le d^p \sum_{j=1}^d \|\delta Z_j\|_{L^p(X)}^p.$$

Proof.

$$\int \frac{\partial \varphi}{\partial \xi_2} k d\xi = \int \varphi q_1 k d\xi \qquad \varphi \in C_b^1.$$

We get by transposition that $\frac{\partial k}{\partial \xi_i}$ exists *in the sense of the theory of distributions* and that

$$\frac{\partial k}{\partial \xi_i} = -q_i k$$

$$\frac{\|\nabla k\|^p}{k^p} = \left(\sum_{i=1}^{q} (q_i)^2\right)^{p/2} \leq d^p \sum |q_i|^p.$$

Taking expectations we get

$$\int \frac{\|\nabla k\|^p}{k^p} k d\xi \leq d^p \sum E\left(|E^g \delta Z_i|^p\right).$$

□

4.4. Lemma. *Let $k \in L^1(\mathbb{R}^d)$, $k > 0$. Assume that*

$$\int \left[\frac{\partial_s k}{k}\right]^p k d\xi < +\infty \quad \textit{for all} \quad p, \quad \left(\textit{where} \quad \partial_s = \frac{\partial}{\partial \xi_s}\right),$$

then for all $\gamma > 0$, $\beta < 1$, then

$$k^\gamma \in \mathcal{H}^\beta.$$

Proof. We denote $k_\beta = (k)^\beta$, $\beta \in]0, 1[$. Then $\|\nabla k_\beta\| = \beta k^{\beta-1} \|\nabla k\|$. We take $\beta = p^{-1}$

$$\|\nabla k_\beta\|^p_{L^p(d\xi)} = \beta^p \int \frac{\|\nabla k\|^p}{k^p} k d\xi < +\infty.$$

Therefore

$$(k)^{1/p} \in W_1^p(\mathbb{R}^d)$$

where W_1^p denotes the usual Sobolev space on $\mathbb{R}^d$. Further by the theory of Sobolev spaces on $\mathbb{R}^d$, the inclusion

4.4.1.
$$W_1^p \subset \mathcal{H}^\beta, \quad \text{if} \quad \beta < \frac{p-d}{p}$$

implies $k^{1/p} \in \mathcal{H}^\beta$.

We remark that $k \in L^1$, k uniformly continuous implies that $k \in L^\infty$. Therefore $k^\gamma \in \mathcal{H}^\beta$ implies $k^{\gamma'} \in \mathcal{H}^\beta$ for $\gamma' > \gamma$. Then fixing γ_0 the proposition

$$k^{\gamma_0} \in \mathcal{H}^\beta \quad \text{will result from}$$
$$k^{1/p} \in \mathcal{H}^\beta \quad \text{for} \quad p > \frac{1}{\gamma_0},$$

which is a consequence of 4.4.1. □

4.5. *Proof of Theorem 4.2.*

We have to suppress the condition 4.2.

Let $v \in C^\infty(\mathbb{R}^d)$ be non-negative, with compact support, and of integral 1: define $v_n(\xi) = n^{-d} v(n^{-1}\xi)$; then the convolution product $\rho * v_n$ is a measure which has a density k_n. The previous estimates show that $\|k_n\|_{\mathcal{H}^\beta} < c$ where c is independent of n. Therefore by the Theorem of Ascoli the weak convergence of $\rho_n \to \rho$ implies the uniform convergence of $k_n \to k_\infty \in \mathcal{H}^\beta$; we deduce that $\rho = k_\infty d\xi$. □

5. Regularity under D_∞ hypothesis

5.1. Theorem. *Assume that $g \in D_\infty(X; \mathbb{R}^d)$ and that $\left(\det g'(x)\right)^{-1} \in L^{\infty-0}$, then the law of g has a density which is C^∞ and which belongs to the Schwartz space $\mathcal{S}(\mathbb{R}^d)$ of functions which are rapidly decreasing along with all their derivatives.*

Proof. Denote by $\mathcal{E}_d$ the set postive integer-valued functions defined on $[1, d]$. If $s \in \mathcal{E}_d$ denote $|s| = \sum_{j=1}^d s(j)$. Denote

$$\partial_s = \frac{\partial^{|s|}}{\partial \xi_1^{s(1)} \ldots \partial \xi_d^{s(d)}}.$$

5.2. Lemma. *For every $s \in \mathcal{E}_d$, there exists*

$$Q_s \in D_\infty(X)$$

such that

$$E\left((\partial_s \varphi)^{\sim}\right) = E(Q_s \tilde{\varphi})$$

for every $\varphi \in C_b^{|s|}(\mathbb{R}^d)$.

Proof. We shall proceed by induction on $|s| = r$. For $r = 1$, we take $Q_s = \delta Z_i$, where $s(i) = 1$.

We have computed δZ_i in 3.1; it results by inspection of this formula that $\delta Z_i \in \mathbb{D}_\infty(X)$.

Assume the lemma true for s, $|s| < r$.

Given s with $|s| = r$, we can write $s = s_1 + s_2$ with $|s_1| = r - 1$, $|s_2| = 1$. To fix notations, assume $s_2(1) = 1$. Then $\partial_s = \partial_{s_1} \partial_1$. Denote

$$\psi = \partial_1 \varphi.$$

We have for every function $\psi \in C_b^\infty(\mathbb{R}^d)$

$$E\left((\partial_{s_1} \psi)^{\sim}\right) = E(Q_{s_1} \tilde{\psi}).$$

Given $\varphi \in C_b^\infty(\mathbb{R}^d)$, we remark that

$$(\partial_1\varphi)\check{} = D_{Z_1}\tilde{\varphi}$$

$$\begin{aligned} E\left((\partial_s\varphi)\check{}\right) &= E\left((\partial_{s_1}\psi)\check{}\right) = E\left(Q_{s_1}\tilde{\psi}\right) \\ &= E\left(Q_{s_1}D_{Z_1}\tilde{\varphi}\right) = E\left(D_{Z_1}\left(Q_{s_1}\tilde{\varphi}\right)\right) - E\left(\tilde{\varphi}D_{Z_1}Q_{s_1}\right) \end{aligned}$$

which leads to the inductive formula

$$Q_s = \delta(Z_1)Q_{s_1} - D_{Z_1}Q_{s_1}.$$

By the inductive hypothesis $Q_{s_1} \in \mathbb{D}_\infty$, therefore $Q_s \in \mathbb{D}_\infty$. □

5.3. *Proof of Theorem 5.1.*

We introduce the characteristic function u of the law of g:

$$u(\xi) = E\big(\exp(i\xi.g)\big).$$

Then denoting by Δ the Laplacian on $\mathbb{R}^d$, we have for all integers k

$$\begin{aligned} |\xi|^{2k}u(\xi) &= E\big(\psi_k(g)\big) \quad \text{where} \quad \psi(\eta) = (-1)^k\Delta_\eta^k\exp(i\xi.\eta)) \\ &= E\left(Q_k\exp(i\xi.g)\right), \end{aligned}$$

where Q_k is a functional on X, determined by the application of Lemma 5.2. This functional is independent of ξ; therefore

$$|u(\xi)| \le |\xi|^{-2k}E(|Q_k|)$$

which means that u is rapidly decreasing. The partial derivatives of u can be obtained from the identities

$$\partial_s\left(|\xi|^{2k}u(\xi)\right) = E\left(\theta_s(g)Q_k\exp(i\xi.g)\right) \quad \text{where} \quad \theta_s(\xi) = \prod_j (i\xi_j)^{s(j)}.$$

By Schwarz' inequality

$$\left|\partial_s\left(|\xi|^{2k}u(\xi)\right)\right| \le \|g\|_{L^{2|s|}}\|Q_s\|_{L^2}.$$

Therefore $u \in \mathcal{S}(\mathbb{R}^d)$; as the Fourier transform is an isomorphism of $\mathcal{S}(\mathbb{R}^d)$ we deduce that the law of g itself $\in \mathcal{S}(\mathbb{R}^d)$. □

6. Pushing down and lifting up through a smooth non-degenerate map

This paragraph is an amplification of 2.2 and 2.3. A lifting up will be made by the methods of differential geometry and pushing down by conditional expectation. We start with the following remark.

The conditional expectation E^g defines

$$E^g : L^{\infty-0}(X) \mapsto L^{\infty-0}(kd\xi).$$

The *range* of a non-degenerate map k will be the open subset of $\mathbb{R}^d$ where $k(\xi) > 0$.

6.1. Direct image of smooth functions

Theorem. *Given a non-degenerate map g with range O, then E^g defines a map from $\mathbb{D}_\infty(X)$ to $C^\infty(O)$.*

More precisely, for all $p < +\infty$ and r there exists p' such that

$$\|k\ E^g(f)\|_{W^{p,r}} \leq c\|f\|_{\mathbb{D}_r^{p'}}.$$

Proof. Given $f \in \mathbb{D}_\infty(X)$, consider the signed measured $d\nu = f d\mu$. We have

$$(g)_* \nu = k u d\xi$$

where $E^g(f) = u$.

Write σ for the covariance matrix, $\gamma = \sigma^{-1}$ and $Z_1 = \sum_k \gamma_{1,k} D g_k$. Then

$$D_{Z_1} \tilde{\varphi} = (\partial_1 \varphi)\tilde{}$$

for every $\varphi \in C_b^1(\mathbb{R}^d)$

$$E\left(f D_{Z_1} \tilde{\varphi}\right) = E\left(\delta(Z_1) f \tilde{\varphi}\right) - E\left(\tilde{\varphi} D_{Z_1} f\right)$$

which implies

$$\int u \frac{\partial \varphi}{\partial \xi_1} k d\xi = \int \varphi E^g\left((\delta Z_1) f - D_{Z_1} f\right) k d\xi$$

which implies

$$\frac{\partial(uk)}{\partial \xi_1} = q_1^f k,$$

where

$$q_1^f = E^g\left((\delta Z_1) - D_{Z_1} f\right).$$

More generally given a multi-index s there exists q_s such that

$$\partial_s(uk) = q_s^f k$$

where $q_s^f \in L^{\infty - 0}(k d\xi)$. Therefore, denoting $O_\varepsilon = \{k(\xi) > \varepsilon\}$

$$\partial_s(uk) \in L^p(O_\varepsilon, d\xi)$$

for all s, which implies $uk \in C^\infty(O)$. As $k \in C^\infty(O)$

$$k^{-1}(uk) = u \in C^\infty(O).$$

□

Remark. Detailed computations of derivatives are carried out in the Appendix.

6.2. Inverse image of distributions

Watanabe introduces the space $D_{-\infty}(X)$ of *distributions* on X defined as

$$D_{-\infty}(X) = \bigcup_{p,r} \Lambda^{-r}\left(L^p(X)\right)$$

where $p \in]1, +\infty]$, $r \in \mathbb{N}$.

Then $T \in D_{-\infty}(X)$ defines a linear form on $D_\infty(X)$ by the formula

$$\langle T, \varphi \rangle = \langle \Lambda^{-r} f, \varphi \rangle = E\left(f \Lambda^{-r} \varphi\right).$$

Conversely every *continuous* linear form on $D_\infty(X)$ has this expression. We denote

$$\|T\|_{D^q_{-r}} = \|\Lambda^{-r} T\|_{L^q} \leq +\infty.$$

We denote by $\mathcal{E}'(O)$ the Schwartz space of distributions with *compact support* contained in 0, then $\mathcal{E}'(O)$ is the dual space of $C^\infty(O)$.

Theorem (S. Watanabe; Inverse image of distributions). *There exists a natural map $(E^g)^*$ from $\mathcal{E}'(O) \to \mathbb{D}_{-\infty}(X)$.*

Proof. Given $T \in \mathcal{E}'(O)$, define a linear map ℓ from $\mathcal{L}$ on D_∞ by

$$\ell(f) = \langle E^g(f), T \rangle.$$

We have to show that ℓ is continuous. This means that there exist k, p, r such that

6.2.1. $$|\ell(f)| < c \|f\|_{\mathbb{D}^p_r}.$$

The continuity of T means that there exists s and a compact $K \subset O$ such that

$$|\langle T, \varphi \rangle| \leq c \sup_{\substack{|s|<r \\ \xi \in K}} |\partial_s \varphi(\xi)|.$$

There exists $\varepsilon > 0$ such that

$$k(\xi) > \varepsilon \quad \text{if} \quad \xi \in K,$$

therefore $k^{-1} \in C^\infty(K)$ and

$$\|\varphi\|_{W^{p,r}} < c \|k\varphi\|_{W^{p,r}} \quad \text{if} \quad \operatorname{supp}(\varphi) \subset K,$$

then by 4.1.1

$$|\ell(f)| < c \|f\|_{\mathbb{D}^p_r}.$$

□

6.2.2. A duality formula

$$\langle E^g(f), T\rangle = \langle f, (E^g)^*T\rangle$$

which means that $(E^g)^*$ is the *adjoint of the conditional expectation*; this fact justifies *a posteriori* our notation.

6.3. Pushed-down Riemannian structure

The Cameron-Martin space with its Hilbert structure provides a "Riemannian structure" on X. We want to get its image under a smooth non-degenerate map. To show the intrinsic character of our construction we shall replace the space $\mathbb{R}^d$ by a smooth abstract manifold V of dimension d.

6.3.1. Manifold-valued smooth maps

We shall denote by $\mathbb{D}_\infty^{\text{loc}}(X; V)$ the space of maps $g : X \to V$ such that there exists an atlas $\varphi_j : O_j \to \mathbb{R}^d$, O_j open in V such that

$$u(\varphi_j \circ g) \in \mathbb{D}_\infty(X; R^d)$$

where u is any smooth map from $\mathbb{R}^d \to \mathbb{R}^d$ with support contained in $\varphi_j(O_j)$. It is sufficient to check this relation for a system u_j of maps such that $O'_j = \{v;\, u_j(\varphi_j(0)) = \varphi_j(v)\}$ constitutes a covering of V.

We denote $g_j = u_j \circ g$. Then $g_j \in D_\infty(X, R^d)$.

We say that g is non-degenerate if $\left(\det g'_j\right)^{-1} \mathbf{1}_{\Omega_j}(g_j) \in L^{\infty-0}$ where $\Omega_j = \varphi_j(O_j)$.

We shall assume throughout this paragraph that g is smooth and non-degenerate. Then by applying 5.1 in a local chart we find that $(g)_* u$ *has a smooth density* k.

We shall denote $V' = \{v \in V, k(v) > 0\}$. Then V' is an open set and therefore a C^∞ manifold. Furthermore by Fang [2], V' *is connected*. We shall suppose in all the following that $V = V'$, which means that $k > 0$ on all V.

6.3.2. Lifting up differential forms

We denote by $\Lambda_0(V)$ the vector space of smooth differentiable forms on V of degree 1, with compact support. Given $\pi \in \Lambda_0$, we take a partition of unity subordinated to the covering O'_j and we write π as a finite sum

$$\pi = \sum_{j=1}^{n} \pi_j \quad \text{where} \quad \operatorname{supp}(\pi_j) \subset O'_j.$$

We also define

$$g^*\pi = \sum_j g_j^* \pi_j$$

where

$$g_j^* \left(\sum_{k=1}^d a_k d\xi^k \right) = \sum_{k=1}^d a_k\big(g(x)\big) Dg^k(x).$$

Then $(g^*\pi)$ is a differential 1-form on X in the sense of Shigekawa.

6.3.3. Pushing down the Riemannian structure

Given $\pi \in \Lambda_0(V)$ we define $g^*\pi$. Then $\|(g^*\pi)(x)\|_H^2$ is a positive function. We denote

$$\|\pi_v\|^2 = E^g\left(\|g^*\pi\|^2\right).$$

Then if g is *non-degenerate*, $\|\pi_v\|^2$ is a non-degenerate quadratic form on $T_v^*(V)$. We therefore get a Riemannian metric on V. Applying 5.1 in a local chart we get that this Riemannian structure is smooth.

6.3.4. Intrinsic Ornstein-Uhlenbeck operator

Given $v_0 \in V$ we denote by n_{v_0} the normal chart of center v_0, that is the map of $T_{v_0}(V)$ into V which associates to each ray of $T_{v_0}(v)$ the geodesic on V tangent to this ray.

We define the *Ornstein-Uhlenbeck operator* $\mathcal{L}_V$ on V-valued maps on X by

$$(\mathcal{L}_V g)(x_0) = \left[\mathcal{L}\left(n_{g(x_0)} \circ g\right)\right]_{x=x_0}.$$

Then $(\mathcal{L}_v g)(x)$ is a vector tangent to V at $g(x)$.

6.3.5. Theorem. *Denote by dv the volume element of the Riemannian metric, by k the density of $(g)_*\mu$ relative to dv, by ∇_V the gradient operator on the Riemannian manifold V, then*

$$\nabla_V(\log k) = 2E^g\left(\mathcal{L}_V(g)\right).$$

Proof. See Airault [1].

7. Absolute continuity under first derivative hypothesis

We shall treat the case of a scalar valued random variable. For $\mathbb{R}^d$-valued random variables a D_1^2 theorem is true; this is a basic result of Bouleau-Hirsch [7].

7.1. Theorem (Nualart-Zakai). *Given $g \in \mathbb{D}_1^1(X)$, denote*

$$A = \{x : Dg(x) \neq 0\},$$

then $(g)_*(\mathbf{1}_A P)$ is absolutely continuous relative to the Lebesgue measure.

Proof. We define a vector field $Z \in L^\infty(X; H)$ on X by

$$Z = \frac{Dg}{1 + \|Dg\|^2}$$

then

$$(Dg \mid Z) = \frac{\|Dg\|^2}{1 + \|Dg\|^2}.$$

Denote $A_\varepsilon = \{x; \|Dg\|(x) > \varepsilon\}$. Then

$$(Dg \mid Z) > \frac{1}{2}\varepsilon^2 \quad \text{on} \quad A_\varepsilon.$$

The inconvenience with the vector field Z is that Z has no kind of smoothness. We regularize it by using the Ornstein-Uhlenbeck semigroup

$$Z^t = P_t Z \qquad (t \geq 0),$$

then Z^t converges in $L^{\infty-0}$ towards Z^0. Furthermore for $t > 0$, $Z^t \in \mathbb{D}_\infty(X; H)$. Therefore δZ^t exists. Given $\varphi \in C^1(R)$, we have

$$E\left((Z^t \mid D\tilde{\varphi})\right) = E\left(\tilde{\varphi}\delta Z^t\right).$$

Remark that

$$D\tilde{\varphi} = \varphi'\big(g(x)\big)Dg(x),$$

then we get

$$E\left(\varphi'\big(g(x)\big)\big(Z^t \mid Dg\big)_x\right) = E\left(\varphi\big(g(x)\big)\delta Z_x^t\right).$$

We introduce

$$\begin{aligned} \alpha_t(\xi) &= E^{g(x)=\xi}\left[\big(Z^t \mid Dg\big)_H\right] \\ \beta_t(\xi) &= E^{g(x)=\xi}\left[\delta Z^t\right] \end{aligned}$$

Denoting by ρ the law of g we get

7.1.1.
$$\begin{cases} \int \varphi'(\xi)\alpha_t(\xi)d\rho(\xi) = \int \beta_t(\xi)\varphi(\xi)d\rho(\xi) \\ \int [\alpha_t(\xi)]^p \, d\rho(\xi) \leq \|Z^t\|_{L^p}^p < +\infty \\ \int [\beta_t(\xi)]^p \, d\rho(\xi) \leq \|Z^t\|_{\mathbb{D}^p}^p < +\infty \end{cases}$$

α_t converges in $L^{\infty-0}(\rho)$ towards α_0. Assume that $\rho_\varepsilon = (g)_*\left(\mathbf{1}_{A_\varepsilon} P\right)$ is not absolutely continuous relatively to the Lebesgue measure. Then there exists a compact set K' of zero Lebesgue measure such that $\rho_\varepsilon(K') > 0$. We therefore have

$$\alpha_0(\xi) \geq \frac{\varepsilon^2}{2} q(\xi) \quad \text{where} \quad q = \frac{d\rho_\varepsilon}{d\rho}.$$

There exists $K \subset K'$, $\rho(K) > 0$ and we fix t_0 such that

$$\alpha_{t_0}(\xi) > \frac{\varepsilon^2}{4} \qquad \xi \in K.$$

We take a sequence u_n of uniformly bounded continuous functions such that

$$0 < u_n < 1, \quad u_n(\xi) \to \mathbf{1}_K(\xi), \quad u_n(\xi) = 0 \quad \text{if } \xi < -R, \quad (R \text{ large enough}).$$

We define

$$\varphi_n(\xi) = \int_{-\infty}^{\xi} u_n(\lambda) d\lambda,$$

then $\varphi_n \in C^1(R)$, $\|\varphi_n\|_{L^\infty} \to 0$. We denote

$$\begin{aligned} A_n &= \int_K \alpha_{t_0}(\xi) u_n(\xi) d\rho(\xi) + \int_{K^c} \alpha_{t_0}(\xi) u_n(\xi) d\rho(\xi) \\ B_n &= \int \varphi_n(\xi) \beta_{t_0}(\xi) dp(\xi), \end{aligned}$$

then we have

$$\liminf(A_n) \geq \frac{\varepsilon^2}{4} \rho(K) \quad \text{and} \quad \lim B_n = 0.$$

Now by 7.1.1., $A_n = B_n$, which is a contradiction.

8. Appendix: Derivations computed by divergence operators

Given a map $g : X \to \mathbb{R}^d$, and z a vector field on $\mathbb{R}^d$, denote by Z its lift to the Wiener space.

Denote by μ the Wiener measure and let $q \in L^1_\mu(X)$; denote

$$\rho = g_*(q\mu); \quad \nu = g_*(\mu); \quad k = \frac{d\rho}{d\nu} = E^g(q).$$

Theorem.

(8.1) $$\delta_\nu(kz) = E^g\big(\delta_\mu(qZ)\big),$$

E^g denoting the conditional expectation under g.

Proof. Given $f \in C^1(M)$ we write $F = f \circ g$; then $[\partial_z f] \circ g = \partial_Z F$.
Therefore

$$\int \delta_\nu(kz) f \, d\nu = \int \partial_{kz} f \, d\nu = \int [\partial_z f] k \, d\nu = \int [\partial_z f] \circ g \; q d\mu$$

$$= \int \partial_Z F \; q d\mu = \int \partial_{qZ} F \, d\mu = \int F \delta_\mu(qZ) \, d\mu = \int f E^g [\delta_\mu(qZ)] \, d\mu.$$

As the equality between the first and the last term is valid $\forall f \in C^1$ this implies (8.1).

Proposition.

$$\begin{aligned}(8.2) \qquad \partial_z k &= -E^g(\delta_\mu(qZ)) + kE^g(\delta_\mu(Z)) \\ &= E^g(\partial_Z q) + [kE^g(\delta_\mu(Z)) - E^g(q\delta_\mu(Z))] \\ &= E^g[\partial_Z q - q\delta(Z)] + kE^g[\delta_\mu(Z)].\end{aligned}$$

Lemma.

(8.3) $$\delta_\nu(kz) = k\delta_\nu(z) - \partial_z k.$$

Proof. 1.3.3.

Proof (of (8.2)). By (8.3)

$$\partial_z k = k\delta_\nu(z) - \delta_\nu(kz).$$

We apply (8.1) twice, for the first term we take $q = 1$ and we get the first identity (8.2); applying (1.3.3) to the first term of the middle equation in (8.2) we prove the second identity. □

We want to transform this result into an expression that can be iterated more easily.

Theorem. *Denote by $\mathcal{A}$ the conditional expectation operator E^g, then*

(8.5) $$\partial_z \mathcal{A} - \mathcal{A}\partial_Z = M_{\mathcal{A}\delta Z}\mathcal{A} - \mathcal{A}M_{\delta Z}$$

where M_φ denotes the operator of multiplication by φ.

Example: Computation of the Hessian. (see Malliavin-Stroock [1]) Denote by ν the law of g and by π its density relative to the volume measure of $\mathbb{R}^d$. We give on $\mathbb{R}^d$ two vector fields z_1, z_2 having a vanishing divergence relative to the Lebesgue measure of $\mathbb{R}^d$.

Theorem.

$$\begin{aligned}-\partial_{z_1}\partial_{z_2} \log \pi &= E^g(\partial_{Z_1}\delta_\mu(Z_2)) \\ &\quad + [E^g(\delta_\mu(Z_1))E^g(\delta_\mu(Z_2)) - E^g(\delta_\mu(Z_1)\delta_\mu(Z_2))] \\ (8.7) \qquad &= E^g[\partial_{Z_1}\delta_\mu(Z_2) - \delta_\mu(Z_1)\delta_\mu(Z_2)] \\ &\quad + E^g(\delta_\mu(Z_1))E^g(\delta_\mu(Z_2)).\end{aligned}$$

Proof.

$$\int \delta_\nu(z) f \, d\nu = \int \left[\partial_z f\right] \, \pi dm = -\int f \partial_z[\log \pi] \, d\nu.$$

The identity between the first and the last term being valid $\forall f \in C^1$, we deduce that

$$\partial_{z_2} \log \pi = -\delta_\nu(z_2) = -E^g\big(\delta_\mu(Z_2)\big) \tag{8.8}$$

where the last equality comes from the application of (8.1) where we have taken $q = 1$. We multiply the two members of (8.8) by ∂_{z_1}; we compute the r.h.s. by (8.2) and get the theorem. □

9. Appendix: Law of a weakly non degenerated map

Theorem. *Given* $g \in \mathbb{D}_2^{\infty-0}(X; R^d)$ *such that* $\det(g') > 0$ *almost surely, then the law of g is absolutely continuous relatively to the Lebesgue measure.*

Proof. We introduce the cutoff function $\psi_\epsilon(x) = v(\epsilon^{-1}\det(g')(x))$, where $v(s)$ is a C^∞ function, equal to 1 for $s \geq 1$, $1 > v(s) > 0$ for $s > \frac{1}{2}$, $v(s) = 0$ for $s \leq \frac{1}{2}$ and such that furthermore $|v'(s)| = c(\delta) v^{1-\delta}(s)$, $\forall \delta > 0$. A building block which permit the construction of such function is $1_{s>0} \exp(-s^{-2})$. We introduce the measure $\mu_\epsilon = \psi_\epsilon \mu$; then $\lim_{\epsilon \to 0} \|\mu_\epsilon - \mu\| = 0$; therefore it is sufficient to prove that $\forall \epsilon > 0$, $\rho_\epsilon = g_* \mu_\epsilon$ is absolutely continuous relatively to the Lebesgue measure. We lift the coordinate vector field $\frac{\partial}{\partial \xi_k}$ into the vector field Z_k defined in 3.1. Define $Z_k^\epsilon = \psi_{\frac{\epsilon}{2}} Z_k$ then $Z_k^\epsilon \in \mathbb{D}_1^{\infty-0}$, therefore $\delta(Z_k^\epsilon)$ exists and belongs to $L^{\infty-0}$. By 1.3.3 we have $\delta_{\mu_\epsilon}(Z_k^\epsilon) = \delta_\mu(Z_k^\epsilon) - D_{Z_k^\epsilon}(\log(\psi_\epsilon))$; then

$$\left[\int |\delta_{\mu_\epsilon}(Z_k^\epsilon)|^p \psi_\epsilon d\mu\right]^{\frac{1}{p}} \leq \|\delta_\mu(Z_k^\epsilon)\|_{L^p(\mu)} + c(1 - \frac{1}{2p}) \|Z_k^\epsilon\|_{L^{2p}} \|g\|_{\mathbb{D}_2^{2p}};$$

then using the method of 4.4 we obtain that μ_ϵ is absolutely continuous relatively to the Lebesgue measure. □

Notes and comments on Part I

For the purposes of quantum theory, Segal developed in the Fifties his algebraic approach to integration; in this context a probability space is a C^*-commutative algebra with unit on which a positive linear form is defined; such a C^*-algebra can be realized as an algebra of random variables on a classical probability space *defined up to isomorphism*. In the Sixties, taking the opposite point of view, Gross [1] looked for a *specific realization* on the Borel σ-field of a fixed Banach space: when specialized to Gaussian random variables, Gross' theory gives the theory of *Abstract Wiener spaces*, which extends Wiener's construction [1] of the *Brownian measure* on the Banach space of continuous function. Gross' methodology works in the framework of vector spaces; it is not applicable to Riemannian path spaces (see Part V). Itô began the abstract theory of Gaussian probability space in [10]: he does not deal with a specific numerical mode; roughly speaking, he considers a Gaussian process indexed by the points of the Cameron-Martin space, a process which is separable in the sense of Doob.

The theory of H-derivation on an abstract Wiener space was initiated in Gross [2]; the corresponding Sobolev-type spaces were developed in Stroock [1] (see also Krée, P. [1]). The construction of Sobolev spaces by means of differentiable vectors of the representation of a group of translations is taken from Malliavin [16].

The Krée-Meyer inequalities were first proved in the L^2 case by Krée M. ([1]): in this case L^p-inequalities become L^2-identities; it is a general fact in Littlewood-Paley theory that L^2-identities sometimes induce L^p inequalities; in Gaussian spaces this induction was realized by P.A. Meyer [2], [3] applying to the OU semigroup Stein's extension [1] of the Littlewood-Paley theory to an abstract semigroup. Krée (M. and P. [1]) and Watanabe [2] deduced from L^p-inequalities the existence of the divergence for smooth vector fields.

The Calderon-Zygmund *rotation method* [1] proves the boundedness in L^p of the Riesz transform in $\mathbb{R}^d$ from the boundedness of the Hilbert transform, using a fibering of $\mathbb{R}^d$ by straight lines parallel to a fixed direction and averaging on directions; fibering the product Gaussian spaces by a family of circles, Pisier [1] gave a rotation-type proof of the Krée-Meyer inequalities which is the approach used in this book. Malliavin-Nualart [2] used Pisier's method for proving the Krée-Meyer inequalities for the case of certain vector-valued functionals.

Bakry [2], [3] constructs an abstract Littlewood-Paley theory which proves Riesz transform inequalities for general semigroups satisfying a *positivity assumption*.

Cruzeiro [1], [3], [4] proved that a smooth vector field Z on X defines a unique flow on X, preserving Wiener measure up to equivalence under the main assumption that $\delta(Z)$ belongs to an exponential class.

An important fact is that the set of smooth functionals D_∞ constitutes an algebra which is stable under C^∞ symbolic calculus; thinking of D_∞ as the Schwartz space $\mathcal{S}(\mathbb{R}^d)$, Watanabe [1] introduced a corresponding class of distributions.

The notion of a non-degenerate map was introduced in Malliavin [10], by a differential calculus based on Itô calculus along the OU process (see Part V). Shigekawa [1], [2] initiated the approach based on the usual differential calculus.

An important problem is to find minimal regularity assumptions for absolute continuity of the laws. In the case of scalar functionals, the hypothesis D_1^1 is shown to be sufficient by Nualart-Zakai [4]; this result has been applied to the law of the maximum of a continuous process (see Florit-Nualart [1] for the Brownian sheet); Nualart's book [9] discusses these results.

Bouleau-Hirsch proved the absolute continuity of non-degenerate $\mathbb{R}^d$-valued functionals in D_1^2 by doing a direct geometric construction of a Federer-type co-area formula (see Chapter V) in the context of Dirichlet forms. It should be emphasized that there does not exist currently an alternative approach to the regularity of laws for $\mathbb{R}^d$-valued functionals with one derivative.

At the opposite extreme to low regularity are higher regularity statements: real analytic functionals or functionals belonging to a Gevrey class; under such assumptions the Gevrey regularity of the laws was proved by Ren [6], Malliavin-Taniguchi [1].

Part II

Quasi-Sure Analysis

In finite dimensions, a function belonging locally to all the Sobolev spaces is a C^∞ function. This fact is a consequence of the Bernstein-Sobolev embedding $H^s(\mathbb{R}^d) \subset C_b(\mathbb{R}^d)$ for $s > d/2$. By contrast, $D_\infty(X)$ is not contained in any space of continuous functions. The quasi-sure analysis is built as a substitute on a Gaussian probability space for the continuity of the functions of D_∞.

A Borel measure on X is of *finite energy* if it defines a continuous linear form on $D_\infty(X)$. A Borel set is said to be *slim* if it does not carry a measure of finite energy. The basic σ-field of quasi-sure analysis will be the Borel σ-field "quotiented" by the slim sets. In classical probability, the basic σ-field is the Borel σ-field "quotiented" by the events of probability zero. As every slim set is of probability zero, quasi-sure analysis will be a refinement of classical probability theory.

The notion of continuity in finite dimensions will be replaced by the notion of *quasi-continuity*, which means continuity outside "almost negligible" sets. The notion of continuity will be formulated relative to the topology of a fixed numerical model. A basic issue is to prove that this definition is independent of the numerical model which has been fixed. The proof of this fact will result from the development in infinite dimensions of a non-linear potential theory. Quasi-sure analysis can be constructed on any abstract Wiener space; again we shall show the independence of this construction from the choice of a Gross radonifying norm.

Quasi-sure analysis makes it possible to build a *co-area formula* describing the disintegration of the Gaussian measure along an $\mathbb{R}^d$-valued, non-degenerate map. In quasi-sure analysis, the "leaves" of such a map exist for all $\xi \in \mathbb{R}^d$. The desired disintegration formula is based on a construction of the *area* for the *finite-codimensional submanifolds* constituted by the leaves. The computations of differential geometry on submanifolds of $\mathbb{R}^d$ can be transferred to finite-codimensional submanifolds of a Gaussian space.

It seems that an exposition of quasi-sure analysis appears here for the first time in book form. In the remainder of this book quasi-sure analysis is used only slightly: **As this Part II is highly technical, it can be skipped on first reading.**

Chapter IV

Foundations of Quasi-Sure Analysis: Hierarchy of Capacities and Precise Gaussian Probability Spaces

Contents: Capacities on a numerical model – Tightness of capacities – Quasi-continuous functions – Tchebycheff inequalities – Redefinitions – Positive generalized functionals, their representation by a Borel measure – Equilibrium potentials – Continuity of capacities – Capacitability of Borel sets – Equilibrium measures – Measures of finite energy – Charge and capacity – Slim sets are null sets for all measures of finite energy – Invariance of capacities under a change of numerical model, precise Gaussian probability space – Quasi-sure analysis on an abstract Wiener space.

The essence of quasi-sure analysis is to work with a smaller class of negligible sets than the usual negligible sets of *almost sure analysis*, the sets of probability zero.

One advantage of *quasi-sure analysis* is the *principle of descent* which will make *non-degenerate conditioning* possible. For instance, conditioning on the value at time 1 of a Brownian motion is non-degenerate. Therefore *pinned* Brownian motion is well defined in quasi-sure analysis; this approach could lead to sharper estimates than other well established methods.

The principle of descent is a consequence of the existence of a continuous *disintegration* of Wiener measure induced by a non-degenerate conditioning. Conditioning by the first coordinate of a random variable with a *smooth density* on $\mathbb{R}^d$ induces a *continuous disintegration.* Quasi-sure analysis will lead to a natural extension of this result to a Gaussian probability space.

We shall first build quasi-sure analysis on a *numerical model*; the two main ingredients will be the *hierarchy of capacities* and *redefinition of a smooth function.*

An important fact will be the *tightness* of capacities on an increasing family of compacts sets.

The notion dual to capacity is the notion of Watanabe's *generalized functionals of finite energy*. Then the following important *positivity result* of Sugita holds true.

A *positive distribution of finite energy is given by a positive Borel measure.*

We shall define the (p, r) potential of a Watanabe generalized functional. Then we shall prove that every Borel set of positive (p, r) capacity has an *equilibrium potential*. Furthermore this equilibrium potential is equal to a non-linear expression involving the (p', r) potential of a positive measure: the *equilibrium measure*.

It will follow that the class of Borel sets of (p, r)-capacity zero coincides with the class of Borel sets which are not charged by any measure of finite energy.

Finally, for a compact K we define its (p, r)-*charge* as the supremum of its mass for measures of (p, r)-energy less than 1. We shall prove the equality of charge with capacity.

The non-linear theory of potential used here is a well established subject in finite dimensions, the passage to the infinite dimensions is possible with essentially a single loss: the maximum principle.

All the previously described theory has been developed on a *numerical model*. To show that quasi-sure analysis is intrinsic we must establish covariance of *numerical quasi-sure analysis* under a change of basis in H; this fact will be established in a section of its own.

Another approach to Gaussian integration is the Gross theory of abstract Wiener space. The theory of capacity will be developed starting with the topology induced by the underlying Banach norm. Then *tightness* of these new capacities will be proved, and finally, the equivalence of quasi-sure analysis in these two frameworks.

1. Hierarchy of capacities on a numerical model

1.1. Notation

We consider the Hilbert space $H = \ell_2(\mathbb{N})$ of square summable real sequences. Then $\mathrm{Seg}(H)$ is the space $\mathbb{R}^{\mathbb{N}}$ with the Gaussian measure μ. The Hermite polynomials define a canonical basis of $L^2(\mathbb{R}^{\mathbb{N}}; \mu)$.

We define a map φ from $\mathbb{R}$ to the circle $\mathbb{T}$ by

$$\xi \to 2\arctan(\xi).$$

We shall consider on $\mathbb{R}^{\mathbb{N}}$ the product metric

1.1.1. $$d(\xi_1, \xi_2) = \sum_{k=2}^{+\infty} k^{-2} d\left(\varphi(\xi_1^k), \varphi(\xi_2^k)\right)$$

where d is the distance on the circle, and the topology is defined by this metric. We define a fundamental system of neighbourhoods of ξ_0 by

$$V_{p,q}(\xi_0) = \left\{\xi;\, \left|\xi^s - \xi_0^s\right| < \frac{1}{q},\quad s \in [1, p]\right\} \quad \text{with} \quad p, q \in \mathbb{N}$$

then

1.1.2. $$\mu(\mathcal{O}) > 0 \quad \text{for every open set} \quad \mathcal{O}.$$

We remark that $\Phi : \mathbb{R}^{\mathbb{N}} \to \mathbb{T}^{\mathbb{N}}$ defined by $\xi^p \to \varphi(\xi^p)$, $p \in \mathbb{N}$, is a homeomorphism of $\mathbb{R}^{\mathbb{N}}$ onto its image.

1.2. Capacities

We fix $p \in]1, +\infty[$ and an integer r. Given an open set $\mathcal{O}$ we define

(i) $$c_{p,r}(\mathcal{O}) = \inf\left\{\|\varphi\|_{\mathbb{D}_r^p};\ \varphi(\xi) \geq 1 \ \text{ a.e. } \ \xi \in \mathcal{O} \right. \\ \left. \text{and} \ \ \varphi(\xi) \geq 0 \quad \mu - \text{a.e.} \quad \xi \in \mathbb{R}^{\mathbb{N}}\right\}.$$

Throughout this chapter we will use on $\mathbb{D}_r^p$ the norm

$$\|\varphi\|_{\mathbb{D}_r^p} = \|\Lambda^{-r}\varphi\|_{L^p}.$$

We remark that $\mathbb{D}_r^p$ has been defined as a subspace of L^p; $\varphi \in \mathbb{D}_r^p$ is defined μ-a.e. and the inequality appearing in (i) can be stated μ-a.e. Furthermore if $0 < r < r'$

$$[\mu(\mathcal{O})]^{1/p} = c_{p,0}(\mathcal{O}) \leq c_{p,r}(\mathcal{O}) \leq c_{p,r'}(\mathcal{O}).$$

Given a subset A of $\mathbb{R}^{\mathbb{N}}$ we define

(ii) $$c_{p,r}(A) = \inf c_{p,r}(\mathcal{O}), \quad \mathcal{O} \quad \text{open}, \quad \mathcal{O} \supset A.$$

When A is open, this definition is consistent with the definition (i).

A property π will be said to hold true (p, r)-quasi-everywhere if

$$c_{p,r}\left(\{x, \pi_x \text{ is not satisfied }\}\right) = 0.$$

We remark that

$$c_{p,r}(A) \leq c_{p,r}(B) \quad \text{if} \quad A \subset B.$$

A subset G of $\mathbb{R}^{\mathbb{N}}$ will be called *slim* if

(iii) $$c_{p,r}(G) = 0$$

for all $p \in]1, +\infty[$ and all integers r.

A property π will be said to hold *quasi-surely* if

$$\{x; \pi_x \quad \text{is not satisfied }\} \quad \text{is slim}.$$

Remark. In finite dimensions, slim implies empty; in fact the embedding $W_r^p(\mathbb{R}^d) \subset C_b(\mathbb{R}^d)$ for $pr > d$ generates the implication $c_{p,r}(A) = 0 \Longrightarrow A$ empty. In infinite dimension, there exist non-empty slim sets.

1.2.1. Proposition.

$$c_{p,r}\left(\bigcup_i G_i\right) \leq \sum_{i=1}^{+\infty} c_{p,r}(G_i).$$

Proof. It is sufficient to prove the lemma for G_i open. Consider $\varphi_i > 0$ such that

$$\varphi_i(\xi) \geq 1 \quad (\text{a.e.} \quad \xi \in \mathcal{O}_i) \quad \text{and} \quad \|\varphi_i\|_{\mathbb{D}_r^p} < c_{p,r}(G_i) + \varepsilon 2^{-i}.$$

Denote $\psi = \sum_{i=1}^{+\infty} \varphi_i$. Then ψ satisfies $\psi > 1$ a.e. $\xi \in \bigcup \mathcal{O}_i$ and the inequality $\|\psi\| \leq \sum \|\varphi_i\|$ proves the lemma. □

1.2.2. Corollary. *Assume that $c_{p,r}(A_i) = 0$, then*

$$c_{p,r}\left(\bigcup_{i=1}^{+\infty} A_i\right) = 0.$$

1.2.3. Corollary. *The union of a countable family of slim sets is slim.*

1.2.4. Corollary (Borel-Cantelli). *If*

$$\sum c_{p,r}(A_k) < +\infty,$$

then outside a set of (p, r)-capacity zero every point belongs to only a finite number of A_k. That is to say, more briefly

$$c_{p,r}\left(\limsup_k A_k\right) = 0.$$

Proof. Denote $G_s = \bigcup_{k>s} A_k$. We have

$$c_{p,r}\left(\bigcap G_s\right) < c_{p,r}(G_s) < \sum_{s=k}^{+\infty} c_{p,r}(A_k).$$

The last term tends towards zero when $k \to \infty$. □

1.2.5. Tchebycheff lemma (weak form). *Given p, r, there exists a constant $M_{p,r}$ such that for every $u \in \mathbb{D}_r^p$ satisfying*

$$u \quad \textit{is lower semi-continuous},$$

then we have

$$c_{p,r}(\{x, u(\xi) > R\}) \leq R^{-1} M_{p,r} \|u\|_{\mathbb{D}_r^p}.$$

Remark. See 2.2 for a stronger statement.

Proof. We remark that

$$\mathcal{O}_R = \{\xi;\, u(\xi) > R\}$$

is open. Denote by ψ a smooth C^∞-function, $0 < \psi < 1$, such that $\psi(t) = 1$ if $t > 1$, $\psi(t) = 0$ if $t < \frac{1}{2}$ and write $\psi_R(t) = \psi(Rt)$. Then, by the chain rule

$$\begin{aligned} \left\| D^{(s)}\left(\psi_R(u)\right)\right\|_{L^p} &\le \tilde{M}_{p,s} R^{-s} \|u\|_{\mathbb{D}^p_s} \\ \|\psi_R(u)\|^p_{L^p} &\le (2^{-1}R)^{-p} \|u\|^p_{L^p}. \end{aligned}$$

We deduce by the Krée-Meyer inequalities that

$$\|\psi_r(u)\|_{\mathbb{D}^p_r} < R^{-1} M_{p,r}.$$

We conclude by remarking that $u(\xi) \ge 1$ if $\xi \in \mathcal{O}_R$. □

1.3. Tightness Theorem. *Consider the quadratic form*

$$q(\xi) = \sum_{k=1}^{+\infty} a_k (\xi^k)^2 \qquad \textit{where}$$

$$\sum a_k < +\infty \quad \textit{and} \quad a_k > 0 \;\; \textit{for all} \;\; k,$$

then

1.3.1. $$q \in \mathbb{D}_\infty$$

1.3.2. $$A_R = \left\{\xi \in \mathbb{R}^{\mathbb{N}};\quad q(\xi) \le R\right\} \quad \textit{is compact}$$

1.3.3. $$c_{p,r}(A^c_R) \to 0 \quad \textit{where} \quad R \to +\infty.$$

Proof. We compute

$$E\big(q(\xi)\big) = \sum a_k < +\infty$$

$$\begin{aligned} E\big(\exp(tq)\big) &= \prod_k E\left(\exp(t a_k \xi_k^2)\right) \\ &= \prod_k \left(1 - \frac{t}{2} a_k\right)^{-1} < +\infty \quad \text{for} \quad t < 2\|a\|^{-1}_{\ell_\infty}. \end{aligned}$$

Therefore $q \in L^{\infty-0}$. Furthermore q is a polynomial of degree 2, and we have

$$\|Dq\|^2_{\ell_2} = 4 \sum a_i^2 \xi_i^2 < +\infty$$

and 1.3.1 is established.

We remark that q is a lower semi-continuous function on $\mathbb{R}^{\mathbb{N}}$. Therefore A_R is closed; then, using the identification $\Phi : \mathbb{R}^{\mathbb{N}} \to \mathbb{T}^{\mathbb{N}}$, $\Phi(A_R)$ is closed in $\mathbb{T}^N$, and therefore compact.

The proof of 1.3.3 results from the lower semi-continuity of q and from the Tchebycheff Lemma 1.2.5.

1.3.4. Corollary (Exhaustion lemma). *Given a subset G of $\mathbb{R}^{\mathrm{N}}$, then*

$$c_{p,r}(A_R \cap G) = 0$$

for all R is equivalent to

$$c_{p,r}(G) = 0.$$

2. Redefinitions

2.1. Quasi-continuous functions

2.1.1. Definition. A function φ defined on $\mathbb{R}^{\mathrm{N}}$ is said to be (p, r)-quasi-continuous if for all ε, it is possible to find an open set $\mathcal{O}_\varepsilon$ such that

$$c_{p,r}(\mathcal{O}_\varepsilon) < \varepsilon$$

and the restriction of φ to $\mathcal{O}_\varepsilon^c$ is continuous.

Taking a sequence ε_n such that $\sum \varepsilon_n < +\infty$ and using 1.2.1 we get that φ is (p, r)-quasi-continuous if and only if there exists a decreasing sequence of open sets Ω_s such that

$$c_{p,r}(\Omega_s) \to 0 \quad \text{and such that}$$
$$\varphi \quad \text{is continuous on the complement of} \quad \Omega_s.$$

2.1.2. Definition. A function φ is said to be ∞-quasi-continuous if it is (p, r)-quasi-continuous for all p, r.

A function φ is ∞-quasi-continuous if and only if there exists a *net* consisting of a decreasing sequence of open sets Ω_n, such that on Ω_n^c φ is continuous and $c_{n,n}(\Omega_n) \to 0$.

We shall call the sequence $\{\Omega_n\}$ a *net* associated to φ.

Remark. We recall that it is possible to construct the theory of the Lebesgue integration starting with the following definition:
A function φ on $[0, 1]$ defines a Lebesgue measurable class if for every $\varepsilon > 0$ it is possible to find an open set $\mathcal{O}$ of Lebesgue measure smaller than ε such that φ restricted to $\mathcal{O}^c$ is continuous.

2.1.3. Essential part of a closed set

Given a closed set K we denote

$$\text{ess}(K) = G^c \cap K$$

where G is the *greatest open set* such that

(i) $$\mu(G \cap K) = 0.$$

Then

$$\mu\big(\mathrm{ess}(K)\big) = \mu(K).$$

We remark that if u is continuous on K and equal to zero μ-a.e. then the restriction of u to $\mathrm{ess}(K)$ is zero.

Given an open set Ω, set $K = \Omega^c$ and $G = \big(\mathrm{ess}(K)\big)^c$. Then

$$G \supset \Omega$$

$$\mu(G) = 1 - \mu\big(\mathrm{ess}(K)\big) = 1 - \mu(K) = \mu(\Omega)$$

which means

$$\mathbf{1}_G = \mathbf{1}_\Omega \quad \mu\text{-a.e.}$$

Therefore, the capacities of G being defined as the infinimum of $\|\psi\|_{p,r}$ such that $\psi \geq \mathbf{1}_G$ μ-a.e., we deduce that

$$c_{p,r}(G) = c_{p,r}(\Omega).$$

2.1.4. Definition. A (p, r)-net will be a decreasing sequence of open sets Ω_k such that

$$\lim_k c_{p,r}(\Omega_k) = 0$$

$$\Omega_k^c \quad \text{is compact}$$

$$\mathrm{ess}(\Omega_k^c) = \Omega_k^c.$$

Then given a (p, r)-quasi-continuous function u, it is possible to find a net on which u restricted to Ω_k^c is continuous; we shall call such net a *continuity net* for u.

2.1.5. Theorem. *Given a (p, r)-quasi-continuous function φ, then for any open set $\mathcal{O}$*

$$\mu\big(\{\xi; \varphi(\xi) \neq 0\} \cap \mathcal{O}\big) = 0$$

implies that

$$c_{p,r}\big(\{\xi; \varphi(\xi) \neq 0\} \cap \mathcal{O}\big) = 0.$$

Proof. Consider a continuity net Ω_s of φ. Define

$$D_s = \Omega_s^c \cap \mathcal{O};$$

then φ is continuous on D_s and is equal to zero almost everywhere. As

$$\mathrm{ess}(D_s) = D_s$$

we deduce that $\varphi(\xi) = 0$ for $\xi \in D_s$.

Denote by F_s the closure of D_s then

$$F_s \subset \Omega_s^c.$$

Define

$$G_s = F_s^c$$

then G_s is open and

$$G_s \subset \Omega_s$$

we deduce that

$$\Gamma = \bigcap G_s \quad \text{satisfies} \quad c_{p,r}(\Gamma) = 0.$$

As D_s is an increasing sequence we have

$$\mathcal{O} = \left(\bigcup_s D_s \right) \cup \Gamma.$$

□

2.1.6. Corollary. *If φ is (p, r)-quasi-continuous*

$$\varphi(\xi) \geq 0 \quad \text{a.e.} \quad \xi \in \mathcal{O},$$

then

$$\varphi(\xi) \geq 0 \quad \text{q.e.} \quad \xi \in \mathcal{O}.$$

Proof. Consider a smooth function v on $\mathbb{R}$, such that v and all its derivatives are bounded, $v(t) > 0$ if $t < 0$, and $v(t) = 0$ if $t \geq 0$. We apply 2.1.3 to $\varphi_1 = v \circ \varphi$.

□

2.1.7. Proposition. *Given $u_1, \ldots u_k, \ldots$ a sequence of quasi-continuous functions, it is possible to find a decreasing sequence of open sets $\mathcal{O}_j$, with their intersection of capacity zero, such that for every k, u_k restricted to $\mathcal{O}_j^c$ is continuous.*

Proof. We associate to u_k a decreasing sequence Ω_k^s of open sets. We proceed by a Cantor diagonalization procedure. We can find *functions* ε_j, $\varepsilon_j(k) < \varepsilon_{j+1}(k)$ such that

$$\sum_k c_{p,r} \left(\Omega_k^{\varepsilon_j(k)} \right) < \frac{1}{j};$$

then we take

$$\mathcal{O}_j = \bigcup_k \Omega_k^{\varepsilon_j(k)}.$$

□

2.2. Tchebycheff inequality theorem. *Given (p, r), there exists $M_{p,r}$ such that for any (p, r)-continuous function f we have*

$$c_{p,r}(A_R) \leq R^{-1} M_{p,r} \| f \|_{\mathbb{D}_r^p}, \quad \text{where}$$

$$A_R = \{\xi : f(\xi) > R\}.$$

Proof. Consider a continuity net Ω_s of f; there exists an open set G_s such that

$$G_s \cap \Omega_s^c = A_R \cap \Omega_s^c.$$

Define ψ_R as in 1.2.5, consider $u_R = \psi_R(f)$, then

$$\|u_R\|_{\mathbb{D}_r^p} \leq R^{-1} M_{p,r},$$

$$G_s \subset A_R \cup \Omega_s$$

implies that

$$\mathbf{1}_{G_s} \leq u_R + v$$

where

$$v \geq \mathbf{1}_{\Omega_s} \quad \text{and} \quad \|v\|_{\mathbb{D}_r^p} < 2 \ \ c_{p,r}(\Omega_s).$$

Therefore, as G_s is open,

$$c_{p,r}(G_s) \leq \|u_R\|_{\mathbb{D}_r^p} + \|v\|_{\mathbb{D}_r^p}.$$

On the other hand

$$A_R \subset G_s \cup \Omega_s$$

implies that

$$c_{p,r}(A_R) \leq \|u_R\|_{\mathbb{D}_r^p} + 3 \ \ c_{p,r}(\Omega_s).$$

Letting $s \to +\infty$ we obtain the theorem. □

2.3. Redefinition

2.3.1. Definition. Given a measurable function ψ, we call a (p,r)*-redefinition of* ψ a function ψ^* such that

$$\psi = \psi^* \quad \text{a.s.}$$
$$\psi^* \quad \text{is} \quad (p,r)\text{-continuous.}$$

2.3.2. *Uniqueness of a redefinition* It follows from 2.1.6 that two redefinitions differ on a set of (p,r)-capacity zero. If we say that two functions which differ on a set of (p,r)-capacity zero are (p,r) equivalent, then there is only one (p,r)-equivalence class containing all the redefinitions of ψ. In this sense we shall say that *the* redefinition of ψ is *essentially unique*.

2.3.3. Theorem. *Every function* $\varphi \in \mathbb{D}_r^p$ *has a* (p,r)*-redefinition, which can be taken to be in the first Baire class.*

Proof. Denote by $\mathcal{A}_s$ the σ-field on $\mathbb{R}^{\mathbb{N}}$ generated by the first s coordinates. We denote by P_t the Ornstein-Uhlenbeck semigroup, then $P_t E^{\mathcal{A}_s} = E^{\mathcal{A}_s} P_t$ and the same commutation holds true for the operator Λ^r. Given $\varphi \in D_r^p(X)$, then by

Krée-Meyer inequalities $\varphi = \Lambda^r \psi$, for some $\psi \in L^p$; denote by φ''_k, ψ_k the conditional expectations relative to $\mathcal{A}_k$; then ψ_k is an L^p martingale converging to ψ; the commutation relation implies that $\varphi''_k = \Lambda^r \psi_k$, therefore convergence of the ψ_k in L^p implies, by the Krée-Meyer inequalities, that

$$\|\varphi - \varphi''_k\|_{D^p_r} \to 0.$$

By extracting a subsequence k_s and writing $\varphi'_s = \varphi''_{k_s}$ we obtain a sequence such that

$$\|\varphi'_s - \varphi\|_{D^p_r} < s^{-2}.$$

We can find a sequence t_s such that $\varphi_s = P_{t_s}\varphi'_s$ satisfies the same strict inequality

$$\|\varphi_s - \varphi\|_{D^p_r} < s^{-2}.$$

By the spectral decomposition of $\mathcal{L}$ in L^2 we have

$$\|\mathcal{L}^r e^{t\mathcal{L}} \varphi_s\|_{L^2} \leq \left[\sup_j j^r e^{-t_s j}\right] \|\varphi'_s\|_{L^2}, \qquad r = 1, 2, \ldots$$

As φ_s is defined on R^{k_s}, using the last inequality with $r > \frac{1}{4}k_s$ the Bernstein-Sobolev embedding Lemma gives that φ_s is continuous.

Consider the open set

$$G_s = \left\{\xi; |\varphi_{s+1}(\xi) - \varphi_s(\xi)| > s^{-1/2}\right\}.$$

Then by 1.2.5 we have that

$$c_{p,r}(G_s) < M_{p,r} s^{-3/2}.$$

Write $\Gamma_q = \bigcup_{s>q} G_s$. Then by 1.2.1 we have

$$c_{p,r}(\Gamma_q) < 2q^{-1/2}$$

and

$$\sum_{s=q}^{+\infty} (\varphi_{s+1} - \varphi_s)$$

converges uniformly on Γ_q^c.

We denote by h_q its sum. Then h_q is a continuous function defined on Γ_q^c. We define φ^* by:

$$\begin{cases} \varphi^*(\xi) = h_q(\xi) + \varphi_q(\xi) & \text{if} \quad \xi \in \Gamma_q^c \\ \varphi^*(\xi) = 0 \quad \text{if} \quad \xi \in \bigcap \Gamma_q \end{cases}.$$

We remark that if $\xi \in \Gamma_q^c$, then $\xi \in \Gamma_{q+1}^c$; our definition is not ambiguous because

$$h_q(\xi) + \varphi_q(\xi) = h_{q+1}(\xi) + \varphi_{q+1}(\xi) \quad \text{if} \quad \xi \in \Gamma_q^c.$$

Furthermore $\varphi^*(\xi)$ is a Borel function (in fact of the first Baire class).

For all q, we have $\varphi^*(\xi) = \varphi(\xi)$ a.s. $\xi \in \Gamma_q^c$. □

2.3.4. Definition. We shall call a *redefinition net* of u a *continuity net* of u^*.

We remark that if Ω_q is a redefinition net of u, then the restriction of u to Ω_q^c is a representative for a.s. equivalence. As $\operatorname{ess}(\Omega_q^c) = \Omega_q^c$ it follows that this representative is *unique*. Therefore a redefinition net of u fully determines u^*.

2.3.5. Theorem. *Given* $\{\varphi_n\}, \varphi_\infty \in \mathbb{D}_r^p$, *assume that* $\|\varphi_n - \varphi_\infty\|_{\mathbb{D}_r^p} \to 0$. *Denote by* $\varphi_n^*, \varphi_\infty^*$ *their* (p,r)*-redefinitions. Then it is possible to find a subsequence* n_j *and a sequence of open sets* $\mathcal{O}_j$ *such that*

$$\varphi_{n_j}^* \ \text{converges uniformly towards} \ \varphi_\infty^* \ \text{on} \ \mathcal{O}_j^c,$$
$$c_{p,r}(\mathcal{O}_j) \to 0 \ \text{and} \ \mathcal{O}_j^c \ \text{compact}.$$

Proof. We remark that $\varphi_n^* - \varphi_\infty^*$ is a redefinition of $\varphi_n - \varphi_\infty$. Then by Tchebycheff's inequality, ε being fixed, and writing

$$A_n'(\varepsilon) = \left\{x;\ \left|\varphi_n^* - \varphi^*\right|(x) > \varepsilon\right\},$$

we have $c_{p,r}\left(A_n'(\varepsilon)\right) \to 0$; we can find open sets $O_n'(\varepsilon) \supset A_n'(\varepsilon)$ such that $c_{p,r}\left(O_n'(\varepsilon)\right) \to 0$.

We construct a subsequence n_j and and a sequence ε_j satisfying $\sum \varepsilon_j < \infty$, such that, denoting $O_j = O_{n_j}'(\varepsilon_j)$, we have:

$$\sum c_{p,r}(O_j) < \infty.$$

The sequence $\mathcal{O}_j = A_j^c \cup \left[\bigcup_{s>j} O_j\right]$, where A_j is defined in 1.3.2, satisfies the conclusion of the theorem. □

2.4. Quasi-sure redefinition

Given $f \in D_\infty(X)$ a quasi-sure redefinition f^* of f will be defined by the two following properties

(i) $f = f^*$ a.e.
(ii) For all (p,r), f^* is (p,r)-quasi-continuous.

2.4.1. Uniqueness

Given two quasi-sure redefinitions u and v of f, then $u - v = w$ is equal to zero a.e. As w is (p,r)-quasi-continuous then 1.4.3 implies that

$$c_{p,r}(\{w \neq 0\}) = 0.$$

Therefore

$$w = 0 \qquad \textit{except on a slim set.}$$

2.4.2. Existence

Given $f \in \mathbb{D}_\infty(\mathbb{R}^{\mathbb{N}})$, then $f \in D_n^n(X)$ for all positive integers n. Therefore we can find a (n, n)-redefinition u_n of f

$$u_n = f \quad \text{a.e..}$$

Denote by H_q a (n, n)-redefinition net of f.

By choosing q_n sufficiently large we will have

$$c_{n,n}\left(H_{q_n}\right) < n^{-3}.$$

Consider

$$O_n = \bigcup_{r \geq n} H_{q_r}$$

then

$$c_{n,n}(O_n) \leq n^{-2}.$$

We define

$$G_n = \left(\text{ess}(O_n^c)\right)^c$$

then G_n is a *decreasing* sequence of open sets. By 2.1.3 we have

$$c_{n,n}(G_n) = c_{n,n}(O_n).$$

Denote

$$\Gamma = \bigcap_r \left(\bigcup_{n>r} G_n \right),$$

then by Borel-Cantelli (1.2.4) Γ is *slim*. Furthermore

$$G_n^c \subset O_n^c \subset H_{q_m}^c \qquad \text{for} \quad m \geq n.$$

As $\text{ess}(G_n^c) = G_n^c$ the identities

$$u_{n+s}(x) = u_n(x) \quad \text{a.e.} \quad x \in G_n^c$$

imply the same equality *everywhere* on G_n^c. Finally, we define f^* by

$$\begin{aligned} f^*(\xi) &= \lim u_n(\xi) \qquad & x \in \Gamma^c \\ f^*(\xi) &= 0 & \text{if} \quad x \in \Gamma. \end{aligned}$$

□

3. Positive generalized functionals

We denote by $D_{-\infty}(\mathbb{R}^{\mathrm{N}})$ the space of *Watanabe distributions* defined in III.6.2.

3.0. Sugita Theorem. *Given* $\ell \in \mathbb{D}_{-\infty}(\mathbb{R}^{\mathrm{N}})$, *satisfying*

$$\ell(f) \geq 0 \quad \textit{for all} \quad f \in \mathbb{D}_{\infty}(\mathbb{R}^{\mathrm{N}}) \quad \textit{such that} \quad f > 0,$$

then there exists a unique positive Borel measure θ *on* $\mathbb{R}^{\mathrm{N}}$, *of finite total mass, such that*

(i)
$$\ell(f) = \int_{\mathbb{R}^{\mathrm{N}}} f^*(\xi) d\theta(\xi).$$

(ii) Furthermore, θ *does not charge slim Borel sets.*

Proof. We shall need the following preliminary result:

(iii) $\mathcal{L}^r P_t$ *is a bounded operator in* L^p, *for* $1 < p < \infty$.

As $\mathcal{L}^r P_t = \left[\mathcal{L} P_{t/r}\right]^r$ it is sufficient to prove (iii) for $r = 1$. Then this operator is equal to $\frac{d}{dt} P_t$; making the change of variables $e^{-t} = \cos\theta$ it is equivalent for $\theta \neq 0$ to take the derivative in θ:

$$\int_X \frac{d}{d\theta} f(x\cos\theta + y\sin\theta)\ \mu(dy) =$$
$$\int_X (-x\sin\theta + y\cos\theta) f(x\cos\theta + y\sin\theta)\ \mu(dy).$$

We have to take the pth-power of this integral and integrate in x; by using Hölder's inequality this leads to

$$\int_{X^2} |-x\sin\theta + y\cos\theta|^p\ |f(x\cos\theta + y\sin\theta)|^p \quad \mu(dx)\mu(dy).$$

We now use the fact that on X^2 the two functions $(x\cos\theta + y\sin\theta)$ and $(-x\sin\theta + y\cos\theta)$ are independent; this independence means that the double integral splits as a product of two simple integrals. □

By the continuity of ℓ there exists (p, r) such that $\|\ell\|_{\mathbb{D}^q_{-r}} < +\infty$ and we have

$$|\ell(f)| \leq \|\ell\|_{D^q_{-r}}\ \|f\|_{\mathbb{D}^p_r}.$$

We use the Ornstein-Uhlenbeck semigroup P_t to suppress the differentiability condition. Define $P_t^*\ell = \ell_t$ by

$$\langle (P_t^*\ell), f\rangle = \langle \ell, P_t f\rangle.$$

As $\|P_t f\|_{\mathbb{D}^p_r} \leq c_t \|f\|_{L^p}$ we obtain the existence of a function

$$u_t \in L^q, \qquad u_t \geq 0,$$

such that

$$\langle P_t^* \ell, f \rangle = \int_{\mathbb{R}^N} f \; u_t \; d\mu.$$

When $t \to 0$ we have $\| P_t f - f \|_{\mathbb{D}_r^p} \to 0$ and therefore, for $f \in \mathbb{D}_r^p$,

$$\int_{\mathbb{R}^N} f \; u_t \; d\mu \to \ell(u).$$

The main argument of the proof will consist in restricting this weak convergence to a compact set by the machinery already used in the tightness Theorem 1.3. Denote by ψ a smooth function defined on R^+, such that $\psi(t) = 0$ if $t > 1$, $\psi(t) = 1$ if $t < \frac{1}{2}$, $0 < \psi < 1$. Denote

$$v_R(\xi) = \psi\left(\frac{1}{R} q(\xi)\right)$$

where q has been defined in 1.3. Then

$$v_R \in \mathbb{D}_\infty \quad \text{and} \quad \|1 - v_R\|_{\mathbb{D}_r^p} \to 0 \quad \text{when} \quad R \to +\infty.$$

Denote

$$\ell_{R,t}(f) = \ell_t(v_R f) = \int_{\mathbb{R}^N} f \; v_R \; u_t \; d\mu.$$

Now if $f \in \mathbb{D}_\infty$ we have

$$v_R \; P_t f \to v_R \; f$$

in $\mathbb{D}_\infty$. Therefore, as $t \to 0$,

(ii) $$\int f \; v_R \; u_t \; d\mu \quad \text{converges for every} \quad f \in \mathbb{D}_\infty.$$

According to 1.3 the support K_R of v_R is compact. Denote by $\mathcal{H}$ the linear span of the Hermite polynomials. Denote by J_R the *restriction operator* to K_R and by $C(K_R)$ the Banach algebra of continuous functions on K_R. We then have

$$j_R(\mathcal{H}) \subset C(K_R) \quad \text{and} \quad \overline{j_R(\mathcal{H})} = C(K_R).$$

As $\mathcal{H} \subset \mathcal{D}_\infty$ we get

$$\int_K H \; v_R \; u_t \; d\mu \quad \text{converges when} \quad t \to 0.$$

As $v_R u_t \geq 0$, we have proved:

3.1. Lemma. *$v_R u_t \, d\mu$ converges weakly to a Borel measure θ_R, supported by K_R.*

3.2. Lemma. *For every open set O we have*

$$\theta_R(O) \leq \|\ell\|_{\mathbb{D}_{-r}^q} c_{p,r}(O).$$

Proof. We can find $\varphi \in \mathbb{D}_r^p$ such that

$$\begin{aligned} \varphi(\xi) &\geq 1 \quad \text{a.e. on } O \\ \|\varphi\|_{\mathbb{D}_{p,r}} &< c_{p,r}(O) + \varepsilon. \end{aligned}$$

Then

$$\lim_{t\to 0} \int \varphi \; v_R \; u_t \; d\mu \leq \|\ell\| \left(c_{p,r}(O) + \varepsilon\right)$$

or

$$\lim_{t\to 0} \int \mathbf{1}_O \; v_R \; u_t \; d\mu \leq \|\ell\| c_{p,r}(O).$$

Consider

$$T(O) = \{\psi \in C(K); \;\; (\text{support of } \psi) \subset O\},$$

then

$$\theta_R(O) = \sup_{q\in T(O)} \int \psi d\theta.$$

As $\psi < \mathbf{1}_O$ we have

$$\int \psi \; v_R \; u_t \; d\mu \leq \int \mathbf{1}_O \; v_R \; u_t \; d\mu.$$

Therefore

$$\theta_R(O) \leq \|\ell\|_{\mathbb{D}^q_{-r}} c_{p,r}(O).$$

□

3.3. Lemma. *Given $f \in D_\infty \cap L^\infty$, denoting by f^* its redefinition, we have*

$$\ell(v_R f) = \int_{K_R} f^* d\theta_R.$$

Proof. We can find a sequence $g_n \in \mathcal{H}$ converging in $\mathbb{D}_r^p$ to f; as g_n is continuous, we can take $g_n^* = g_n$. Using 2.3.5 we can find a subsequence n_j and open sets O_j such that

$$g_{n_j} \text{ converges towards } f^* \text{ uniformly on } O_j^c$$

and

$$c_{p,r}(O_j) \to 0.$$

Then by Lemma 3.2

$$\theta_R(O_j) = \varepsilon_j \to 0.$$

Therefore

$$\int f^* d\theta_R = \lim \int_{O_j} g_{n_j} d\theta_R + \delta_j$$

where $|\delta_j| \leq \varepsilon_j \|f^*\|_{L^\infty}$,

$$\int_{O_j^c} g_n d\theta_R = \int g_n d\theta_R + c\varepsilon_j$$

where $c = \sup_n \|g_n\|_{\mathbb{D}_r^p}$,

$$\left| \int g_n \, u_t \, d\mu - \int g_n \, u_{t'} \, d\mu \right| \le c \| u_t - u_{t'} \|_{D^q_{-r}}.$$

Therefore the integrals

$$\int g_n \, u_t \, d\mu \to \int g_n \, d\theta_R$$

uniformly relatively to n. □

3.4. Lemma. *The measure θ_R converges in total variation norm $\| \; \|_M$ towards θ. Furthermore we have for every $f \in L^\infty \cap \mathbb{D}_\infty$*

$$\ell(f) = \int f^* \, d\theta.$$

Proof. We have for $R' < R$

$$\|\theta_R - \theta_{R'}\|_M = \lim \int (v_R - v_{R'}) \, u_t \, d\mu.$$

Set $R'' = \frac{1}{2} R$. Then

$$(v_R - v_{R'}) \, (\xi) = 0 \quad \text{if} \quad v_{R''}(\xi) > 0.$$

Therefore

$$|v_R - v_{R'}| < 2 \, (1 - v_{R''})$$

$$\|\theta_R - \theta_{R'}\|_M < 2c \| 1 - v_{R''} \|_{\mathbb{D}^p_r} \to 0 \quad \text{when} \quad R' \to +\infty.$$

Then, as f^* is bounded,

$$\ell(f) = \lim \int f^* \, d\theta_R = \int f^* \, d\theta.$$

□

3.5. *Proof of the theorem*

We have to eliminate the condition $f \in L^\infty$. Denote by g any odd, smooth truncation function g satisfying on $\xi > 0$, $g(\xi) = \xi$ if $\xi < 1$, $g(\xi) = 2$ if $\xi > 2$, $0 \le g(\xi) \le 2$.

Denote $g_M(t) = Mh \, (t/M)$ and $f_M = g_M(f)$. Then

$$f_M \in D_\infty \cap L^\infty.$$

Furthermore

$$g_M(f^*)$$

is *quasi-surely continuous* and therefore is *the* quasi-sure redefinition of f_M. We have by 3.4

$$\ell(f_M) = \int g_M(f^*)d\theta$$

when $M \to \infty$, $f_M \to f$ in $\mathbb{D}_\infty$. Therefore

(i) $$\ell(f) = \lim_{M\to\infty} \int g_M(f^*)d\theta.$$

If $f \in \mathbb{D}_\infty$, then $f^2 \in \mathbb{D}_\infty$. Apply the last equality to $f^2 \in \mathbb{D}_\infty$. We get

$$\ell(f^2) = \lim_{M\to\infty} \int g_M\left((f^*)^2\right) d\theta.$$

The fact that the r.h.s. has a finite limit implies that $f^* \in L^2(\theta)$. Therefore going back to (i) we get

$$\lim_M \int g_M(f^*)d\theta = \int f^* d\theta.$$

□

4. Equilibrium potentials

4.1. Definition. Given a subset A of $\mathbb{R}^N$, we define

$$F_{p,r}(A) = \left\{u \in \mathbb{D}_r^p; \quad u^* \geq 1 \ \ (p,r)\text{-quasi-everywhere on } A\right\}.$$

4.1.1. Lemma. *$F_{p,r}(A)$ is a closed convex subset of $\mathbb{D}_r^p$.*

Proof. Given a sequence $u_s \in F_{p,r}$ converging strongly in $\mathbb{D}_r^p$ towards u_∞, by 2.3.5 we can find a subsequence such that u_s^* converges towards u_∞^* (p,r)-quasi-everywhere; therefore $u_\infty^* \geq 1$ (p,r)-q.e. □

4.1.2. As $\mathbb{D}_r^p$ is isomorphic to an L^p space, there exists a *unique element* $\varphi \in F_{p,r}(A)$ which minimizes the norm $\|\varphi\|_{\mathbb{D}_r^p}$. We denote this element by φ_A and we shall call φ_A the *equilibrium potential* of A.

4.2. Proposition. *We have*

$$\varphi_A = \Lambda^r f \quad \text{with} \quad f \geq 0 \quad \text{and} \quad \varphi_A > 0 \quad \text{everywhere.}$$

Proof. Introduce $\Lambda^{-r}\varphi_A = f \in L^p$. Write $f^+ = \sup(f,0)$, then

$$\Lambda^r f^+ \geq \Lambda^r f \quad \text{almost everywhere}$$

implies by 2.1.6 the same inequality quasi-everywhere. Therefore $\Lambda^r f^+ \in F_{p,r}(A)$; therefore by the extremality of φ_A we must have

$$\|f^+\|_{L^p} \geq \|f\|_{L^p}$$

which implies $f \geq 0$. □

4.2.1. Corollary. *For all* $\xi \in \mathbb{R}^N$, $\varphi_A(\xi) \leq +\infty$ *is well defined; the function* φ_A *is lower semi-continuous.*

Proof. The expression of $(\Lambda^r f)(\xi)$ is given by the integral of a positive function depending upon the parameter $\xi \in \mathbb{R}^N$; the Fatou lemma proves the corollary. □

4.3. Lemma. *Assume A is an open set, then*

$$\|\varphi_A\|_{\mathbb{D}_r^p} = c_{p,r}(A).$$

Proof. We remark that we have used in the Definition 1.2 the class $\Gamma(A)$ of functions φ satisfying

- $\varphi(\xi) \geq 1$ almost everywhere on A
- $\varphi(\xi) \geq 0$ almost everywhere,

then granted 2.1.6, the first equality is true quasi-everywhere. Therefore

$$\|\varphi_A\|_{\mathbb{D}_r^p} \leq c_{p,r}(A).$$

But as $\varphi_A \in \Gamma(A)$ we get the equality. □

4.4. Theorem. *For any Borel set A, we have*

$$\|\varphi_A\|_{\mathbb{D}_r^p} = c_{p,r}(A).$$

Proof. Given an open set O, $O \supset A$, we have

$$F_{p,r}(O) \subset F_{p,r}(A)$$

the infinimum taken over the class on the left is therefore greater than the infinimum taken over the class on the right. Using 4.3 we get

$$c_{p,r}(O) \geq \|\varphi_A\|_{\mathbb{D}_r^p};$$

this inequality holds for every $O \supset A$. Therefore we get

$$c_{p,r}(A) \geq \|\varphi_A\|_{\mathbb{D}_r}.$$

To prove the reverse inequality we introduce the redefinition φ_A^* of φ_A. Given $\varepsilon > 0$ we can find an open set Ω such that

$$c_{p,r}(\Omega) < \varepsilon \quad \text{and such that}$$
$$\varphi_A^* \quad \text{is continuous on} \quad \Omega^c \quad \text{and} \quad \varphi_A^* \geq 1 \quad \text{on} \quad \Omega^c \cap A.$$

Introduce $G_1 = \{\xi \in \mathbb{R}^N;\ \varphi_A(\xi) > 1 - \varepsilon\}$. Then by 4.2.1 G_1 is *open*; then by 2.1.6 $\varphi_A^* > 1 - \varepsilon$ quasi-everywhere. Introduce $G = G_1 \cup \Omega$. Then G is open and $G \supset A$.

Denote by φ_Ω the equilibrium potential of Ω. Then

$$\varphi_A + \varphi_\Omega > (1-\varepsilon) \quad \text{almost everywhere on} \quad G.$$

Therefore

$$\begin{aligned} c_{p,r}(A) \leq c_{p,r}(G) &\leq \frac{1}{1-\varepsilon} \|\varphi_A + \varphi_\Omega\|_{\mathbb{D}_r^p} \\ &\leq \frac{1}{1-\varepsilon} \left[\|\varphi_A\|_{\mathbb{D}_r^p} + \|\varphi_\Omega\|_{\mathbb{D}_r^p} \right]. \end{aligned}$$

By Lemma 4.3 $\|\varphi_\Omega\|_{\mathbb{D}_r^p} = c_{p,r}(\Omega) < \varepsilon$.

As ε is arbitrary we get

$$c_{p,r}(A) \leq \|\varphi_A\|_{\mathbb{D}_r^p}.$$

□

5. Continuity of capacities

5.1. Theorem. *For an increasing sequence A_n we have*

$$c_{p,r}\left(\bigcup A_n\right) = \lim c_{p,r}(A_n).$$

Proof. Denote by φ_n the equilibrium potential of A_n. Then

$$\|\varphi_n\|_{\mathbb{D}_r^p} = c_{p,r}(A_n).$$

As φ_n are bounded in norm, it is possible to find a convex combination

$$\psi_s = \sum \alpha_s^j \varphi_j \quad \text{with} \quad \lim_s \alpha_s^j = 0$$

such that ψ_s converges *strongly* in $\mathbb{D}_r^p$ to ψ_∞.

By 2.3.5 we can find a subsequence such that

$$\lim \psi_{s_r}^*(\xi) = \psi_\infty^*(\xi)$$

quasi-everywhere.

Granted the fact that

$$\psi_s^*(\xi) \geq 1 - \sum_{j<k} \alpha_s^j \quad \text{quasi-everywhere on } A_k,$$

we get

$$\psi_\infty^* \in F_{p,r}\left(\bigcup A_k\right).$$

Using Theorem 4.4, we obtain

$$\|\psi_\infty\|_{\mathbb{D}_r^p} \geq c_{p,r}\left(\bigcup A_k\right).$$

As

$$\|\psi_\infty\|_{\mathbb{D}_r^p} \leq \sup \|\varphi_n\|_{\mathbb{D}_r^p},$$
$$c_{p,r}\left(\bigcup_k A_k\right) \leq \lim c_{p,r}(A_k).$$

As capacity is an increasing function, the reverse inequality is trivial. □

5.2. Proposition. *For any decreasing sequence of compact sets K_n, we have*

$$c_{p,r}\left(\bigcap K_n\right) = \lim c_{p,r}(K_n).$$

Proof. Write $K_\infty = \bigcap K_n$. Given ε, we can find O open, $O \supset K_\infty$, such that $c_{p,r}(O) \leq c_{p,r}(K_\infty) + \varepsilon$.

On the other hand, there exists n_0 such that $O \supset K_{n_0}$. Therefore,

$$c_{p,r}(K_\infty) + \varepsilon > c_{p,r}(K_{n_0}).$$

□

6. Capacitability of a Borel set

6.1. Theorem. *For any Borel set A, $\varepsilon > 0$ being given, it is possible to find $K \subset A$, K compact such that*

$$c_{p,r}(K) \geq c_{p,r}(A) - \varepsilon.$$

Proof. By tightness we can find a compact set B such that $c_{p,r}(B^c) < \varepsilon/2$. We take B as our basic space on which to apply Choquet's theorem. We define for all $Z \subset B$ an increasing function of sets $I(Z)$ by

$$I(Z) = c_{p,r}(Z),$$

then I satisfies the hypothesis (Doob [1], page 750) where we take for $\mathcal{H}$ the class of compact subsets of B; therefore we can find H compact such that

$$H \subset B \cap A$$
$$I(H) > I(B \cap A) - \varepsilon/2.$$

Finally

$$c_{p,r}(A) < I(B \cap A) + c_{p,r}(B^c) \leq c_{p,r}(H) + \varepsilon.$$

□

7. Equilibrium measure

7.1. Potential of a measure

We denote by $p_t(\xi, d\eta)$ the transition probability of the Ornstein-Uhlenbeck process. A Borel measure ρ will be said to be *admissible* if for all $t > 0$,

$$P_t^*(\rho)(d\eta) = \int p_t(\xi, d\eta)d\rho(\xi)$$

is absolutely continuous relative to the Gaussian measure μ. Then the r-potential of an admissible measure is the *function* U_r^θ defined by

$$U_r^\theta(\eta)d\mu(\eta) = \frac{1}{\Gamma(r/2)} \int_0^{+\infty} P_t^*(\rho)(d\eta)t^{r/2}e^{-t}\frac{dt}{t}.$$

7.2. Theorem. *Denote by φ_A the equilibrium potential of A. There exists an unique measure θ supported on $\bar{A}$, such that*

$$\theta \in \mathbb{D}_{-r}^{p'} \qquad (p' \text{ denoting the conjugate exponent of } p)$$

$$\varphi_A = \Lambda^r (U_r^\theta)^{p'-1}$$

$$c_{p,r}(A) = \|U_r^\theta\|_{L^{p'}}^{p'-1} = \|\theta\|_{\mathbb{D}_{-r}^{p'}}^{p'-1} = \left[\int \varphi_A^* d\theta\right]^{1/p}.$$

We shall call θ the equilibrium measure of A.

Proof. If $v > 0$ q.e. on A, $v \in \mathbb{D}_r^p$

$$\varphi_A + \lambda v \in F_{r,p}(A) \quad \text{for} \quad \lambda > 0.$$

Therefore

$$\frac{d}{d\lambda}\|\varphi_A + \lambda v\|_{\mathbb{D}_r^p}^p \geq 0 \quad \text{for} \quad \lambda = 0.$$

We have

$$\|\varphi_A + \lambda v\|_{\mathbb{D}_r^p}^p = \|\Lambda^{-r}(\varphi_A + \lambda v)\|_{L^p}^p.$$

We recall that if $\xi, \eta \in R$

$$\left(\frac{d}{d\lambda}|\xi + \lambda\eta|\right)_{\lambda=0} = \eta \quad \text{signum } (\xi)$$

$$\left(\frac{d}{d\lambda}\left|\Lambda^{-r}(\varphi_A + \lambda v)\right|^p\right)_{\lambda=0} = p\left|\Lambda^{-r}\varphi_A\right|^{p-1} \left(\text{signum } (\Lambda^{-r}\varphi_A)\right)\Lambda^{-r}v.$$

If $v > 0$ everywhere, $v \in \mathbb{D}_r^p$, we have *a fortiori*

$$\xi(v) = E\left(\left|\Lambda^{-r}\varphi_A\right| \left(\text{signum } (\Lambda^{-r}\varphi_A)\right)\Lambda^{-r}v\right) > 0.$$

This means that ξ is a positive linear form on $\mathbb{D}_r^p$. By the Sugita theorem there exists a positive Borel measure θ such that

(i) $$\xi(v) = \int v^* d\theta.$$

As $\int v^* d\theta \geq 0$ for all $v > 0$ on $\bar{A}$, this implies

$$\text{support}(\theta) \subset \bar{A}.$$

7.2.1. Lemma.

$$\int v^* d\theta = E\left(\left(\Lambda^{-r} v\right) U_r^\theta\right).$$

Proof. In fact this identity is true when v is an Hermite polynomial; the two sides are continuous linear forms in v for the $\mathbb{D}_r^p$ norm; then the identity extends by continuity. □

7.2.2. *End of the proof of Theorem 7.2.*

$$\xi(v) = \int \Lambda^{-r} v \, U_r^\theta \, d\mu$$

we deduce then

$$|\Lambda^{-r}\varphi_A|^{p-1} \quad \text{signum}\left(\Lambda^{-r}\varphi_A\right) = U_r^\theta.$$

As the r.h.s. is positive we deduce that signum $\left(\Lambda^{-r}\varphi_A\right) = 1$. Therefore

$$\left(\Lambda^{-r}\varphi_A\right)^{p-1} = U_r^\theta$$

or

$$\begin{aligned}
\varphi_A &= \Lambda^r (U_r^\theta)^{p'-1} \\
\|\varphi_A\|_{\mathbb{D}_r^p}^p &= \|\Lambda^{-r}\varphi_A\|_{L^p}^p = E\left([U_r^\theta]^{p'}\right) = \|\theta\|_{\mathbb{D}_{-r}^{p'}}^{p'}.
\end{aligned}$$

By Lemma 7.2.1

$$\int \varphi_A^* d\theta = E\left((U_r^\theta)^{\frac{p}{p-1}}\right).$$

□

7.3. Corollary. *Given a compact set K, such that $c_{p,r}(K) > 0$ then there exists a Borel measure σ, $\sigma \in \mathbb{D}_{-r}^{p'}$, such that $\sigma(K) > 0$.*

Proof. Take for σ the equilibrium measure θ of K. Then θ is supported by K, and

$$\theta(K) = \int d\theta > 0 \qquad (\text{otherwise} \quad \varphi_K = 0).$$

□

7.4. Theorem. *A Borel set A satisfies*

$$c_{p,r}(A) = 0 \quad \textit{if and only if}$$
$$\sigma(A) = 0 \quad \textit{for all measures} \quad \sigma \in \mathbb{D}_{-r}^{p'}.$$

Proof. If $c_{p,r}(A) > 0$ then by 6.1 there exists a compact $K \subseteq A$ such that $c_{p,r}(K) > 0$.

We now use 7.3 and find $\sigma \in D^{p'}_{-r}$ such that

$$0 < \sigma(K) \leq \sigma(A).$$

□

Proof of the converse implication

Given θ a measure $\theta \in D^{p'}_{-r}$, the linear form

$$\ell_1(H) = \int H d\theta \qquad H \in \mathcal{H}$$

extends by continuity to a positive linear form ℓ defined on $\mathbb{D}^p_r$. Then this linear form can be represented according Sugita's theorem by a positive Borel measure $\tilde{\theta}$. As the restrictions of Hermite polynomials are dense in $C(K')$ for all compact K', we get $\theta = \tilde{\theta}$.

According to Sugita's theorem, $\tilde{\theta}$ does not charge the sets of (p, r)-capacity zero. □

7.5. Measures of finite energy

A Borel measure σ is of finite energy if $\sigma \in \mathbb{D}_{-\infty}$. This means that there exists (p, r) and c such that for

$$|\int H d\sigma| < c\|H\|_{\mathbb{D}^p_r} \quad \text{for every smooth cylindrical function} \quad H.$$

In Itô's construction of quasi-sure analysis (Itô [14]), the measures of finite energy play the same role as the hierarchy of capacities in our approach. The equivalence of these two approaches results from

7.5.1. Theorem. *A compact set K is slim if and only if*

$$\theta(K) = 0 \quad \text{for all measures of finite energy.}$$

Proof. Theorem 7.4. □

8. Charge and capacity

8.1. The simplex Δ^p_r

We denote

$$\Delta^p_r = \left\{\rho \in \mathbb{D}^{p'}_{-r}; \quad \rho \text{ is a positive measure,} \quad \|\rho\|_{\mathbb{D}^{p'}_{-r}} \leq 1\right\}.$$

We remark that if v is a Borel measurable function, $0 \leq v \leq 1$, then $\rho \in \Delta^p_r$ implies $v\rho \in \Delta^p_r$. In fact

$$P_t^*(vd\rho) \leq P_t^*(d\rho).$$

8.2. The charge

We define, for every Borel set A, its *charge*

$$\gamma(A) = \sup \rho(A) \quad \text{where} \quad \rho \in \Delta_r^p.$$

We have

$$0 \leq \gamma(A) \leq 1.$$

8.3. Theorem.

$$\gamma(A) = c_{p,r}(A).$$

Proof. Given A, denote by φ_A its equilibrium potential. Then

$$\varphi_A \geq 1 \quad (p,r)\text{-q.e. on } K; \quad \text{which means:}$$
$$G := \{\xi \in K; \varphi_A(\xi) < 1\} \quad \text{satisfies } c_{p,r}(G) = 0.$$

Given any measure $\rho \in \Delta_r^p$, by 7.4 we have $\rho(G) = 0$; hence

$$\varphi_A \geq 1 \quad \rho\text{-a.e. on } K; \quad \text{therefore}$$

$$\rho(K) \leq \int \varphi_A^* d\rho \leq \|\varphi_A\|_{\mathbb{D}_r^p} \|\rho\|_{\mathbb{D}_{-r}^{p'}} \leq \|\varphi_A\|_{\mathbb{D}_r^p} = c_{p,r}(K).$$

Taking the supremum over $\rho \in \Delta_r^p$ we get

$$\gamma(A) \leq c_{p,r}(A).$$

□

The remaining part of the proof requires the following theorem.

8.4. Theorem. *Given a compact subset K of $\mathbb{R}^n$, denote by φ_K its equilibrium potential and by θ_K its equilibrium measure, then*

$$\varphi_K(\xi) = 1 \qquad \theta_K\text{-a.e.}$$

Remark on the finite-dimensional case. We can replace $\mathbb{R}^{\mathbb{N}}$ by $\mathbb{R}^n$ in the first seven paragraphs, then all the previous results hold true; this last statement is obvious; nevertheless it could be proved "formally" using the identification, of $\mathbb{D}_r^p(\mathbb{R}^n)$ with a closed subspace of $\mathbb{D}_r^p(\mathbb{R}^{\mathbb{N}})$ given by the inverse image associated to $\pi_n : \mathbb{R}^{\mathbb{N}} \to \mathbb{R}^n$.

8.5. *Proof of Theorem 8.4.*

We shall denote by $d\xi$ the Lebesgue measure on $\mathbb{R}^n$. Then the transition probability of the Ornstein-Uhlenbeck semigroup can be written as

$$p_t(\xi, d\eta) = \tilde{p}_t(\xi, \eta)\mu(d\eta)$$

where

$$\mu(d\eta) = (2\pi)^{-n/2} \exp\left(-\frac{1}{2}\|\eta\|^2\right) d\eta.$$

Then, for every $t > 0$, $\tilde{p}_t(\xi, \eta)$ is a continuous function of (ξ, η), symmetric in (ξ, η). Then

$$U_r^\theta(\eta) = \int_{\mathbb{R}^n \times \mathbb{R}^+} \tilde{p}_t(\xi, \eta)\theta(d\xi)e^{-t} \frac{t^{r/2}}{\Gamma(r/2)} \frac{dt}{t}.$$

This integral defines U_r^θ *everywhere* as a function taking values in $]0, +\infty]$. By Fatou's lemma, U_r^θ is a *lower semi-continuous function*. In the same way

$$\varphi_K(\xi) = \int_{\mathbb{R}^n \times \mathbb{R}^+} \tilde{p}_t(\xi, \eta) \left[U_r^\theta(\eta)\right]^{p'-1} \mu(d\eta)e^{-t} \frac{t^{r/2}}{\Gamma(r/2)} \frac{dt}{t}$$

is a lower semi-continuous function defined on $\mathbb{R}^n$. Choose $\delta > 1$ and denote

$$\mathcal{O}_\delta = \{\xi; \varphi_K(\xi) > \delta\}.$$

Now, for every $v \in C^\infty$ with compact support in $\mathcal{O}_\delta$ we have

$$(\varphi_K + tv) \in F_{p,r}(K) \quad \text{for } t \text{ small enough;}$$

therefore

$$I(v) := \frac{d}{dt} \|\varphi_K + tv\|_{\mathbb{D}_r^p}^p = 0.$$

By definition of θ we have

$$I(v) = p \int v d\theta = 0.$$

As this holds true for every v with compact support in $\mathcal{O}_\delta$ this implies that

$$\theta(\mathcal{O}_\delta) = 0 \quad \text{or}$$

$$\theta\left(\varphi_K^{-1}(]1, +\infty])\right) = 0.$$

Since, on K, we have $\varphi_K \geq 1$, (p, r)-q.e., *a fortiori* θ-a.e. we get

$$\varphi_K(\xi) = 1 \qquad \theta\text{-a.e.}$$

□

8.6. *Proof of Theorem 8.3 in finite dimension*

There exists an increasing sequence of compact subsets K_n, such that

$$c_{p,r}(K_n) \to c_{p,r}(A).$$

Denote by θ_n the equilibrium measure of K_n; we have

$$\theta_n(A) \geq \theta_n(K_n) = \int \varphi_{K_n} d\theta_n = \left(c_{p,r}(K_n)\right)^p .$$

As

$$\left(\|\theta_n\|_{\mathbb{D}^{p'}_{-r}}\right)^{-1} \theta_n \in \Delta^p_r ,$$

we deduce

$$\gamma(A) \geq \|\theta_n\|^{-1}_{\mathbb{D}^{p'}_{-r}} \left(c_{p,r}(K_n)\right)^p .$$

But we know that

$$\|\theta_n\|_{\mathbb{D}^{p'}_{-r}} = \left(c_{p,r}(K_n)\right)^{p-1} ;$$

finally

$$\gamma(A) \geq c_{p,r}(K_n)$$

and

$$\gamma(A) \geq c_{p,r}(A).$$

□

8.7. Cylindrical approximation of a compact set

We denote by π_m the natural projection of $\mathbb{R}^{\mathbb{N}} \to \mathbb{R}^m$. Given a compact set K we shall call a *cylindrical neighbourhood* of K any open set Ω containing K of the form $\pi_m^{-1}(\mathcal{O})$ where $\mathcal{O}$ is an open set of $\mathbb{R}^m$.

8.7.1. Lemma. *We can construct a decreasing sequence Δ_n of cylindrical neighbourhoods of K, such that for any open set $\mathcal{O} \supset K$, there exists n_0 such that $\mathcal{O} \subset \Omega_{n_0}$.*

Proof. We consider on the infinite-dimensional torus $\mathbb{T}^{\mathbb{N}}$ the following fundamental family of neighbourhoods of the unit element

$$U_n = \left\{\theta; \quad |\theta_i| < \frac{1}{n} \qquad i \in [1, n]\right\}.$$

We have constructed a continuous injection Φ of $\mathbb{R}^{\mathbb{N}}$ into $\mathbb{T}^n$. Then

$$\Phi(K) \quad \text{is a compact} \quad K'.$$

We define

$$\Omega_n = \Phi^{-1}(K' + U_n).$$

□

8.8. *Proof of Theorem 8.3 for a compact set*

We have by 8.7.1 and 5.1

$$c_{p,r}(K) = \lim c_{p,r}(\Omega_n).$$

Potential theory on Ω_n reduces to the finite-dimensional case; therefore

$$\gamma(\Omega_n) = c_{p,r}(\Omega_n).$$

We can find $\theta_n \in \Delta_{-r}^{p'}$ such that, $\mathrm{supp}(\theta_n) \subset \Omega_n$ and

$$\theta_n(\Omega_n) \geq \left(1 - \frac{1}{n}\right) c_{p,r}(\Omega_n).$$

We can extract a subsequence θ_{n_j} converging to σ in the weak topology of $\mathbb{D}_{-r}^{p'}$; then $\sigma \in \Delta_{-r}^{p'}$. As $1 \in \mathbb{D}_r^p$ we have

$$\sigma(1) = c_{p,r}(K).$$

As $\mathrm{supp}(\sigma) = \bigcap \mathrm{supp}\, \theta_n$ we get

$$\sigma(K) = c_{p,r}(K).$$

□

8.9. *Proof of Theorem 8.3.*

We go from compact sets to Borel sets as in Section 8.6. □

9. Precise Gaussian probability spaces

We resume the same point of view as developed in Chapter I, Section 5. An *irreducible Gaussian probability space* is the data $(\Omega, \mathcal{A}, P;\ H)$ of a probability space $(\Omega, \mathcal{A}, P)$ together with a closed subspace $H \subset L^2(\Omega, \mathcal{A}, P)$ such that all functions in H are Gaussian variables and such that the σ-field generated by H is equal to $\mathcal{A}$. An intrinsic property of $(\Omega, \mathcal{A}, P, H)$ will be a property which does not depend upon the choice of a numerical model representing $(\Omega, \mathcal{A}, P, H)$.

More precisely, given two orthonormal bases of H $e_{k'}$, $e'_{k'}$ there exists an infinite orthogonal matrix such that

$$e'_s = \sum_s \alpha_s^k e_k.$$

We consider the following random variables on $\mathbb{R}^{\mathbb{N}}$

$$J_s(\xi) = \sum \alpha_s^k \xi_k.$$

Then $J_s \in \mathbb{D}_\infty$, therefore by 2.3.3, J_s possesses an ∞-redefinition J_s^*. Let

$$J^* : \mathbb{R}^{\mathbb{N}} \to \mathbb{R}^{\mathbb{N}}$$

be defined by

$$\xi \to \left(J_s^*(\xi)\right), \qquad s \in \mathbb{N}.$$

Our purpose is to show that "J^* preserves quasi-sure analysis". This fact is fully stated in:

9.1. Theorem. *We choose a redefinition J^* of J, then*

(9.1.1.) $$\|u \circ J\|_{\mathbb{D}_r^p} = \|u\|_{\mathbb{D}_r^p}.$$

9.1.2. *For every measure $\theta \in \mathbb{D}_{-r}^{p'}$, $(J^*)_*\theta$ is a well-defined measure, independent of the choice of J^* and*

$$\|(J^*)_*\theta\|_{\mathbb{D}_{-r}^{p'}} = \|\theta\|_{\mathbb{D}_{-r}^{p'}}.$$

9.1.3. *Given $u \in \mathbb{D}_r^p(X)$ the redefinition of $u \circ J$ is $u^* \circ J^*$.*
9.1.4. *Denote by ${}^tJ$ the map defined by the transposed matrix, then*

$$J^* \circ ({}^tJ)^* = ({}^tJ)^* \circ J^* = \text{identity}.$$

9.1.5. *For any Borel set A, we have*

$$c_{p,r}(A) = c_{p,r}\left((J^*)^{-1}(A)\right).$$

Proof of 9.1.1. The map J preserves the L^p-space, the Cameron-Martin representation and therefore the space $\mathbb{D}_r^p$. □

9.2. Lemma. *There exists a sequence $\mathcal{O}_s$ of open sets such that*

$$c_{p,r}(\mathcal{O}_j) \to 0 \quad \textit{and}$$

$$J^* : \mathcal{O}_j^c \to \mathbb{R}^N \quad \textit{is continuous}.$$

Proof. We choose for $\mathcal{O}_j$ the sequence constructed in 2.1.7 relative to the sequence $J_1, \ldots J_s \ldots \in \mathbb{D}_r^p$. Then on $\mathcal{O}_j^c$ all the components of J^* are continuous; therefore by the definition of the product topology, J^* is continuous. □

9.3. Lemma. *If $u \in \mathbb{D}_r^p$ and u is continuous, then denoting*

$$v = u \circ J,$$

we have

$$v^* = u \circ J^*.$$

Proof. In fact by 9.1, $u \circ J \in \mathbb{D}_r^p$; therefore its redefinition is well defined. By 9.2, $u \circ J^*$ satisfies the properties of a redefinition. □

9.4. *Proof of 9.1.2.*

We choose a redefinition J^*; then we have an associated net $\mathcal{O}_j \supset \mathcal{O}_{j+1}$. Given $\theta \in \mathbb{D}_{-r}^{p'}$ we denote $1_{\mathcal{O}_j}\theta = \theta_j$. Then, by Sugita's theorem, $\|\theta_j - \theta\| \to 0$ for the total variation norm. As J^* restricted to $\mathcal{O}_j$ is continuous, we deduce that $(J^*)_*\theta_j$ is a well defined measure. We define

$$(J^*)_*\theta = \lim_j (J^*)_*\theta_j.$$

Any two redefinitions differ on a set of (p, r)-capacity zero. Therefore they are equal θ-a.e. Therefore the image of θ does not depend upon the choice of a redefinition. We denote $\sigma = (J^*)_*\theta$.

For every *continuous* $u \in \mathbb{D}_r^p$ we have

$$\int u\left(J^*(\xi)\right) d\theta(\xi) = \int u(\eta) d\sigma(\eta).$$

By 9.1.1

$$\langle (u \circ J)^*, \theta \rangle = \langle u, \sigma \rangle.$$

The l.h.s. is continuous for the norm of $\mathbb{D}_r^p$, therefore the equality extends by continuity to all the functions in $\mathbb{D}_r^p$; then σ is a continuous linear form on $\mathbb{D}_r^p$; the map $(J^*)_*$ is the transpose of the map $(J)^*$ which is an isometry on $\mathbb{D}_r^p$. □

9.5. Lemma. *For any Borel set A we have*

$$c_{p,r}\left((J^*)^{-1}(A)\right) \leq c_{p,r}(A) \leq c_{p,r}\left(J^*(A)\right).$$

Proof. We shall first prove the right hand inequality; we can limit ourselves to the case where A is a compact set K. Denote by Ω_s the redefinition net of J^*, $K_s = K \cap \Omega_s^c$, then

$$c_{p,r}(K_s) \to c_{p,r}(K).$$

Denote by

$$\Gamma(K) = \Delta_r^p \cap \{\rho; \quad \text{supp}(\rho) \subset K\};$$

then the direct image by J^* of K_s is *compact*; therefore by 9.1.2

$$(J^*)_* : \Gamma(K_s) \to \Gamma\left(J^*(K_s)\right).$$

We have by 8.3

$$c_{p,r}(K_s) = \text{supp}\rho(1) \qquad \rho \in \Gamma(K_s);$$

therefore as the direct image conserves the mass

$$c_{p,r}\left(J^*(K_s)\right) \geq \text{supp}\rho(1) = c_{p,r}(K_s).$$

□

The left hand inequality is proved by taking $B = (J^*)^{-1}A$ in the right hand inequality

$$c_{p,r}(B) \leq c_{p,r}\left(J^*(B)\right) \quad \text{or} \quad c_{p,r}\left((J^*)^{-1}(A)\right) \leq c_{p,r}(A).$$

□

9.6. *Proof of 9.1.3 and 9.1.4.*

Given $u \in \mathbb{D}_r^p$ denote by Ω_s one of its redefinition nets. Then by 9.5

$$c_{p,r}\left((J^*)^{-1}(\Omega_s)\right) \to 0.$$

Therefore there exists a decreasing sequence of open sets $\mathcal{O}'_{n'}$, $\mathcal{O}'_{n'} \supset (J^*)^{-1}(\Omega_n)$ satisfying $c_{p,r}(\mathcal{O}'_n) \to 0$. Denote by $\mathcal{O}''_n$ a redefinition net of J^* and write $\mathcal{O}_n = \mathcal{O}'_n \cup \mathcal{O}''_n$; then

$$c_{p,r}(\mathcal{O}_n) \to 0 \quad \text{and on} \quad \mathcal{O}_n^c \quad u^* \circ J^* \quad \textit{is continuous}.$$

Therefore $u^* \circ J^*$ is (p,r)-quasi-continuous. As $u \circ J = u^* \circ J^*$ μ-a.e. we deduce that $u^* \circ J^*$ is the redefinition of $u \circ J$. □

We have ${}^tJ \circ J = J \circ {}^tJ$ μ-a.e. = identity, then by 9.1.3, $({}^tJ \circ J)^* = {}^tJ^* \circ J^* =$ identity *quasi-surely*.

9.7. *Proof of 9.1.5.*

We have ${}^tJ^*\left(J^*(\xi)\right) = \xi$ if $\xi \in S^c$ where S is *slim*. Then $J^*(\xi) \in A \cap S^c$ implies that $\xi \in {}^tJ^*(A \cap S^c)$. Therefore

$$\text{(i)} \qquad (J^*)^{-1}(A \cap S^c) \supset {}^tJ^*(A \cap S^c).$$

By 9.5

$$c_{p,r}\left((J^*)^{-1}(A \cap S^c)\right) \leq c_{p,r}(A \cap S^c) \leq c_{p,r}\left({}^tJ^*(A \cap S^c)\right).$$

The inclusion (i) implies the inequality in the reverse order; therefore we get

$$c_{p,r}\left((J^*)^{-1}(A \cap S^c)\right) = c_{p,r}(A \cap S^c)$$

$$c_{p,r}\left((J^*)^{-1}(A \cap S^c)\right) \leq c_{p,r}\left((J^*)^{-1}(A)\right)$$
$$\leq c_{p,r}\left((J^*)^{-1}(A \cap S^c)\right) + \mathrm{cap}_{p,r}\left((J^*)^{-1}(S)\right).$$

Since by 9.5 $c_{p,r}\left((J^*)^{-1}(S)\right) = 0$, we get the identity

$$c_{p,r}\left((J^*)^{-1}(A)\right) = c_{p,r}(A \cap S^c).$$

As $c_{p,r}(S) = 0$, the r.h.s. is equal to $c_{p,r}(A)$. □

Remark. We can modify a redefinition outside a slim set S. Given a choice of J^*, we choose an uncountable Borel slim set S_0; we define

$$q(\xi) = J^*(\xi) \qquad \xi \notin S_0, \qquad\qquad q(\xi) = u(\xi) \qquad \xi \in S_0,$$

where u is a Borel map which maps S_0 onto $\mathbb{R}^{\mathbb{N}}$. Then q is a redefinition and

$$c_{p,r}\big(q(A)\big) = 1 \quad \text{for all Borel sets } A, \qquad A \supset S_0.$$

Therefore the analogue of statement 9.1.5 written for the direct image cannot hold true for *all choices* of a redefinition J^*.

Remark. It is easy to construct "well-defined" redefinitions of J that preserve the capacity *by direct image.* From a functorial point of view, such constructions are not useful.

9.8. Precise Gaussian probability space

Very briefly, *a precise Gaussian probability space* is the "object" which is defined by quasi-sure analysis on a numerical model, *up to an* isomorphism *defined by a change of basis*.

We shall formulate in this section a more elaborate definition obtained by a procedure which is *more formal than illuminating*.

Given a Gaussian probability space $(\Omega, \mathcal{A}, P, H)$ we denote by $\mathcal{U}$ the space of all isometric maps of H onto ℓ_2. We can identify $\mathcal{U}$ with the family of all orthonormal bases of H.

Given $u \in \mathcal{U}$ we denote by Y_u the numerical model associated to u : $Y_u \approx \mathbb{R}^{\mathbb{N}}$. We denote by $\mathcal{A}_u$ the σ-field on Y_u generated by the Borel set of Y_u and the slim sets. Given u, u' we have defined in 9.1 a measurable map $q_{u,u'}$ $Y_u \to Y_{u'}$ such that $q_{u,u'}^{-1}(\mathcal{A}_{u'}) = \mathcal{A}_u$.

We denote by $\mathcal{E}_u$ the measures on Y_u of finite energy. Then $(q_{u,u'})_*\mathcal{E}_u = \mathcal{E}_{u'}$ and so on. We consider the space

$$Z = \operatorname{map}(\mathcal{U} \to \mathbb{R}^{\mathbb{N}}).$$

We define a separable subspace $X \subset Z$ by

9.8.1. $\qquad x = x(u) \quad \text{where} \quad x(u) = q_{u',u}x(u') \quad \text{for all} \quad u, u' \in \mathcal{U}.$

If we specify the value at one point u_0 of the function $x(\cdot)$ then all its other values will be defined by

$$x(u) = q_{u_0,u}\,(x(u_0))\,.$$

If we proceed in this way we give a special meaning to the basis u_0. The formulation 9.8.1 is symmetric in the choice of basis.

(i)
$$\operatorname{cap}^{(2)}\big(\Phi(A)\big) \geq \operatorname{cap}^{(1)}(A).$$

(ii) By 10.1 we can find an increasing sequence of compact sets K_n of B such that $\operatorname{cap}^{(1)}(K_n^c) \to 0$. Set $\tilde{K}_n = \Phi(K_n)$. Then $\tilde{K}_n$ is compact in $\mathbb{R}^{\mathbb{N}}$. Therefore $\tilde{K}_n^c = \mathcal{O}_n$ is open in $\mathbb{R}^{\mathbb{N}}$. We have $\Phi(K_n^c) \subset \mathcal{O}_n$ and $\mu\left(\mathcal{O}_n \cap \Phi(K_n^c)\right) = 0$; this means that the conditions

$$\begin{array}{rcccl} u(\xi) & \geq & 1 & \text{a.e.} & \xi \in \mathcal{O}_n \\ u \circ \Phi & \geq & 1 & \text{a.e.} & \xi \in K_n^c \end{array}$$

are *identical* conditions. Therefore

$$\operatorname{cap}^{(1)}(K_n^c) = \operatorname{cap}^{(2)}(\tilde{K}_n^c).$$

As the l.h.s. tends to zero for $n \to +\infty$, we get

$$\lim \operatorname{cap}^{(2)}(\tilde{K}_n^c) = 0$$

which proves 10.3.1. □

Given a compact subset K of B, Φ restricted to K is a continuous injective map. Therefore Φ is a homeomorphism of K with its image $\tilde{K}$; $(\Phi)_*$ realizes a bijection between the simplex of probability measures on K and $\tilde{K}$; as $(\Phi)_*$ preserves the $\mathbb{D}_{-p}^{r'}$ norms, we deduce that the charges are equal:

$$\gamma(K) = \gamma(\tilde{K}).$$

By 3.2, (which is true for $\operatorname{cap}^{(1)}$)

$$\operatorname{cap}^{(1)}(K) \geq \gamma(K)$$

and by 8.4

$$\gamma(\tilde{K}) = \operatorname{cap}^{(2)}(\tilde{K});$$

therefore

$$\operatorname{cap}^{(1)}(K) \geq \operatorname{cap}^{(2)}(\tilde{K}).$$

We have proved the converse inequality by (i). Therefore

$$\operatorname{cap}^{(1)}(K) = \operatorname{cap}^{(2)}\big(\Phi(K)\big) \quad \text{for every compact } \ K \subset B.$$

Given now a Borel set A in B, denote $\tilde{A} = \Phi(A)$. Then $\tilde{A}$ is a Borel set in $\mathbb{R}^{\mathbb{N}}$. We can find, by 6.1, a sequence of compact subsets K_s' of $\mathbb{R}^{\mathbb{N}}$ such that

$$\operatorname{cap}^{(2)}(K_s') \to \operatorname{cap}^{(2)}(\tilde{A}), \qquad K_s' \subset \tilde{A}.$$

Consider the sequence of compact sets K_n defined in (ii). Set

$$\tilde{K}_{s,n} = K_s' \cap \tilde{K}_n \quad \text{with} \quad \tilde{K}_n = \Phi(K_n).$$

Then the restriction of $\psi = \Phi^{-1}$ to $\tilde{K}_n$ is a homeomorphism. Therefore $\tilde{K}_{s,n}$ is the image of a compact subset $K_{s,n}$ of B and

$$\mathrm{cap}^{(2)}(\tilde{K}_{s,n}) = \mathrm{cap}^{(1)}(\tilde{K}_{s,n}).$$

We have

$$\lim_n \mathrm{cap}^{(2)}(\tilde{K}_{s,n}) = \mathrm{cap}^{(2)}(K'_s) \to \mathrm{cap}^{(2)}(\tilde{A}).$$

As $\tilde{K}_{s,n} \subset A$ we have also

$$\mathrm{cap}^{(1)}(K_{s,n}) \leq \mathrm{cap}^{(1)}(A).$$

Therefore

$$\begin{aligned} \mathrm{cap}^{(1)}(A) &\geq \lim_s \lim_n \mathrm{cap}^{(2)}(K_{s,n}) = \mathrm{cap}^{(2)}(\tilde{A}) \quad \text{or} \\ \mathrm{cap}^{(1)}(A) &\geq \mathrm{cap}^{(2)}\big(\Phi(A)\big). \end{aligned}$$

Combined with the inequality (i) we get the equality

$$\mathrm{cap}^{(2)}\big(\Phi(A)\big) = \mathrm{cap}^{(1)}(A).$$

Consider now a function $u \in \mathbb{D}_\infty(\mathbb{R}^N)$. Denote by K'_s a redefinition net of u on $\mathbb{R}^{\mathrm{N}}$. Define K_n and $\tilde{K}_{s,n} = K'_s \cap \tilde{K}_n$ as previously. Then, on $\tilde{K}_{s,n}$, ψ is an isomorphism to its image $K_{s,n}$ which is therefore compact. Define

$$v = u^* \circ \Phi.$$

Then v is continuous on $K_{s,n}$

$$v = (u \circ \Phi) \quad \text{a.s.,}$$

and finally

$$\mathrm{cap}^{(1)}(K_{s,n}) = \mathrm{cap}^{(2)}(K_{s,n}).$$

The r.h.s. tends to 1 when $s, n \to +\infty$; we can therefore find a sequence $n = q(s)$ such that $G_s = K_{s,q(s)}$ will satisfy

$$\lim \mathrm{cap}^{(1)}(G_s) = 1.$$

We have, by 1.2.1.,

$$1 = \mathrm{cap}^{(1)}(B) \leq \mathrm{cap}^{(1)}(G_s) + \mathrm{cap}^{(1)}(G_s^c).$$

Finally $\mathrm{cap}^{(1)}(G_s^c) \to 0$ and v is a redefinition on B of $u \circ \Phi$. □

10.4. Remark. For some applications it could be better to use quasi sure analysis on an abstract Wiener space instead of quasi-sure analysis on a numerical model. The *redefinition* associated to the first case will enjoy stronger properties of continuity.

For instance, in the case of Brownian motion, it could be useful (see Airault-Malliavin [1]) to construct quasi-sure analysis using the following norm:

$$f_{\alpha,q}(x) = \left(\int_0^1 \int_0^1 \left\| x(\tau) - x(\tau') \right\|^{2q} \frac{d\tau d\tau'}{|\tau - \tau'|^{1+2q\alpha}} \right)^{1/2q}$$

where q is a natural integer and where $\alpha < \frac{1}{2}$.

Chapter V

Differential Geometry on a Precise Gaussian Probability Space

Contents: Regular disintegration corresponding to a non-degenerate functional; principle of descent – Partial function associated to a splitting – Finite codimensional projection of a slim set is slim – Implicit function theorem in finite codimension – Differential forms of degree p – Non-normalized conditional expectation of differential forms – Currents; currents along a fibre of a non-degenerate map – Submanifolds; their defining functions – Gauss map of a submanifold, its approximate continuity – Hausdorff area measure of a submanifold – Functional characterization of the area measure – Coarea formula along a non-degenerate map – The operator δ; its commutation with the non-normalized conditional expectation – Oriented submanifold – Stokes' theorem.

Given an $\mathbb{R}^d$-valued random variable with a smooth law, it is an elementary fact that any marginal conditional law is immediately computable. If we replace *marginal conditioning* by a *non-linear conditioning* given by a smooth map of $\mathbb{R}^d$ into $\mathbb{R}^p$ again the conditional law is very explicitly computable; its intrinsic expression is given by the *Federer co-area formula*.

A natural question is the study of the conditional laws of Wiener measure by a non-degenerate map. Then Watanabe's distribution theory combined with Sugita's theorem implies that a *non-degenerate functional* generates a *regular disintegration* of Wiener measure into a continuous family of conditional laws which are all measures of finite energy.

The next objective is to obtain a co-area formula. We need a geometric study of the level sets of a non-degenerate functional; this will lead us to the notion of a *finite codimensional submanifold* of X. The study of submanifolds will need some implicit function theory which will be provided by the *splitting lemma.* A geometric co-area measure will be defined on submanifolds and the co-area formula will be obtained. A Stokes' formula for the integration of differential forms on submanifolds will be a consequence of the co-area formula.

1. Regular disintegration corresponding to a non-degenerate functional

1.1. Theorem. *Given a non-degenerate $g \in D_\infty(X; R^d)$, then we have $(g)_*\mu = k d\xi$ where $k \in C^\infty(\mathbb{R}^d)$. Denote by $\mathcal{O}$ the range of g, in the sense of III.6. Then there exist p, r and a continuous map $\mathcal{C}$ of $\mathcal{O}$ into $\mathbb{D}^p_{-r}$ such that $\mathcal{C}(\xi) = \theta_\xi$ is a Borel measure and such that for every $\psi \in C^\infty(\mathbb{R}^d)$, $u \in \mathbb{D}_\infty(X)$*

$$\int_X u(x)\psi\big(g(x)\big)d\mu(x) = \int_{\mathbb{R}^d} \psi(\xi)k(\xi)d\xi \left[\int u^*(x)\theta_\xi(d\xi)\right].$$

Proof. By the Sobolev imbedding, the Dirac mass δ_ξ is a continuous linear form on $W^{p',r'}(\mathbb{R}^d)$ for $p'r' > d$. By III.6.1, there exists (p, r) such that

$$(g)_*(u\mu) \in W^{p',r'}_{\text{loc}}(\mathcal{O}) \quad \text{for} \quad u \in \mathbb{D}^p_r.$$

We go back to III.6.2; we consider the Watanabe map:

$$(E^g)^* : \mathcal{E}'(0) \to \mathbb{D}_{-\infty}(X).$$

We take for δ_ξ the Dirac mass at $\xi \in \mathcal{O}$; then

$$(E^g)^*\delta_\xi = \ell_\xi \in \mathbb{D}^p_{-r}(X) \subset \mathbb{D}_{-\infty}(X).$$

If $u \in D_\infty(X)$, we have

$$\ell_\xi(u) = \langle (E^g)^*\delta_\xi, u\rangle = \langle \delta_\xi, E^g u\rangle = (E^g u)(\xi).$$

Therefore if $u > 0$, $\ell_\xi(u) \geq 0$. Using Sugita's theorem, there exists a measure of finite energy θ_ξ such that

$$\ell_\xi(u) = \int_X u^*(x)\theta_\xi(dx).$$

For fixed u,

$$\ell_\xi(u) = (E^g u)(\xi)$$

is a continuous function of ξ which implies the weak continuity of $\mathcal{O} \to \mathbb{D}^p_{-r}(X)$ defined by $\xi \to (E^g)^*\delta_\xi$. □

1.2. Corollary (Principle of descent). *Consider on X a property (π) which is true quasi-surely, then under every conditioning $g(x) = \xi$, where ξ belongs to the range of g, the property (π) is true almost surely.*

Proof. The measure θ_ξ is the conditional law under the conditioning $g(x) = \xi$.

$$S = \{\xi;\, (\pi) \quad \text{is not satisfied in} \quad \xi\}.$$

Then since S is slim, $\theta_\xi(S) = 0$. □

Remark. In Section 6 we will realize a geometric construction of the measures $\mathcal{O}_\xi$.

2. Splitting of a precise Gaussian probability space

2.1. Partial functions

2.1.1. Notations. Given a separable Hilbert space H, we consider an orthogonal direct sum decomposition

$$H = V_1 \oplus V_2.$$

We denote $X = \mathrm{Seg}(H)$, $Y_i = \mathrm{Seg}(V_i)$ $(i = 1, 2)$, then we have

$$X = Y_1 \times Y_2 \quad \text{or} \quad x = (y_1, y_2).$$

We denote $\Pi_{y_i} : x \to y_i$, $i = 1, 2$.

Given $f \in L^2(X)$ we can consider its partial function $f_{y_1^0} : Y_2 \to \mathbb{R}$ defined by

$$f_{y_1^0}(y_2) = f\left(y_1^0, y_2\right);$$

then, by the Fubini theorem, the map $\Phi : y_1^0 \mapsto f_{y_1^0}$ defines a map

$$Y_1 \to L^2(Y_2).$$

We shall call Φ the *partial function map*. By the Fubini theorem, Φ is an isometric isomorphism

$$\begin{aligned} \Phi : L^2(Y) &\mapsto L^2\left(Y_1; L^2(Y_2)\right) \qquad \text{or simply} \\ L^2(Y) &\subset L^2\left(Y_1; L^2(Y_2)\right). \end{aligned}$$

For smooth functionals we find the following analogue.

2.2. Theorem. *Given a splitting $H = V_1 \oplus V_2$, $X = Y_1 \times Y_2$ then the partial function map realizes an injection of*

$$\mathbb{D}_r^p(X) \quad \textit{into} \quad \mathbb{D}_{r_1}^p\left(Y_1, \mathbb{D}_{r_2}^p(Y_2)\right)$$

where $r_1 + r_2 = r$ or

(i) $$\mathbb{D}_r^p(X) \subset \mathbb{D}_{r_1}^p\left(Y_1; \mathbb{D}_{r_2}^p(Y_2)\right).$$

Remark. The r.h.s. of this inclusion has to be understood as the space of smooth functionals on Y_1, with values in the Banach space $D_{r_2}^p(Y_2)$.

In this book we have developed differential calculus for Hilbert-space-valued functionals. A parallel theory has been developed for L^p-valued functionals, for which the validity of the Krée-Meyer inequalities has been shown (Malliavin – Nualart [1]). We want to avoid this theory and we shall prove the weaker statement

(ii) $$\mathbb{D}_r^p(X) \subset \mathbb{D}_{r_1}^p\left(Y_1; \mathbb{D}_{r_2}^2(Y_2)\right) \quad \text{where} \quad p \geq 2 \quad \text{and} \quad r_1 + r_2 = r.$$

Proof. We choose a numerical model of X associated to an orthonormal basis of H, adapted to the decomposition $V_1 \oplus V_2$. Then every Hermite polynomial on X is the sum of products of Hermite polynomials on Y_1 and Y_2. At the level of derivatives we get the splitting

$$D_h = D_{h_1} + D_{h_2} \quad \text{if} \quad h = h_1 + h_2, \qquad h_i \in V_i$$

$$D_h^r = \sum_s \binom{r}{s} D_{h_1}^s D_{h_2}^{r-s}$$

$$\|D^2 f\|_{H.S.}^2 = \sum_s \left[\binom{r}{s}\right]^2 \|D_1^s D_2^{r-s}\|_{H.S.}^2;$$

therefore

$$\|D_1^{r_1} D_2^{r_2} f\|_{H.S.} \le \|D^r f\|_{H.S.} \quad \text{if} \quad r_1 + r_2 \le r.$$

□

2.3. Theorem. *Consider a direct sum decomposition* $H = Z \oplus Z^\perp$ *where* $\dim Z < +\infty$. *Denote* $X = \mathrm{Seg}(H)$, $Y = \mathrm{Seg}(Z^\perp)$, $\pi_Y : X \to Y$. *Then given a slim set* S *on* X, *we have*

$$\Pi_Y(S) \quad \textit{is a slim set of} \quad Y$$

$$\Pi_Y^{-1}(\Pi_Y(S)) \quad \textit{is a slim set of} \quad X.$$

Proof. We remark that $\mathrm{Seg}(Z) = Z$.

Using the Sobolev embedding theorem on $\mathbb{R}^d$ we have, denoting by $C(Z)$ the space of continuous functions on Z,

$$\mathbb{D}_s^p(Z) \subset C(Z) \quad \text{if} \quad ps > d.$$

Furthermore, by localisation, we know that for every M given, there exists γ_M such that

(i) $$\|u_y\|_{C(\Delta'_M)} \le \gamma_M \|u_y\|_{\mathbb{D}_s^p(Z)}$$

where $\Delta'_M = \{z; |z| < M\}$. Denote $\Delta_M = \Pi_Z^{-1}(\Delta'_M)$; set

$$S_M = \Delta_M \cap S.$$

Given n, we consider an open set $\mathcal{O}$ in X such that $\mathcal{O} \supset S_M$ and $\mathrm{cap}_{2n,2n}(\mathcal{O}) < \varepsilon$. Set $\mathcal{O}_M = \Delta_M \cap \mathcal{O}$; there exists $u \in \mathbb{D}_{2n}^{2n}(X)$ such that

$$u(x) > 1 \quad \text{a.e. on} \quad \mathcal{O}_M \quad \text{and} \quad \|u\|_{D_n^n} < \varepsilon.$$

As Π_Y is an open continuous mapping, then $\mathcal{O}_1 = \Pi_Y(\mathcal{O})$ is an open set containing $S_1 = \Pi_Y(S)$.

Consider the partial function φ_y defined on Z by $\varphi_y(z) = u(y, z)$ and the map $\chi : y \to \varphi_y$. Then by 2.2

$$\chi \in \mathbb{D}_n^n\left(Y; \mathbb{D}_n^2(Z)\right).$$

Denote

$$p := \|\chi\|^2_{\mathbb{D}_n^2(Z)};$$

then

$$p \in \mathbb{D}_n^n(Y);$$

furthermore

$$\|p\|_{\mathbb{D}_n^n(Y)} < \varepsilon.$$

As $u(x) > 1$ a.e. on $\mathcal{O}$ we have by (i) that $p > \gamma_M^{-1}$ a.e. on $\mathcal{O}_1$ a.e. relative to the measure $(\Pi_Y)_*(\mathbf{1}_0\mu_X)$, which is equivalent to the measure $\mathbf{1}_{\mathcal{O}_1}\mu_Y$. □

3. Implicit function theorem in finite codimension

3.1. Notations

We shall consider a *finite-dimensional subspace* Z of H; then

$$H = Z^\perp \oplus Z.$$

We remark that $\mathrm{Seg}(Z) = Z$; we are therefore splitting

$$X = Y \times Z \qquad \text{where} \quad Y = \mathrm{Seg}(Z^\perp).$$

We shall consider a non-degenerate map $g \in \mathbb{D}_\infty(X; Z)$. Then

$$g(x) = g(y, z).$$

We denote by $\frac{\partial g}{\partial z}(x)$ the *linear endomorphism* of Z defined by

$$\frac{\partial g}{\partial z}(x) \cdot h = (D_h g)(x) \quad \text{for every} \quad h \in Z.$$

Furthermore by 2.1

$$g_y \in \mathbb{D}_s^p\left(Y; \mathbb{D}_r^2(Z; Z)\right)$$

for every integer $p \geq 2$.

3.2. Theorem. *There exists a decreasing sequence of open sets Ω_q of Y whose intersection is slim in Y, such that the partial function*

$$y \to g_y^*$$

is a continuous map of Ω_q^c into $C^\infty(Z; Z)$.

Remark. We denote by $C^\infty(Z; Z)$ the space of C^∞ maps of Z into Z, the topology being defined by the uniform convergence of all derivatives on compact sets.

Proof. We consider the space

$$D^2_\infty(Z;Z) = \bigcap_r \mathbb{D}^2_r(Z;Z).$$

Then by the Sobolev embedding theorem, we have a continuous injection

$$j : \mathbb{D}^2_\infty(Z;Z) \to C^\infty(Z;Z).$$

We can find a net $\mathcal{O}^r_q$ on Y associated to a redefinition $g^{(r)}_y$; $g^{(r)}_y$ is a $\mathbb{D}^2_r(Z;Z)$-valued continuous function on $(\mathcal{O}^r_q)^c$. We have furthermore,

$$c_{q,q}(\mathcal{O}^r_q) \mapsto 0 \qquad r \quad \text{fixed}, \quad q \to +\infty.$$

We can therefore find a function $q_n(r)$ such that, for every fixed n,

$$\sum_r c_{n,n}\left(\mathcal{O}^r_{q_n(r)}\right) < \frac{1}{n};$$

we take

(i) $$\Omega_n = \left(\text{ess}\left(\bigcup_r \mathcal{O}^r_{q_n(r)}\right)\right)^c;$$

then on Ω^c_n the redefinition g^* is completely defined by the class of g for the μ-a.e. convergence. It does not depend on r. Furthermore it is continuous for all the $\mathbb{D}^2_r$ norms. This implies the $C^\infty(Z;Z)$ continuity. □

3.3. Implicit function theorem. *Consider the redefinition net Ω_n defined in 3.2. Fix a point $z_0 \in Z$ and consider the function*

$$q(y) = \left|\det\left(\frac{\partial(g_y)^*}{\partial z}(z_0)\right)\right|;$$

then q is continuous on Ω^c_n.

Assume that there exist n and δ such that

$$K(n,\delta) = \text{ess}\left(\{x \in \Omega^c_n;\ \ q(y) \geq \delta\}\right)$$

is non-empty. Then it is possible to find an open neighbourhood U of z_0 such that, for $y \in K(n,\delta)$, the restriction of $(g_y)^*$ is a C^∞-diffeomorphism of U onto its image.

3.4. Lemma. *A closed subset F of $C^\infty(Z,Z)$ is compact if and only if it is bounded.*

Remark. A subset Q of $C^\infty(Z;Z)$ is said to be bounded if for every compact L of Z

$$\sup_{u\in Q, z\in L} \|D^s u(z)\|_{Z^{\otimes(s+1)}} < +\infty.$$

Proof. The image $p_s : F \times L \to C\left(L; Z^{\otimes(s+1)}\right)$ defined by

$$(q, z) \to D_q^{(s)}(z)$$

is continuous and therefore is bounded if $(F \times L)$ is compact.

Conversely if $D^s u$ is bounded then by *Ascoli's theorem* $D^{(s-1)}u$ will be in a relatively compact set in $C(Z; Z^{\otimes S})$. □

3.5. Lemma. *Denote by F a compact subset of $C^\infty(Z, Z)$. Fix $z_0 \in Z$, $\delta > 0$. Set*

$$F_\delta = \{v \in F; \quad |\det(Dv)| \geq \delta\};$$

then F_δ is a compact subset of $C^\infty(Z; Z)$. Furthermore it is possible to find an open neighbourhood U of z_0 in Z such that for every $v \in F_\delta$ the restriction v_U of v to U is a diffeomorphism onto its image.

Proof. The function $q(v) = \det(Dv)$ is a continuous function on $C^\infty(Z; Z)$; therefore F_δ is compact.

Consider the ball $B(z_0, 1)$. Then by 3.2 there exists $M < +\infty$ such that for $k = 0, 1, 2$ we have $\|D^{(k)}u\| \leq M$ on $B(z_0, 1)$ for all $u \in F_\delta$; then the eigenvalues of ${}^t(Du)(Du)$ are less than M^2. Therefore the condition $\det(Du) > \delta$ implies that all the eigenvalues are greater than $\delta' := \delta M^{-d+1}$ (where $d = \dim(Z)$).

Define $r = \inf(1, \delta'/2M)$, then spectrum $(Du)(z)\left({}^t Du(z)\right) > \delta'/2$ if $z \in B(z_0, r)$. Therefore u restricted to $B(z_0, r)$ is a local diffeomorphism.

We have

$$\left\|u(z_1) - u(z_2) - u'(z_1)(z_1 - z_2)\right\| < \frac{M}{2} \|z_1 - z_2\|^2;$$

we define $r' = \inf(r, \delta/2M)$; then the restriction of u to $B(z_0, z_1)$ is an injective local diffeomorphism, and therefore a diffeomorphism. □

3.6. *Proof of the implicit function theorem*

As $q(y) := (g_y)^*$ is continuous on $K(n, \delta)$, the image $q\left(K(n, \delta)\right)$ is compact in $C^\infty(Z; Z)$; Lemma 3.5 gives the result. □

4. Currents and differential forms

4.1. Exterior algebra of an L^2-space

Given an abstract probability space $(\Gamma, \mathcal{A}, P)$ we consider the Hilbert space $L^2(\Gamma)$; then

$$\begin{aligned} L^2(\Gamma) \otimes L^2(\Gamma) &= L^2(\Gamma^2) \\ \left(L^2(\Gamma)\right)^{\otimes n} &= L^2(\Gamma^n). \end{aligned}$$

We write

$$\left(L^2(\Gamma)\right)^{\wedge n} = \left\{\psi \in L^2(\Gamma^n); \quad \psi \text{ antisymmetric}\right\}.$$

Antisymmetry means that for every permutation $\sigma \in \mathcal{O}_n$, the symmetric group, we have

$$\psi^\sigma(\gamma_1 \dots \gamma_n) := \psi\left(\gamma_{\sigma(1)}, \dots, \gamma_{\sigma(n)}\right) = \varepsilon(\sigma)\psi(\gamma_1, \dots, \gamma_n)$$

where $\varepsilon(\sigma)$ is the signature of the permutation σ.

Furthermore $\left(L^2(\Gamma)\right)^{\wedge n}$ is a closed subspace of $L^2(\Gamma^n)$; therefore it is a Hilbert space.

Given $\psi_1 \in L^2(\Gamma^{n_1})$, $\psi_2 \in L^2(\Gamma^{n_2})$ we define $\psi_1 \otimes \psi_2 \in L^2\left(\Gamma^{n_1+n_2}\right)$ by

$$\begin{aligned}(\psi_1 \otimes \psi_2)\left(\gamma_1, \dots, \gamma_n, \gamma_{n_1+1}, \gamma_{n_1+n_2}\right) = \\ \psi_1\left(\gamma_1, \dots, \gamma_{n_1}\right)\psi_2\left(\gamma_{n_1+1}, \gamma_{n_1+n_2}\right)\end{aligned}$$

$$\psi_1 \wedge \psi_2 = \frac{1}{(n_1+n_2)!}\sum_{\sigma \in \mathcal{O}_n} (\psi_1 \otimes \psi_2)^\sigma .$$

Given $f_1 \dots f_n \in L^2(\Gamma)$ we have

$$(f_1 \wedge \dots \wedge f_n)(\gamma_1 \dots \gamma_n) = \frac{1}{n!}\sum \varepsilon(\sigma) f_1\left(\gamma_{\sigma(1)}\right) \dots f_n\left(\gamma_{\sigma(n)}\right).$$

Given an element $\psi \in \left(L^2(\Gamma)\right)^{\wedge n}$ we define a n-linear antisymmetric form on $L^2(\Gamma)$ by

$$\begin{aligned}\psi(f_1, f_2, f_n) = (\psi \mid f_1 \wedge \dots \wedge f_n) := \\ \int_{\Gamma^n} \psi(\gamma_1, \dots, \gamma_n) f_1(\gamma_1) \dots f_n(\gamma_n)\Gamma^{\otimes}(d\gamma_1 \dots d\gamma_n).\end{aligned}$$

We have

$$\begin{aligned}(f_1 \wedge \dots \wedge f_n \mid g_1 \wedge \dots \wedge g_n)_{L^2(\Gamma^n)} = \\ \frac{1}{n!}\det\begin{pmatrix} (h_1 \mid h'_1)_H & (h_1 \mid h'_n)_H \\ (h_n \mid h'_1)_H & (h_n \mid h'_n)_H \end{pmatrix}.\end{aligned}$$

Remark. As it is clear that $L^2(\Gamma)$ is itself an abstract Hilbert space we can apply to $L^2(\Gamma)$ either the construction of paragraph 4.1 or the construction of paragraph 4.2; therefore these two paragraphs seem inconsistent! In fact the tensor product and exterior product are defined in algebra as solutions of a universal problem; they are unique up to a *canonical isomorphism*. The two constructions proposed for $\left(L^2(\Gamma)\right)^\wedge$ are *canonically isomorphic.*

4.3. Differential forms on X

Given a Gaussian probability space on X we shall call a differential form on X of degree n an element of $\mathbb{D}_\infty(X; H^{\wedge n})$.

For instance given $\varphi \in D_\infty(X)$ we shall write

$$d\varphi = D\varphi.$$

Then $d\varphi \in D_\infty(X; H)$ is a differential form of degree 1. We have

$$\|d\varphi_1 \wedge \ldots \wedge d\varphi_n\|^2_{H^{\wedge n}} = \frac{1}{n!} \det\left((D\varphi_i \mid D\varphi_j)\right).$$

We recognize on the r.h.s. the determinant of the covariance matrix introduced in Chapter III.

Coboundary. Given $\pi \in \mathbb{D}_\infty(X; H^{\wedge n})$ we define $d\pi \in \mathbb{D}_\infty\left(X; H^{\wedge(n+1)}\right)$ by

$$(d\pi \mid h_1 \wedge \ldots \wedge h_n \wedge h_{n+1}) = \frac{1}{(n+1)!} \sum_\sigma \varepsilon(\sigma) D_{h_{\sigma(1)}} \left(\pi \mid h_{\sigma(2)} \wedge \ldots h_{\sigma(n+1)}\right).$$

4.4. Currents on X

We call a *current* on X of degree n any element $\mathcal{C} \in \mathbb{D}_{-\infty}(X; H^{\wedge n})$. We have a *duality* between currents and differential forms defined by

$$\langle \Pi, \mathcal{C} \rangle = \lim_{t \to 0} E\left[\left(\Pi \mid (P_t^* \mathcal{C})\right)_{H^{\wedge n}}\right]$$

where P_t^* is the dual of the Ornstein-Uhlenbeck operator.

4.5. Pushing down differential forms, lifting up currents

We consider in this paragraph a non-degenerate map $g \in \mathbb{D}_\infty(X; \mathbb{R}^d)$. We denote by $\mathcal{O}$ the range of g which is an open connected subset of $\mathbb{R}^d$.

Given a function $u \in \mathbb{D}_\infty(X)$, we denote by $\tilde{E}^g(u)$ its *non-normalized conditional expectation* defined by

$$\tilde{E}^g(u) = kE^g(u)$$

where $kd\xi$ is the law of g. Then

$$E\left(u\psi(g)\right) = \int_{\mathbb{R}^d} \psi(\xi)\tilde{E}^g(u) d\xi.$$

4.5.1. Pushing down

Given $\pi \in \mathbb{D}_\infty(X; K^{\wedge n})$ we define a differential form

$$\tilde{E}^g(\pi) \in C^\infty\left(\mathcal{O}; (\mathbb{R}^d)^{\wedge n}\right)$$

by the formula

$$\tilde{E}^g(\pi) = \sum_{s \in \beta_{n,d}} \rho_s d\xi_{i_1} \wedge d\xi_{i_2} \wedge d\xi_{i_n}$$

(where $\beta_{n,d}$ denotes the subsets of $[1, d]$ of cardinality n) where ρ_s is given by the formula

$$\rho_s = \tilde{E}^g \left((\pi \mid dg_{i_1} \wedge \ldots \wedge dg_{i_n})_{H^{\wedge n}} \right).$$

The operator $\tilde{E}^g$ on differential forms is defined by the combination of the following two operations

(i) scalar product in the space in $H^{\wedge n}$.
(ii) non-normalized conditional expectation on the scalar functions generated by this projection.

4.5.2. Inverse image of a differential form

Given a differential form θ of degree n on the range of g:

$$\theta = \sum a_{i_1,\ldots,i_n}(\xi) d\xi_{i_1} \wedge \ldots d\xi_{i_n}.$$

We define its inverse image under g, which will be denoted by $g^{\#}\theta$ as the following differential form on X

$$g^{\#}\theta = \sum a_{i_1,\ldots,i_n}(g(x)) dg_{i_1} \wedge \ldots \; dg_{i_n};$$

then we have

4.5.3. Proposition.

$$d\left(g^{\#}(\theta)\right) = g^{\#}(d\theta).$$

Proof. We verify this formula for differential forms of degree zero: given a function ψ on the range of g we have to show

$$d(\psi \circ g) = g^*(d\psi)$$

or

$$D(\psi \circ g) = \sum_{i=1}^{d} \frac{\partial \psi}{\partial \xi_i} dg_i.$$

For a form θ we have

$$\begin{aligned} d\theta &= \sum da_{i_1\ldots i_n} \wedge d\xi_{i_1} \wedge \ldots \wedge d\xi_{i_n} \\ g^{\#}(d\theta) &= \sum g^{\#}\left(da_{i_1,\ldots,i_n}\right) \wedge dg_{i_1} \ldots \wedge dg_{i_n}. \end{aligned}$$

We have previously shown

$$g^{\#}\left(da_{i_1,\ldots,i_n}\right) = d\left(g^{\#}\left(a_{i_1,\ldots,i_n}\right)\right).$$

□

4.5.4. Theorem.

$$E\left((\pi \mid g^{\#}\theta)\right) = \int_{\mathbb{R}^d} \left(\tilde{E}^g \pi \mid \theta\right) d\xi .$$

Remark. This formula shows that at the level of differential forms the operator $\tilde{E}^g$ plays the same role that the non-normalized conditional expectation does at the level of functions.

Proof. We have

$$\begin{aligned} \theta &= \sum a_{i_1 \ldots i_n} d\xi_{i_1} \wedge \ldots \wedge d\xi_{i_n} \\ g^{\#}(\theta) &= \sum g^{\#}\left(a_{i_1 \ldots i_n}\right) dg_{i_1} \wedge \ldots \wedge dg_{i_n} \\ \left(\pi \mid g^{\#}(\theta)\right) &= \sum g^{\#}\left(a_{i_1, \ldots, i_n}\right) \left(\pi \mid dg_{i_1} \wedge \ldots \wedge dg_{i_n}\right) \\ E\left((\pi \mid g^{\#}\theta)\right) &= \sum \int a_{i_1, \ldots, i_n} E^g \left(\pi \mid dg_{i_1} \wedge \ldots \wedge dg_{i_n}\right) d\rho . \end{aligned}$$

□

4.5.5. Lifting up

We denote by $\mathcal{E}'\left(\mathbb{R}^d; (\mathbb{R}^d)^{\wedge n}\right)$ the dual of $C^\infty\left(\mathbb{R}^d; (\mathbb{R}^d)^{\wedge n}\right)$. We shall call this space the *space of currents* of degree n with compact support. Then we define the *lifting up* of a current $\mathcal{C}$

$$\langle \tilde{E}^g(\pi), \mathcal{C} \rangle = \langle \pi, (\tilde{E}^g)^* \mathcal{C} \rangle$$

where $\pi \in \mathbb{D}_\infty(X; H^{\wedge n})$, $\mathcal{C} \in \mathcal{E}'\left(\mathbb{R}^d; (\mathbb{R}^d)^{\wedge n}\right)$.

If

$$\mathcal{C} = \sum_s \alpha_s d\xi_{i_1} \wedge \ldots d\xi_{i_s} \quad \text{with} \quad \alpha_s \in \mathcal{E}'(\mathbb{R}^d)$$

then

$$(\tilde{E}^g)^* \mathcal{C} = \sum_s \left((\tilde{E}^g)^* \alpha_s\right) dg_{i_1} \wedge \ldots \wedge dg_{i_s} .$$

Remark. The Watanabe-lifting-up of the distribution $\alpha_s \in \mathcal{E}'(\mathbb{R}^d)$ is equal to $k\left(\tilde{E}^g\right)\alpha_s$; this change of definition is taylored to the proof of the Stokes' formula in 7.6.

4.5.6. Fundamental current along a fiber

Given $\xi_0 \in \mathcal{O}$ we denote by θ_{ξ_0} the measure of finite energy such that

$$\tilde{E}^{g(x)=\xi_0}(u) = \int u^*(x) \theta_{\xi_0}(dx).$$

We associate to ξ the current of degree d on $\mathcal{O}$ defined by

$$C_{\xi_0} = \delta_{\xi_0} d\xi_1 \wedge \ldots \wedge d\xi_d.$$

Then using 4.5.2 we have

$$(\tilde{E}^g)^* \left(C_{\xi_0}\right) = \theta_{\xi_0} dg_1 \wedge \ldots \wedge dg_d.$$

5. Submanifolds of finite codimension

5.1. Definition of a submanifold. We shall call a submanifold of codimension d any Borel subset V of X such that there exists $g \in \mathbb{D}_\infty(X; \mathbb{R}^d)$, $\xi_0 \in$ Range(g), S_1, S_2 two slim sets such that

$$(g^*)^{-1}(\xi_0) \cup S_1 = V \cup S_2.$$

We shall call g a *defining function* of V. Of course V possesses many defining functions. The relation between these defining functions is made explicit in the next paragraph.

5.2. Relations between defining functions

5.2.1. Proposition. *Consider two non-degenerate functions $f, g \in \mathbb{D}_\infty(X; \mathbb{R}^d)$ such that $(f^*)^{-1}(0)$ and $(g^*)^{-1}(0)$ are equal up to slim sets.*

Then it is possible to construct:

(i) a redefinition net K_s^c of f and g;
(ii) a finite decomposition

$$K_s = \bigcup_{j=1}^{r,s} A_j; \quad A_j \text{ being compact.}$$

(ii) a splitting of the Gaussian space

$$X = Y_j \oplus Z_j \quad \text{where} \quad \dim(Z_j) = d$$

(we shall denote by π_j the projection of X onto $Z_{s,r}$).
(iv) A neighbourhood $U_j \ni 0$ in $\mathbb{R}^d$ and
(v) a diffeomorphism $\varphi_{j,y}$ depending smoothly on the parameter $y \in \Pi_0(A_j)$ such that

$$\varphi_{j,y}\left(f^*(y,z)\right) = g^*(y,z), \quad (y,z) \in A_j, \quad z \in U_j$$

(vi) $D_h \varphi_{j,y}(0) = 0$ for all $h \in Y_j$.

Proof. The proof will depend upon the following lemmas.

5.2.2. Lemma. *Consider the Hilbert space ℓ^2 and a compact subset $K \subset \ell^2$ contained in the unit ball of ℓ^2, but not containing zero. Then we can find a finite number of open sets $\mathcal{O}_s$ such that*

$$K = \bigcup K_s, \qquad K_s := K \cap \mathcal{O}_s$$

and such that for every s we can find a basis vector e_s such that

$$(k_s \mid e_s) \neq 0 \quad \textit{for every} \quad k_s \in K_s .$$

Proof. Set

$$\Omega_j = \{k \in K; (k \mid e_j) \neq 0\};$$

then Ω_j is open; as K does not contain zero, we have that Ω_j is an open covering of K, therefore we can extract a finite covering. □

5.2.3. Lemma. *Given $f, g \in \mathbb{D}_\infty(X; \mathbb{R}^d)$, non-degenerate, we can find a common redefinition net Ω_s for f and g, and for any s, a finite number of subspaces $Z_{1,s}, \ldots, Z_{r,s}$ of H, of dimension d, such that $P_{Z_{r,s}}(df^q(X))$ and $P_{Z_{r,s}}(dg^q(x))$ are d linearly independent vectors for $x \in A_j$, where $\Omega_s^c = \bigcup_{j=1}^{r_s} A_j$, A_j being compact.*

Proof. Given an orthonormal basis of $H : e_1 \ldots e_n, \ldots$ we consider for every finite subset $S \subset \mathbb{N}$, $\mathrm{Card}(S) = d$, $e_S = e_{i_1} \wedge e_{i_2} \wedge e_{i_d}$. Then

$$\{e_S\} \quad \text{is an orthonormal basis of} \quad H^{\wedge s}.$$

We construct a redefinition net Ω_s such that on $\Omega_{s'}^c$, $f, Df, D^2 f \ldots g, Dg, D^2 g \ldots$ are continuous.

We denote $a = |\det(f')|$, $b = |\det(g')|$, then $a, a^{-1} \in \mathbb{D}_\infty$; therefore the sequence $G_n^c := \left\{x; (a^*)^{-1}\left(\left[\frac{1}{n}, n\right]\right)^c\right\}$ have a slim intersection. We can therefore choose ε_s and the redefinition net Ω_s such that

$$a^*(\Omega_s^c) \quad \text{and} \quad b^*(\Omega_s^c) \subset \left[\varepsilon_s, \varepsilon_s^{-1}\right].$$

On Ω_s^c the $H^{\wedge d}$-valued continuous functions $z = df_1 \wedge \ldots \wedge df_d$ and $dg_1 \wedge \ldots \wedge dg_d$ have a compact range. We apply Lemma 5.2.2. We can find $S_1 \ldots S_{r_s}$ such that

$$\left|\left(z(x) \mid e_{S_j}\right)_{H^{\wedge d}}\right| > 0 \quad \text{for} \quad x \in K'_{j,s'}$$

with $K'_{j,s}$ compact, $\Omega_s^c = \bigcup_{j=1}^{r_s} K'_{j,s'}$.

Denote by $V'_{j,s}$ the subspace of H generated by e_q, $q \in S_j$, then the images of $df_1(x), \ldots, df_d(x), dg_1(x) \ldots dg_d(x)$ by $P_{V'_{j,s}}$ are linearly independent. Given x_0, there exists a subspace $Z_{j,s}$ of $V_{j,s}$ of dimension d such that $V_{j,s}$ is transverse both to $u_1(x_0) \ldots u_d(x_0)$ and to $u_{d+1}(x_0)$, $u_{2d}(x_0)$.

This transversality will hold true on an open neighbourhood of x in $K'_{j,s}$. By compactness we can cover $K'_{j,s}$ by a finite number of such neighbourhoods. □

5.2.4. *Proof of 5.2.1*

Choose a point $x_0 \in A_j$. Denote by u (resp. v) the linear map of $\mathbb{R}^d$ into Z_j defined by sending the k-coordinate vector into $\Pi_j\,(df_k(x_0))$ (resp. into $\pi_j\,(dg_k(x_0))$). Define for $x \in A_j$

$$\tilde{f}(x) = u \circ f(x) \qquad (\text{resp} \quad \tilde{g} = v \circ g).$$

We now use the splitting lemma: consider

$$\tilde{f}_y(z) = \tilde{f}(y, z), \qquad \tilde{g}_y(z) = \tilde{g}(y, z);$$

then $\tilde{f}_y, g_y$ are local diffeomorphisms at the origin of Z_j. Denote by $\tilde{g}_y^{-1}$ the inverse diffeomorphism of g. Finally we define φ_y by

$$\varphi_y = \tilde{g}_y^{-1} \circ v \circ u^{-1} \circ \tilde{f}_y.$$

We have $\varphi_y(0) = 0$ for all y. Therefore by differentiation in y we get (vi). □

5.3. The Gauss map and the tangent space

5.3.1. Theorem. *Given a submanifold M of codimension d and two defining functions f and g, we have that there exists a function k such that*

$$(df_1 \wedge \ldots \wedge df_d)\,(x) = k(x) dg_1(x) \wedge \ldots \wedge dg_d(x) \quad \textit{quasi-surely} \quad x \in M.$$

Proof. By differentiating 5.2.1.(v) and using 5.2.1.(vi) we get:

$$\big(k(x)\big)^{-1} = \text{ Jacobian matrix of } \varphi_{j,y}(0) \text{ for } x = (y, z), x \in M.$$

□

5.3.2. Definition of the Gauss map

We define a two-valued map γ_M of M into $H^{\wedge d}$ by

$$\gamma_M(x) := \frac{df_1(x) \wedge \ldots \wedge df_d(x)}{\|df_1(x) \wedge \ldots \wedge df_d(x)\|_{H^{\wedge d}}}.$$

Then γ_M is defined outside a slim set and up to its sign does not depend on the choice of the defining function f. The choice of the sign corresponds to the choice of orientation.

5.3.3. Definition of the tangent space. Given a submanifold M of codimension d then the subspace

$$T_m(M) = \text{orthogonal complement in } H \text{ of the span of } df_1(m), \ldots, df_d(m)$$

does not depend upon the choice of the defining function; it is called the *tangent space* to $m \in M$.

5.3.4. Metric on the Grassmannian

We denote by $\mathcal{G}_\alpha$ the set of all subspaces of dimension d of H. Given a subspace V we associate to this subspace the two points of $H^{\wedge d}$ defined by

$$\pm e_1 \wedge e_2 \dots \wedge e_d$$

where $e_1 \dots e_d$ is an orthonormal basis of V. This map does not depend upon the choice of the orthonormal basis. Therefore we get a two-valued natural map

$$u : \mathcal{G}_d \to H^{\wedge d};$$

we define

$$\text{distance}(V, V') = \text{distance}\big(u(V), u(V')\big).$$

5.3.5. Proposition. *Given a submanifold M we can find an increasing family of compact sets $M_s \subset M$ such that*

$$\left(M - \bigcup_s M_s\right) \quad \textit{is slim}$$

and such that

$$x \to T_m(M)$$

is a continuous map of M_s into the d-dimensional subspace of H.

Proof of the theorem

We take a defining function f and a redefinition net K_s^c of f such that $K_s \cap M$ is compact.

5.3.6. Definition. We shall call such a family M_s a *regular exhaustion of M*.

6. Area of a submanifold

6.1. Notation

Consider a regular exhaustion M_s of the manifold M. Define

$$M_s(\varepsilon) := \\ \big\{x \in X;\ \ x = m + z \text{ with } m \in M_s,\ \ z \in (T_m\,(M_s))^\perp \text{ and } \|z\| \le \varepsilon\big\};$$

then $M_s(\varepsilon)$ is compact.

We denote by γ_d the volume of the unit ball of $\mathbb{R}^d$.

6.2. Theorem. *Given a bounded continuous function ψ on X, then*

$$\gamma_d^{-1} \lim \varepsilon^{-d} E\left(\psi(x)\mathbf{1}_{M_s(\varepsilon)}(x)\right) \quad \textit{exists};$$

this limit defines a Borel measure a_s with its support in M_s; furthermore

$$\lim_s a_s \quad \textit{exists for the total variation norm,}$$
$$\lim a_s := a \quad \textit{will be called the area measure of} \quad M.$$

The non-normalized conditional law for the measure $\mathbf{1}_{M_s(\varepsilon)}\mu$ under the conditioning $f^* = \xi_0$ is:

$$\theta_s = \mathbf{1}_{M_s}\left[\left(\det(f')\right)^*\right]^{-1} a.$$

Proof. The non-normalized conditional law is defined by

$$\int \psi(x)\theta(dx) = \lim_{\varepsilon\to 0} \gamma_d^{-1}\varepsilon^{-d} E\left(\psi\mathbf{1}\{\|f\| < \varepsilon\}\right).$$

If $x \in M_s$

$$f^*(x) = (D_z f)^*(m) + o(\|z\|);$$

therefore

$$\|f(x)\| < \varepsilon \quad \text{is equivalent} \quad \|(D_z f)(m)\| < \varepsilon.$$

Fix m and denote by v the volume in $\mathbb{R}^d$, we get

$$\gamma_d^{-1} \lim \varepsilon^{-d}\left(\{z;\ \|(D_z f)^*(m)\| < \varepsilon\}\right) = \left[\det(f')^*\right]^{-1};$$

therefore taking $\psi_1(x) = \left[\det(f')^*\right]\psi(x)$ we get

$$\int \psi_1(x)\theta(dx) = \lim_{\varepsilon\to 0} \gamma_d^{-1}\varepsilon^{-d} E\left(\psi\mathbf{1}_{M(\varepsilon)}\right).$$

As $\left(\det(f')^*\right)^{-1} \in \mathbb{D}_\infty$ this proves the desired convergence. The various uniform convergences needed are consequences of the exhaustion procedure $\{M_s\}$. □

6.3. Co-area formula

6.3.1. Theorem. *Given a non-degenerate map $g \in \mathbb{D}_\infty(X; \mathbb{R}^d)$, we have for every function $u \in \mathbb{D}_\infty(X)$, $\psi \in C_b(\mathbb{R}^d)$*

$$E\left(\psi(g)\det(g')u\right) = \int_{\mathbb{R}^d} \psi(\xi)\left[\int_{(g^*)^{-1}(\xi)} u^* da\right] d\xi$$

where da is the area measure on the manifold $(g^*)^{-1}(\xi)$.

6.3.2. Corollary. *The law of g has a density given by*

$$k(\xi) = \int_{(g^*)^{-1}(\xi)} \left[\det(g')^*\right]^{-1} da.$$

6.3.3. Corollary. *For $u \in \mathbb{D}_\infty(X)$ we have*

$$\tilde{E}^{g=\xi}(u) = \int_{(g^*)^{-1}(\xi)} \left[(\det g')^*\right]^{-1} u^* da$$

6.3.4. *Proof of Theorem 6.3.1*

Denote by θ the non-normalized conditional law for the measure μ under the conditioning $g^* = \xi_0$; then by Section 1 we have that θ is a measure of finite energy; therefore $\lim_{s\to\infty} \theta(M_s^c) = 0$; computing $\mathbf{1}_{M_s}\theta$ by Theorem 6.2 we obtain 6.3.1. □

7. Stokes' formula

7.1. The operator δ

Given $\pi \in \mathbb{D}_\infty(X; H^{\wedge n})$ we formally define $\delta\pi$ by the duality formula

$$E\big((\pi \mid d\sigma)\big) = E\big((\delta\pi \mid \sigma)\big) \quad \text{where} \quad \sigma \in \mathbb{D}_\infty\left(X; H^{\wedge(n-1)}\right).$$

7.2. Theorem. *If $\pi \in \mathbb{D}_\infty\left(X; H^{\wedge n}\right)$, then $\delta\pi$ is well defined by the formula 7.1 and $\delta\pi \in \mathbb{D}_\infty\left(X; H^{\wedge(n-1)}\right)$.*

Proof. We choose an orthonormal basis of $H : e_1, \ldots, e_n, \ldots$. Then a differential form π will be called *basically finite* if it can be written as a finite sum

$$\pi = \sum a_i \, d\eta_{i_1} \wedge \ldots \wedge d\eta_{i_n}$$

where $d\eta_i$ denotes the linear form on H associated to the vector e_i. Then by integration by parts

$$\delta\pi = \sum \left(\partial_{i_1}^* a_i\right) d\eta_{i_2} \wedge \ldots \wedge d\eta_{i_n} - \left(\partial_{i_2}^* a_i\right) d\eta_{i_1} \wedge d\eta_{i_2} \wedge d\eta_{i_n} + \ldots$$

This formula shows that $\delta\pi$ is well defined for basically finite forms. We have

$$\|\pi\|_{\mathbb{D}_r^2}^2 = \sum \|a_i\|_{\mathbb{D}_r^2}^2$$

$$\|\partial_{i_1}^* a_i\|_{\mathbb{D}_2^2} \leq \|a_i\|_{\mathbb{D}_{r+1}^2}^2 \quad \text{(by Krée-Meyer inequality)}$$

$$\|\delta\pi\|_{\mathbb{D}_r^2}^2 \leq n^2 \|\pi\|_{\mathbb{D}_r^2}.$$

As the basically finite forms are dense in $\mathbb{D}_\infty^2$ this proves the theorem for the $\mathbb{D}_\infty^2$ norms.

The case of $\mathbb{D}_\infty^p$ can be treated in an analogous way. □

7.3. Commutation of δ with $\tilde{E}^g$

Given a smooth differential form θ on the range of g we define $\delta_{\mathbb{R}^d}\theta$ relative to the Lebesgue measure:

$$\int_{\mathbb{R}^d} (\delta_{\mathbb{R}^d}\theta \mid \sigma)\, \delta\xi = \int_{\mathbb{R}^d} (\theta \mid d\sigma) d\xi.$$

Theorem. *We have for $\pi \in \mathbb{D}_\infty(X; H^{\wedge n})$*

$$\tilde{E}^g(\delta\pi) = \delta_{\mathbb{R}^d}\left(\tilde{E}^g(\pi)\right).$$

Remark. This formula is parallel to the formula III.2.4. In Chapter III we were working with the usual conditional expectation; then the duality formula on $\mathbb{R}^d$ had to be formulated using the law of g instead of Lebesgue measure.

Proof. Given a smooth differential form θ on the range of g we have

$$\int \left(\tilde{E}^g(\delta\pi) \mid \theta\right) d\xi = E\left(\delta\pi \mid g^\#(\theta)\right) = E\left((\pi \mid dg^\#(\theta))\right);$$

then by the commutation $dg^\# = g^\# d$, we get

$$\begin{aligned} &= E\left((\pi \mid g^\#(d\theta))\right) = \int_{\mathbb{R}^d} \left(\tilde{E}^g\pi \mid d\theta\right) d\xi \\ &= \int_{\mathbb{R}^d} \left(\delta_{\mathbb{R}^d}\tilde{E}^g\pi \mid \theta\right) d\xi. \end{aligned}$$

□

7.4. Integration of a differential form on an oriented manifold

Given a manifold M of finite codimension we have defined its Gauss map by

$$\gamma_M = \pm\frac{df_1 \wedge \ldots \wedge df_d}{|df_1 \wedge \ldots \wedge df_g|}.$$

An orientation of M will be the choice of a sign in the r.h.s. To an oriented manifold is associated a Gauss map γ_M^+ which is a single valued map from M into $H^{\wedge d}$. Furthermore, we impose the condition that *γ_M^+ must be ∞-quasi-continuous.*

Given a differential form π of degree d we define its integral on the oriented manifold M by

$$\int_M \pi = \int_M (\pi \mid \gamma_M^+)\, da$$

where a is the area measure of M.

7.5. Coarea formula revisited

Proposition. *Given* $g \in \mathbb{D}_\infty(X; \mathbb{R}^d)$ *and* $\pi \in \mathbb{D}_\infty(X; H^{\wedge d})$, *then*

$$\tilde{E}^g(\pi) = \left(\int_{(g^*)^{-1}(\xi)} \pi\right) d\xi_1 \wedge \ldots \wedge d\xi_d.$$

Remark. We choose on $(g^*)^{-1}(\xi)$ the orientation induced by $dg_1 \wedge \ldots \wedge dg_d$.

Proof.

$$\tilde{E}^g(\pi) = \tilde{E}^g\left((\pi \mid dg_1 \wedge \ldots \wedge dg_d)\right) d\xi_1 \wedge \ldots \wedge d\xi_d.$$

By the co-area formula, we have

$$\begin{aligned}\tilde{E}^g\left((\pi \mid dg_1 \wedge \ldots \wedge dg_d)\right) &= \int_{(g^*)^{-1}(\xi)} (\pi \mid dg_1 \wedge \ldots \wedge dg_d) \frac{da}{|dg_1 \wedge \ldots \wedge dg_d} \\ &= \int_{(g^*)^{-1}(\xi)} \left(\pi \mid \gamma_M^+\right) da.\end{aligned}$$

□

7.6. Theorem (Stokes' formula). *Consider a non-degenerate map* $f \in \mathbb{D}_\infty(X; R^{d+1})$, $f = f_0, f_1, \ldots, f_d$, *let* $g \in \mathbb{D}_\infty(X; R^d)$ *be defined by* $g_k = f_k$, $k \in [1, d]$ *then g is non-degenerate.*

Given $\pi \in \mathbb{D}_\infty\left(X; H^{\wedge(d+1)}\right)$ *then*

$$\int_{f^{-1}(\xi'')} \pi - \int_{f^{-1}(\xi')} \pi = \int_{g^{-1}(\eta)\cap\{t'<\xi_0<t''\}} \delta\pi,$$

where $\xi = (t, \eta)$, $t \in \mathbb{R}$, $\eta \in \mathbb{R}^d$, $\xi' = (t', \eta)$, $\xi'' = (t'', \eta)$ *and where the orientations on the manifolds on the left are induced by f and the orientation on the manifolds on the right is induced by g.*

Remark. We have abbreviated the notation omitting to make explicit the redefinitions: $f^{-1}(\xi'')$ must be read as $(f^*)^{-1}(\xi'')$, and so on.

Proof. Given a Lipschitz function $u(t)$ of a scalar variable t, set

$$\tilde{u} \circ f_0.$$

We denote by γ_g^+ the Gauss map associated to the oriented manifold $g^{-1}(\eta)$. Then by 7.5

$$\left(\int_{g^{-1}(\eta)} \left(\delta\pi \mid \gamma_g^p\right) \tilde{u}\, da^g\right) = \tilde{E}^g(\tilde{u}\delta\pi)\ \xi_1 \wedge \ldots \wedge d\xi_d.$$

By consideration of degree,

$$\tilde{E}^g(\tilde{u}\pi) = 0;$$

therefore

$$0 = \delta\left(\tilde{E}^g(\tilde{u}\pi)\right) = \tilde{E}^g\left(\delta(\tilde{u}\pi)\right).$$

Finally we get

$$\delta(\tilde{u}\pi) = \tilde{u}\delta\pi + \tilde{u}' i(df_0)\pi,$$

where we have denoted by $i(df_o)$ the interior product defined by the following duality identity:

$$(i(df_0)\pi \mid \rho) = (\pi \mid df_0 \wedge \rho)$$

where $\rho \in D_\infty(X; H^{\wedge d})$.

Denote by u_n the sequence of functions with compact support defined by $u_n' = v_n$ where

$$v_n(t) = \frac{1}{2n}\left[\mathbf{1}_{[t'-\frac{1}{n}, t'+\frac{1}{n}]} - \mathbf{1}_{[t''-\frac{1}{n}, t''+\frac{1}{n}]}\right];$$

then

$$\begin{aligned}
&\lim \tilde{E}^g\,(\tilde{u}_n \delta\pi) \\
&= \lim \int_{g^{-1}(\eta)} \tilde{v}_n\,(\pi \mid df_0 \wedge \ldots \wedge df_d)\,\frac{da^g}{|df_1 \wedge \ldots \wedge df_d|} \\
&= \tilde{E}^{f_0 = t''} \int_{g^{-1}(\eta)} (\pi \mid df_0 \wedge \ldots \wedge df_d)\,\frac{da^g}{|df_1 \wedge \ldots \wedge df_d} - E^{f_0 = t'} \ldots
\end{aligned}$$

We use now the co-area formula for g

$$\begin{aligned}
&= \tilde{E}^{f_0 = t''}\,\tilde{E}^{g = \eta}\,[(\pi \mid df_0 \wedge \ldots \wedge df_d)] - \tilde{E}^{f_0 = t'} \ldots \\
&= \tilde{E}^{f = (t'', \eta)}\,[(\pi \mid df_0 \wedge \ldots \wedge df_d)] - \tilde{E}^{f = (t', \eta)} \ldots
\end{aligned}$$

Finally we use the co-area formula for f. Then we get

$$\begin{aligned}
&= \int_{f^{-1}(t'', \eta)} (\pi \mid df_0 \wedge \ldots \wedge df_d)\,\frac{da^f}{|df_0 \wedge \ldots \wedge df_d|} - \ldots \\
&= \int_{f^{-1}(t'', \eta)} \left(\pi \mid \gamma_f^+\right) da^f - \ldots
\end{aligned}$$

□

Notes and comments on Part II

Nonlinear potential theory for the study of Sobolev spaces in finite dimensions is a classical subject: see Adams [1], Havin-Mazya [1]. Its introduction to infinite dimensions was initiated in Malliavin [17] and an implicit function theorem was obtained.

The construction of capacities needs an underlying topology; this topology is intrinsic in the case of an abstract Wiener space; in the case of Gaussian spaces this topology depends upon the choice of a numerical model; the fact that quasi-sure analysis does not depend upon this choice was proved by Albeverio-Fukushima-Hansen-Ma-Röckner [1], Shigekawa [14] and Gong-Ma [1]; the tightness of capacities is fundamental to this proof; this tightness was underlined independently in Airault-Malliavin [1] and Sugita [3]. Sugita also proved a realization of positive distributions as integrals with respect to a Borel measure of finite energy.

Another way to demonstrate the independence of the topology is to provide a definition of capacities free from topology: the capacities associated to L^2-type norms have been characterized by Ren [3] and Hirsch-Song [1] in terms of hitting probabilities by OU sheets.

Itô [13] defines the capacities associated to an L^p-type space by duality, the basic concept there being the Borel measures of finite energy.

The study of quasi-sure properties of Brownian motion started with Fukushima [1], [2], [3], [5], Kaneko [1], Takeda [1]. The quasi-sure convergence of smooth martingales was proved in Ren [5], Malliavin-Nualart [2], Kazumi [1].

Airault-Malliavin [1] constructed the regular disintegration of the Gaussian measure under a non-degenerate map by means of a co-area formula, as presented above. In this work, tightness of capacities is the fundamental tool. Lescot [2] proved the existence of a pseudo-regular disintegration along $\mathbb{R}^d$-valued maps when the covariance matrix is quasi-surely invertible.

Fang [1], [2] showed that the essential image of a non-degenerate map is an open connected subset of $\mathbb{R}^d$, a result that can be considered as an intermediate value theorem.

Is it true that the direct image of the set of critical points for a scalar-valued degenerate map is of Hausdorff dimension zero? Negative answers are given in Katznelson-Malliavin [2], positive answers in Lescot [3]. A basic tool is the characterization of compact subsets of D_∞ obtained in Da Prato-Malliavin-Nualart [1].

The existence of a regular disintegration implies the *descent principle* which says that a property true quasi-surely holds true almost surely under conditioning by a non-degenerate map; this principle was used in Malliavin-Malliavin [5] to transfer properties from path groups to loop groups. The large deviation theory for the Brownian bridge can be approached through a quantitative transfer principle: Fang [3], [7], [9], Nobuo Yoshida [1].

A theory of finite-codimensional submanifolds of a Gaussian space can be developed: the notions of the second fundamental form, curvature and Ricci curvature can be defined and computed: Getzler [2], Airault [3], Airault-Van Biesen [1], Kazumi-Shigekawa [3], Van Biesen [1].

A geometric theory of independent σ-fields is constructed by Nualart-Üstünel-Zakai [2], [3].

A general theory of differential forms on a Gaussian space has been constructed in Kusuoka [7], [8].

Notions of the quasi-sure diffeomorphism are developed by Kusuoka [8], Shigekawa [13]. Stokes' formulae have been considered in Goodman [1], Skorokhod [1]. Asymptotic expansions with remainder term estimation in D_∞ lead to sharp estimates of heat kernels for small time (Watanabe [3]).

It is possible to treat anticipative problems using quasi-sure analysis by solving the adapted problem after restricting the solution to the finite-codimensional manifold which describes the anticipativity; procedures of this sort have been implemented by Buckdahn-Föllmer [1] for anticipative Girsanov transformations and by Nualart-Malliavin [2] for anticipative Stratonovich stochastic differential equations. This methodology has also been applied to the computation of Lyapunov exponents of stochastic dynamical systems (Arnold-Imkeller [1]) and to financial probability (Imkeller [3]).

Part III

Stochastic Integrals

The numerical model of Chapter I is associated to the Hilbert space ℓ^2 of square integrable sequences. Another remarkable Hilbert space is the space of square-integrable functions on a space with a measure of finite mass and without atoms. This Hilbert space has a sort of "continuous basis". The associated Gaussian probability space is called the *probability space of white noise*. The chaos decomposition over the white noise has a canonical expression, using what might be described as "Hermite polynomials with continuous indices", defined in terms of symmetric kernels. Those kernels are "integrated against the white noise", integrals which are divergent in absolute value, but become convergent by some kind of "continuous renormalization" of their Riemann-Lebesgue sums; those renormalized integrals are the *Wiener-Itô multiple stochastic integrals*.

The next issue is to compute stochastic integrals of random kernels; the possibility of proceeding to the required renormalization is guaranteed by the existence of first derivatives in the sense of Chapter I; the stochastic integral is equal to the divergence considered in then Chapter III. These two deep results establish a structural link between the theory of white noise stochastic integrals and the differential calculus considered in Part I.

An important case of white noise is the white noise associated to the Hilbert space of square integrable function on [0, 1] for the Lebesgue measure; this generates the *probability space of Brownian motion*. The multiplication operator by the indicator function $\mathbf{1}_{[0,t]}$ defines a one parameter family of Hilbert space projection operators Π_t; lifting this family Π_t to the probability space level we get the family of conditional expectations $E^{\mathcal{N}_t}$ relative to the *Itô filtration* $\mathcal{N}_t$. The family Π_t can be considered as a "homotopy operator" linking continuously the identity with the operator 0; at the level of the probability space, the homotopy is realized by the conditional expectations $E^{\mathcal{N}_*}$. It is well known in elementary differential calculus that a homotopy operator generates a Poincaré Lemma thereby making effective the computation of a differential form through its coboundary; on the probability space of the Brownian motion, the Poincaré Lemma is a formula with which to compute a functional from the knowledge of its gradient; iterating this formula, a stochastic Taylor formula is obtained. Classically, the study of Brownian martingales is done through the Itô calculus which becomes linked via this stochastic Taylor formula with the differential calculus of Part I.

Chapter VI. White Noise Stochastic Integrals as Divergences

Contents: Gaussian probability space over an L^2 space – Construction of $\mathrm{Seg}(L^2(\Gamma))$ by Wiener-Itô multiple integrals – Fock space, naturality of Wiener – Itô multiple integrals – Stroock-Taylor formula – Nualart-Pardoux construction of the Skorokhod stochastic integral – The Gaveau-Trauber theorem identifying the Skorokhod integral with a divergence – Energy estimate for the Skorokhod integral – Existence of traces and of Stratonovich integrals.

The theory of divergence developed in Chapters II and III is essentially of a global nature. It is based on estimates for the Ornstein-Uhlenbeck operator.

The purpose of this chapter is to develop a constructive schema of a *local nature*, so as to compute the divergence by a limiting procedure. We want to avoid, at least formally, the introduction of DZ.

The localisation will be achieved by a splitting of the Cameron-Martin space H into small pieces. This splitting will be obtained by a continuous spectral decomposition of H.

This chapter aims to present this methodology free from any filtrations; the development of the theory of several parameters processes is outside the scope of this book.

1. White noise

1.1 Definition. We shall denote by (Γ, B, ρ) an abstract measure space. We shall assume

(i) $\rho(\Gamma) < +\infty$.
(ii) $L^2(\Gamma, B, \rho)$ is *separable,*
(iii) *the measure* ρ *has no atoms.*

(which means that every $B \in B$ can be written $B = B_1 \cup B_2$ with $\rho(B_1) = \rho(B_1) = \frac{1}{2}\rho(B)$).

One example is $[0, 1]$ with the Lebesgue measure. Furthermore, this example is universal.

The probability space

$$X = \text{Seg}\left(L^2(\Gamma, B, \rho)\right)$$

will be called the probability space of *white noise* over Γ.

In the following we shall use the notation $L^2(\Gamma)$ for $L^2(\Gamma, B, \rho)$ and similar abbreviations.

1.2. Differential calculus

Given $h \in L^2(\Gamma)$ and given $u \in D_1^2(X)$, then

$$(Du) \in L^2\left(X; L^2(\Gamma)\right).$$

As $L^2\left(X; L^2(Z)\right) \simeq L^2(X \times Z)$, there exists $q(x, \gamma)$ defined $\mu \otimes \rho$ a.e. such that

$$(D_h u)(x) = \int_\Gamma q(x, \gamma) h(\gamma) \rho(d\gamma).$$

We shall write $q(x, \gamma) = (D_\gamma u)(x)$. With this notation we have

$$D_h u = \int_\Gamma (D_\gamma u) h(\gamma) d\gamma \quad \text{where} \quad Du \in L^2(\Gamma \times X).$$

In the same way, if $u \in D_2^2(X)$, then there exists $D^{(2)}_{\cdots} u \in L^2(\Gamma^2 \times X)$ such that

$$D_{h_1} D_{h_2} u = \int_{\Gamma^2} \left(D^{(2)}_{\gamma_1, \gamma_2} u\right) h_1(\gamma_1) h_2(\gamma_2) d\rho(\gamma_1) d\rho(\gamma_2).$$

Furthermore

$$\|D^2 u\|_{L^2} = \int_{\Gamma^2 \times X} \left|\left(D^2_{\gamma_1, \gamma_2} u\right)(x)\right|^2 d\rho(\gamma_1) d\rho(\gamma_2) d\mu(x)$$

and

$$D^{(2)}_{\gamma_1, \gamma_2} u = D^{(2)}_{\gamma_2, \gamma_1} u.$$

The splitting of Γ in "small pieces" appears formally in these notations. The notation D_γ is reminiscent of the notation of a partial derivative in finite dimension.

2. Construction of Seg $\left(L^2(\Gamma, B; \rho)\right)$ by Itô-Wiener stochastic integrals

The machinery which produced Seg(H) involved all the bases of H, each basis giving rise to a *numerical model*. Finally the intrinsic properties of Seg(H) are those which are common to *all the numerical models*.

A fixed numerical model enjoys all the properties of Seg(H) but the contrary is false; for instance a fixed numerical model has a canonical vector space structure which Seg(H) does not have.

The Hilbert space $L^2(\Gamma, B, \rho)$ has a "natural basis" which is indexed by "the points of Γ". This *formal* point of view which appeared at the end of the last paragraph will become more meaningful when we introduce finite approximations to B.

2.1. Resolution of $L^2(\Gamma)$

We recall that the data of a finite σ-field $\hat{\mathcal{B}}$ is equivalent to the data of a finite partition of $\Gamma : a_1 \dots a_s \in B$ (where $\cup a_s = \Gamma$, $a_s \cap a_\ell = \emptyset$ if $s \neq \ell$). The σ-field generated by the a_j will have 2^s elements and will be equal to $\hat{\mathcal{B}}$. The elements a_s will be called the *atoms* of $\hat{B}$.

A *resolution* of Γ will be given by the data of an increasing family of finite σ-fields $B_1 \subset B_2 \subset B_n \dots \subset \mathcal{B}$. We will assume that

$$\bigcup_n L^2(\Gamma; B_n; \rho) \quad \text{is dense in} \quad L^2(\Gamma, \mathcal{B}; \rho).$$

We shall denote by Γ_n the *set of atoms* of $\mathcal{B}_n$. Then Γ_n is a finite set. Furthermore the inclusion relation in Γ induces a natural surjection $\Gamma_{n+1} \to \Gamma_n$.

We will assume that all the atoms have a strictly positive measure.

Resolution of Seg $(L^2(\Gamma))$

Given a resolution of $L^2(\Gamma)$, the corresponding resolution of Seg $(L^2(\Gamma))$ will be the data of a countable sequence of Gaussian variables

$$\{G^n_\delta\}, \qquad \delta \in \Gamma_n$$

satisfying 2.2.1, 2.2.2, and 2.2.3.

2.2.1.
$$E\left(|G^n_\delta|^2\right) = \rho(\delta)$$

2.2.2.
$$\sum_{\delta \subset \delta'} G^n_\delta = G^{n-1}_{\delta'} \quad \text{where} \quad \delta' \in \Gamma_{n-1} \quad \text{(consistency condition)}$$

2.2.3. For fixed n, $\{G^n_\delta\}$, $\delta \in \Gamma_n$ are independent.

2.2.4. Recursive construction of the resolution of Seg $(L^2(\Gamma))$

We choose a family of *independent* Gaussian variables

$$G^1_\delta, \qquad \delta \in \Gamma_1,$$

which satisfy 2.2.1.

Then we choose $D^2_{\delta'}$ according to the following procedure. *Under the conditioning*

$$\sum_{\delta' \subset \delta} G^2_{\delta'} = G^1_\delta \qquad \delta \in \Gamma_1$$

we choose

$$\left\{G^2_{\delta'}\right\}, \quad \delta' \in \{\delta'; \delta' \subset \delta\}$$

independent and satisfying 2.2.1, and so on.

2.3. Wiener's stochastic integral

Given $f \in L^2(\Gamma)$, we associate to f the martingale

$$f_n = E^{B_n}(f),$$

then

$$f_n \in L^2(\Gamma_n).$$

We define

$$\int f_n(\gamma) dx(\gamma) = \sum_{\delta \in \Gamma_n} f_n(\delta) G^n_\delta .$$

As f_n is *constant* on the atom δ, $f_n(\delta)$ is a well defined number.

We have the energy identity

2.3.1. $$E\left(\left|\int f_n(\gamma) dx(\gamma)\right|^2\right) = \sum_{\delta \in \Gamma_n} |f_n(\delta)|^2 \rho(\delta) = \|f_n\|^2_{L^2}$$

$$E\left(\left|\int f_n(\gamma) dx(\gamma) - \int f_{n+p}(\gamma) dx(\gamma)\right|^2\right) = \|f_n - f_{n+p}\|^2_{L^2},$$

the right hand side tends to zero as $n \to +\infty$. This proves that

2.3.2. $$I_n = \int f_n(\gamma) dx(\gamma) \quad \text{converges in} \quad L^2(X).$$

We define

2.3.3. $$\int f(\gamma) dx(\gamma) = \lim_{L^2} \int f_n(\gamma) dx(\gamma).$$

We can strengthen this statement:

2.3.4. $$I_n = \int f_n(\gamma) dx(\gamma) \quad \textit{converges almost surely.}$$

In fact, denote by $\mathcal{N}^n$ the σ-field generated by $\{G_\delta\}$, $\delta \in \Gamma_n$. Then almost sure convergence will result from the fact that

2.3.5. $$I_n \quad \textit{is a} \quad \mathcal{N}^n \; \textit{martingale}.$$

In fact

$$E^{\mathcal{N}^n}(I_{n+1}) = \sum_{\delta \in \Gamma_n} A_\delta$$

where

$$A_\delta = E^{\mathcal{N}^n}\left(\sum_{\delta' \subset \delta} f_{n+1}(\delta') G_{\delta'}^{n+1}\right) = f_n(\delta) G_\delta^n .$$

□

2.3.6. Proposition. *Given $f, h \in L^2(\Gamma)$, $\int f dx$ is a centered Gaussian variable*

$$E\left(\left(\int f dx\right)\left(\int h dx\right)\right) = (f \mid h)_{L^2(\Gamma)} .$$

Proof. By direct computation

$$E\left(\left(\int f_n(\gamma) dx(\gamma)\right)\left(\int h_n(\gamma) dx(\gamma)\right)\right) = (f_n \mid h_n)_{L^2(\Gamma)} .$$

Then 2.3.6 is obtained by 2.3.2. □

2.3.7. Proposition. *Given $B_1, B_2 \ldots B_s \in \mathcal{B}$ disjoint, denote by $V(B_k)$ the vector space of random variables of the form $\int f_k dx$ with $f_k(\gamma) = 0$ if $\gamma \notin B_k$, then the $\{V(B_k)\}$ are independent.*

Proof. For Gaussian variables, independence follows from the orthogonality relation which follows from 2.3.6. □

2.3.8. Notation

Given $f \in L^2(\Gamma)$ we write

$$\tilde{W}(f) = \int f(\gamma) dx(\gamma).$$

The correspondence $f \to \tilde{W}(f)$ defines a map $L^2(\Gamma) \to L^2(X)$. This map realizes *a constructive version* of the correspondence $f \to W(f)$ defined in Chapter I. By abuse of notation we drop the tilde and denote $\tilde{W}$ as W.

2.4. Construction of the Itô-Wiener stochastic integral of degree p

2.4.1. Definition. We consider the product space

$$\Gamma^p = \left(\Gamma^p, \mathcal{B} \otimes \ldots \otimes \mathcal{B}, \rho \otimes \ldots \otimes \rho\right).$$

Given a resolution Γ_n of $L^2(\Gamma)$, we define a resolution Γ_n^p of $L^2(\Gamma^p)$ by

$$\Gamma_n^p = \left\{\delta \in \mathcal{B}^{\otimes p}; \delta = \delta_1 \times \delta_2 \ldots \times \delta_p \quad \text{with} \quad \delta_k \in \Gamma_n\right\}.$$

It is clear that the $\delta \in \Gamma_n^p$ constitute a partition of Γ^p; they generate a finite σ-field $\mathcal{B}_n^p$. Given $f \in K^2(\Gamma^p)$ the conditional expectation $E^{\mathcal{B}_n^p} f = f_n$ defines a martingale.

We shall denote by $L_s^2(\Gamma^p)$ the closed subspace of *symmetric functions* of $L^2(\Gamma^p)$.

Given $f \in L_s^2(\Gamma^p)$ we define the family

$$S_n(f) = \sum_{\delta \in \Gamma_n^p} \mathcal{E}(\delta) f_n(\delta) w(\delta)$$

where

$$w(\delta) = \mathcal{E}(\delta) \prod_{k=1}^{p} w(\delta_k).$$

$$\begin{aligned} \mathcal{E}(\delta) &= 0 \quad \text{if there exists} \quad j, k, j \neq k \quad \text{with} \quad \delta_j = \delta_k \\ \mathcal{E}(\delta) &= 1 \quad \text{in the other cases.} \end{aligned}$$

This means that the atoms δ on which the summation takes place have to be "diagonal-free".

2.4.2. Theorem. *$S_n(f)$ converges in $L^2(X)$.*

Proof. We denote by Σ_p the symmetric group on p elements. Given $\delta \in \Gamma_n^p$ and $\sigma \in \Sigma_p$ we denote by δ^σ the atom $\delta_{\sigma(1)} \times \ldots \times \delta_{\sigma(n)}$. Then

$$w(\delta) = w(\delta^\sigma).$$

□

We say that $\delta' \sim \delta$ if there exists σ such that $\delta^\sigma = \delta$. We have the orthogonality lemma.

2.4.3. Lemma.

$$\begin{aligned} E\big(w(\delta)w(\delta')\big) &= \rho^{\otimes p}(\delta) \quad \textit{if} \quad \delta \sim \delta', \\ E\big(w(\delta)w(\delta')\big) &= 0 \quad \textit{otherwise}. \end{aligned}$$

Proof. The $w(\delta_k)$ are independent by the diagonal-free assumption and 2.3.7. Therefore

$$\begin{aligned} E\big(w(\delta)\big) &= \prod_k E\big(w(\delta_k)\big) = 0 \quad \text{and} \\ E\big((w(\delta))^2\big) &= \prod_k E\left([w(\delta_k)]^2\right) = \prod_k \rho(\delta_k). \end{aligned}$$

By using an appropriate permutation we are reduced to the case

$$\begin{aligned} \delta &= \delta_1 \times \ldots \times \delta_{k_0} \times \delta_{k_0+1} \times \ldots \times \delta_p \\ \delta' &= \delta_1 \times \ldots \times \delta_{k_0} \times \delta'_{k_0+1} \times \ldots \times \delta'_p \end{aligned}$$

where $A_{k_0} = \left(\bigcup_s \delta_{k_0+s}\right)$ and $A'_{k_0} = \bigcup_s \delta'_{k_0+s}$ are disjoint. Conditioning on $V\left(\left(A_{k_0} \cup A'_k\right)^c\right) = \mathcal{A}$, we have

$$E\big(w(\delta)w(\delta')\big) = E\left(\prod_{k \le k_0} (w(\delta_k))^2 \, E^{\mathcal{A}}(w'w'')\right)$$

where w' is $V(A_{k_0})$ measurable and w'' is $V(A'_{k_0})$ measurable. This implies their independence and therefore their orthogonality. □

2.4.4. Lemma. *Given $f \in L^2_s(\Gamma^p)$, then*

$$E\left(|S_n(f)|^2\right) \le p! \| f_n \|^2_{L^2}.$$

Proof. We write Γ_n^p as the union of equivalence classes

$$\Gamma_n^p = \cup \Omega_r,$$

then

$$S_n(f) = \sum_{\delta \in \Gamma_n^p} \mathcal{E}(\delta) f_n(\delta) w(\delta) = \sum_r \sum_{\delta \in \Omega_r} f_n(\delta) w(\delta) \mathcal{E}(\delta).$$

By the symmetry of f, $f_n(\delta) = f_n(\delta')$ if $\delta, \delta' \in \Omega_r$. Picking a representative δ^r in Ω_r we have

$$S_n(f) = \sum_r p! \mathcal{E}(\delta^r) w(\delta^r) f_n(\delta^r).$$

Using the orthogonality relation 2.4.3

$$E\left(|S_n(f)|^2\right) = (p!)^2 \sum \mathcal{E}(\delta^r) \rho^{\otimes p}(\delta_r) \left(f(\delta_r)\right)^2.$$

On the other hand

$$\| f_n \|^2_{L^2} = \sum_{\delta \in \Gamma_n^p} \rho^{\otimes p}(\delta) |f(\delta)|^2 = \sum_r p! \rho^{\otimes p}(\delta^r) |f(\delta^r)|^2.$$

□

2.4.5. Lemma. *Denote by $L^2_{s,s}(\Gamma_n^p)$ the space of functions in $L^2_s(\Gamma_n)$ such that $\mathcal{E}(\delta) = 0$ implies $f(\delta) = 0$. Then*

$$S_{n+q}(f) = S_n(f) \quad \textit{for all} \quad q > 0 \quad \textit{and}$$

$$E\left(|S_n(f)|^2\right) = p! \| f \|^2_{L^2}.$$

Proof. Write

$$A(\delta) = \left\{\delta' \in \Gamma^p_{n+1}; \delta' \subset \delta\right\},$$

then

$$\sum_{\delta' \subset \delta} w(\delta') = w(\delta).$$

The second equality is obtained by going back to the proof of 2.4.4 and taking into account the fact that it is possible to omit $\mathcal{E}(\delta)$ in the definition of $S_n(f)$. □

2.4.6. *Proof of 2.4.2.*

The sequence of operators $U_n f \to S_n(f)$ are uniformly bounded operators from $L^2_s(\Gamma^p)$ into $L^2(X)$ by 2.4.3. By 2.4.5, U_n converges on $\bigcup_n L^2_{s,s}(I_n)$, which is dense in $L^2_s(\Gamma)$. Therefore it converges on $L^2_s(\Gamma)$.

2.4.7. Definition. We write

$$\lim S_n(f) = \int f dx^{\otimes p} = \int_{\Gamma^p} f(\gamma_1 \dots \gamma_p) dx(\gamma_1) dx(\gamma_2) dx(\gamma_p)$$

then the right hand side will be called the *Itô-Wiener stochastic integral.*

2.5. Properties of the Itô-Wiener stochastic integral

2.5.1. Theorem. *Given $f \in L^2_s(\Gamma^p)$, define*

$$F(x) = \int f dx^{\otimes p}$$

then

$$D_{\gamma_0} F = G$$

where

$$G = p \int g dx^{\otimes(p-1)}$$

and where

$$g\left(\gamma_1 \dots \gamma_{p-1}\right) = f\left(\gamma_1 \dots \gamma_{p-1}, \gamma_0\right).$$

Proof. We take $f \in L^2_{s,s}(\Gamma^p_n)$, then

$$\int f dx^{\otimes p} = S_n(f) = \sum_{\delta \in \Gamma^p_h} w(\delta) f(\delta).$$

Given $\delta = \delta_1 \times \dots \times \delta_p$, we have

$$D_{\gamma_0} w(\delta) = \mathbf{1}_{\delta_1}(\gamma_0) w\left(\delta_2 \times \dots \times \delta_p\right) + \dots + \mathbf{1}_{\delta_p}(\gamma_0) w\left(\delta_1 \times \dots \times \delta_{p-1}\right).$$

Denote by Σ_p the symmetric group on p objects. Consider

$$u = \mathbf{1}_{\delta_1}\mathbf{1}_{\delta_2}\mathbf{1}_{\delta_p}.$$

We symmetrize u in order to define the stochastic integral; its symmetrization is

$$\tilde{u} = \frac{1}{p!}\sum_{\sigma\in\Sigma_p} u^\sigma.$$

$$D_{\gamma_0}\left(\int \tilde{u}dx^{\otimes p}\right) = p\int u\left(\gamma_1 \dots \gamma_{p-1}, \gamma_0\right) dx^{\otimes(p-1)}.$$

We remark in fact that the right hand side is symmetric in $\left(\gamma_1, \dots, \gamma_{p-1}\right)$. This proves the theorem on $\bigcup_n L^2_{s,s}(\Gamma_n^p)$ which extends to $L^2_s(\Gamma)$ by continuity. □

2.5.2. Lemma. *Given $p \neq q$, $f \in L^2_s(\Gamma^p)$, $g \in L^2_s(\Gamma^q)$, consider*

$$F = \int f dx^{\otimes p}, \qquad G = \int g dx^{\otimes q}, \qquad \textit{then}$$

$$E(FG) = 0.$$

Proof. We can assume $p < q$. By continuity it is sufficient to prove the theorem for $f, g \in \bigcup_n L^2_{s,s}(\Gamma_n^*)$.

The proof will then result from the identity

$$E\left(w(\delta)w(\delta')\right) = 0, \quad \text{if} \quad \mathcal{E}(\delta) = \mathcal{E}(\delta') = 1 \quad \text{and} \quad \delta \in \Gamma_n^p, \quad \delta' \in \Gamma_n^q.$$

In fact, under the action of Σ_p, the equivalence class of Γ_n^p can be identified with $Z \in \mathcal{P}(\Gamma_n)$ with $\operatorname{card}(Z) = p$.

Given $Z \in \mathcal{P}(\Gamma_n)$, $Z' \in \mathcal{P}(\Gamma_n)$ with $\operatorname{card}(Z) = p$, $\operatorname{card}(Z') = q$, we take $\delta_0 \in Z' \cap Z^c$. Then $V(\delta_0)$ and $V(\delta_0^c)$ are independent σ-fields. Write $\mathcal{A} = V(\delta_0)$, then

$$E\left(w(\delta)w(\delta')\right) = E\left(w(\delta)w(\delta'')E^{\mathcal{A}}\left(w(\delta_0)\right)\right) = 0$$

where δ'' is obtained from δ' by deletion of δ_0.

2.5.3. Theorem. *The map*

$$f \to \frac{1}{(r!)^{1/2}}\int f dx^{\otimes p}$$

defines a surjective isometry of $L^2_s(\Gamma^p)$ onto the chaos $\mathcal{C}_p$ of Seg$\left(L^2(\Gamma)\right)$.

Proof. The isometry is proved first for $f \in L^2_{s,s}(\Gamma_n^p)$ by taking up the proof of 2.4.4, erasing the factor $\mathcal{E}(\delta)$ which is then superfluous and then extending by continuity.

To conclude, we shall use an induction on p. If $p = 1$ the chaos $\mathcal{C}_1$ of order 1 is precisely isomorphic to $L^2(\Gamma)$. We make the assumption that the theorem is true for $p < r$. Then given $\varphi \in \mathcal{C}_r$, $D_\gamma\varphi \in \mathcal{C}_{r-1}$. Therefore, by the inductive hypothesis, there exists $f_\gamma \in L^2_s(\Gamma^{r-1})$ such that

$$\int f_\gamma(\gamma_1 \dots \gamma_{r-1})\, dx^{\otimes(r-1)} = D_\gamma \varphi.$$

We denote by $k \in L_s^2(\Gamma^r)$ the symmetrization of the function $f_\gamma(\gamma_1 \dots \gamma_{r-1})$. Define

$$\psi = \int k dx^{\otimes r},$$

then by 2.5.1

$$D_\gamma \psi = D_\gamma \varphi.$$

As $E(\psi) = E(\varphi) = 0$ this implies that $\varphi = \psi$. We have proved the surjectivity onto the chaos of order r. Finally the fact that the multiple integrals of order r belong to the chaos of order $\le r$ results from the fact that its D^{r+1} derivative vanishes identically. The fact that it belongs to the chaos of order r is then proved by induction using the orthogonality relations 2.5.2. □

2.5.4. Corollary. *Given* $f, g \in L^2(\Gamma)$

$$E\left(\left(\int f dx\right)\left(\int g dx\right)\right) = \int f(\gamma) f(\gamma) \rho(d\gamma).$$

Proof. We polarize 2.5.3 written with $r = 1$. This identity results also from 2.3.6. □

2.5.5. Theorem. *Given* $\varphi_1, \varphi_2 \in L^2(\Gamma)$, *write*

$$F(\gamma_1, \gamma_2) = \frac{1}{2}(\varphi_1(\gamma_1)\varphi_2(\gamma_2) + \varphi_1(\gamma_2)\varphi_2(\gamma_1))$$

then

$$\left(\int_\Gamma \varphi_1(\gamma) dx(\gamma)\right)\left(\int_\Gamma \varphi_2(\gamma) dx(\gamma)\right) = \int_{\Gamma^2} F(\gamma_1, \gamma_2)\, dx(\gamma_1) \otimes dx(\gamma_2) + \int_\Gamma \varphi_1 \varphi_2 d\rho.$$

Proof. By bilinearity and continuity we are reduced to proving the statement when $\varphi_1 = \mathbf{1}_{A_1}$, $\varphi_2 = \mathbf{1}_{A_2}$. Using (*in advance*) Theorem 3.2 we will choose a resolution of $L^2(\Gamma)$ which will make the computation easier. We take for $\mathcal{B}_1$ the σ-field generated by A_1, A_2. Then all the atoms δ which will appear in $\mathcal{B}_n$ are either contained in or disjoint from A_i $(i = 1, 2)$. In degree two, $\delta = \delta_1 \times \delta_2$ will satisfy $\mathcal{E}(\delta) = 1$ if and only if $\delta_1 \ne \delta_2$. With this notation, the l.h.s. equals

$$\left(\sum_{\substack{\delta_1 \subset A_1 \\ \delta_1 \in \Gamma_n}} w(\delta_1)\right)\left(\sum_{\substack{\delta_2 \subset A_2 \\ \delta_2 \in \Gamma_n}} w(\delta_2)\right) = \sum_{\delta \in \Gamma_n^2} \mathcal{E}(\delta) w(\delta_1) w(\delta_2) E^{\mathcal{B}_n}(F) + \sum_{\substack{\delta_1 \in \Gamma_n \\ \delta_1 \subset A_1 \cap A_2}} [w(\delta_1)]^2 .$$

Letting $n \to \infty$, the proof will result from the following *contraction theorem*.

2.5.6. Wiener contraction theorem. *Denote*

$$f_n = \sum_{\delta \subset A} [w(\delta)]^2, \quad \delta \in \Gamma_n^\rho,$$

then, when $n \to +\infty$, f_n converges towards the constant $\rho(A)$.

Proof.

$$\begin{aligned} E(f_n) &= \sum_{\delta \subset A} \rho(\delta) = \rho(A) \\ f_n - \rho(A) &= \sum_{\delta \subset A} \left[[w(\delta)]^2 - \rho(\delta) \right]. \end{aligned}$$

We remark that $\left([w(\delta)]^2 - \rho(\delta)\right)$ and $\left([w(\delta')]^2 - \rho(\delta')\right)$ are independent centered random variables. Therefore

$$\begin{aligned} E\left(|f_n - \rho(A)|^2\right) &= \sum_{\delta \subset A} \left(E[w(\delta)]^2 - \rho(\delta)\right)^2 \\ &= 2 \sum_{\delta \subset A} (\rho(\delta))^2 \end{aligned}$$

where the last inequality is obtained by the Gaussian hypothesis.

As $\sum \rho(\delta) = \rho(A)$ and $\rho(\delta) \to 0$ we get

$$E\left(|f_n - \rho(A)|^2\right) \to 0.$$

3. The Fock space of $L^2(\Gamma, \mathcal{B}; \rho)$

We want to describe the effect of a change of *resolution*; this description will be used for defining the intrinsic properties of the white noise. This development is parallel to the method used in Chapter I:

Abstract Hilbert space $H \to$ choice of a basis of $H \to$ numerical model $\to$ effect of the change of basis $\to$ Seg(H).

The role of the choice of a *basis* will now be played by the choice of a *resolution*.

3.1. Probability space associated to a resolution

Given a resolution $B_1 \subset \ldots \subset B_n \subset \ldots$ of $L^2(\Gamma)$, we constructed in 2.2 a resolution of Seg $\left(L^2(\Gamma)\right)$. We want to give this resolution the status of a probability space. Denote

$$\Gamma_{\cdot} = \bigcup_n \Gamma_n.$$

We shall define a probability measure on $\mathbb{R}^{\Gamma_{\cdot}}$ by introducing $\Gamma_\alpha = \bigcup_{n=1}^{\alpha} \Gamma_n$ (α finite integer).

The marginal law θ_α will be the Gaussian measure which satisfies the consistency condition 2.2.2. We get thus a coherent system of marginal law on $\mathbb{R}^{\Gamma_{\cdot}}$. We thus obtain a probability measure θ and the *probability space associated* to the resolution; we shall sometimes write Y for $(\mathbb{R}^{\Gamma_{\cdot}}, \theta)$.

3.2. Theorem. *Consider two resolutions* $\{B_n\}$, $\{B'_n\}$ *of* $(L^2(\Gamma))$*; denote by* Y *and* Y' *the two associated probability spaces.*

Then there exists a canonical isomorphism φ *of the probability space* Y *onto* Y' *such that given* $f \in L_s^2(\Gamma^p)$, *associating to* f

$$I(y) = \int f dy^{\otimes p}, \qquad I'(y') = \int f dy'^{\otimes p}$$

then we have

$$I = I' \circ \varphi.$$

Proof. Given $\delta' \in \bigcup_q \Gamma'_q = \Gamma'_{\cdot}$ we consider the sequence $u_{\delta',n} = E^{B_n}(\mathbf{1}_{\delta'})$. The martingale $\{u_{\delta',n}\}$ will converge in $L^2(\Gamma_{\cdot})$. We get by linearity and completion a map $L^2(\mathbb{R}^{\Gamma'_{\cdot}}; \theta') \to L^2(\mathbb{R}^{\Gamma_{\cdot}}; \theta)$ which will induce a map $\varphi : Y \to Y'$.

Consider now the two following stochastic integrals of degree 2 (to facilitate the notation)

$$\begin{aligned} I &= \int f(\gamma_1, \gamma_2)\, dy(\gamma_1) \otimes dy(\gamma_2), \\ I' &= \int f(\gamma_1, \gamma_2)\, dy'(\gamma_1) \otimes dy'(\gamma_2). \end{aligned}$$

The integral I' has been defined as the limit of $S'_n(f)$ where

$$S'_n(f) = \sum_{\delta' \in \Gamma_n'^2} \mathcal{E}(\delta') f'_n(\delta') w'(\delta'_1) w'(\delta'_2)$$

where $f'_n = E^{B'_n}(f)$. We introduce

$$v'_n(\gamma_1, \gamma_2) = \sum_{\delta' \in \Gamma_n'^2} \mathcal{E}(\delta') f'_n(\delta') \mathbf{1}_{\delta'_1}(\gamma_1) \mathbf{1}_{\delta'_2}(\gamma_2).$$

We have $v'_n \in L_s^2(\Gamma^2)$. Therefore

$$I_n(y) = \int v'_n(\gamma_1, \gamma_2) dy(\gamma_1) \otimes dy(\gamma_2)$$

is well defined.

The theorem is now a consequence of the following lemma.

3.2.1. Lemma.

$$I_n = \left(S'_n(f)\right) \circ \varphi.$$

Proof. For fixed n, as p tends to infinity

$$\begin{aligned} I_n &= \lim S_p(v'_n) \\ &= \lim_p \sum_{\delta \in \Gamma_p^2, \delta' \in \Gamma_n'^2} \mathcal{E}(\delta)\mathcal{E}(\delta') f'_n(\delta') E^{B_p}\left(\mathbf{1}_{\delta'_1}\right) E^{B_p}\left(\mathbf{1}_{\delta'_2}\right) w(\delta_1) w(\delta_2). \end{aligned}$$

Finally we remark

$$\lim_{p\to\infty} \sum_{\delta_1 \in \Gamma_p} \mathcal{E}(\delta_1) E^{B_p}\left(\mathbf{1}_{\delta'_n}\right) w(\delta_1) = \varphi^*\left(w(\delta'_1)\right).$$

□

3.3. Definition of the Fock space. Given $L^2(\Gamma, B; \rho)$ we shall denote by $\mathcal{R}$ the family of all its resolutions. Given $\mathcal{B} \in \mathcal{R}$ we have constructed a probability space $Y_{\mathcal{B}}$ associated to $\mathcal{B}$. Given $\mathcal{B}, \mathcal{B}' \in \mathcal{R}$, there exists a canonical isomorphism

$$\psi_{\mathcal{B}',\mathcal{B}} : Y_{\mathcal{B}} \to Y_{\mathcal{B}'}.$$

The *Fock space* will consist of the collection of all $Y_{\mathcal{B}'}$, $\psi_{\mathcal{B},\mathcal{B}'}$ where $\mathcal{B}, \mathcal{B}' \in \mathcal{R}$.

A property of the Fock space is a property which is invariant under the isomorphisms $\psi_{\mathcal{B}',\mathcal{B}} (\mathcal{B}, \mathcal{B}' \in \mathcal{R})$.

The realization of the chaos by multiple Itô-Wiener integrals is a property of the Fock space.

In the case of Seg $\left(L^2(\Gamma, B; \rho)\right)$ we have the notion of chaos but no canonical representation of the chaos.

We shall denote the Fock space by Foc $\left(L^2(\Gamma, B, \rho)\right)$ or Foc $\left(L^2(\Gamma)\right)$ for short.

Foc $\left(L^2(\Gamma)\right)$ enjoys all the properties of Seg $\left(L^2(\Gamma)\right)$.

Foc $\left(L^2(\Gamma)\right)$ possesses furthermore the *theory of Itô-Wiener multiple integrals*.

A Fock space is therefore a Segal space with an *enriched structure*.

4. Stroock-Taylor formula

4.1. Theorem (Stroock-Taylor formula). *Given* $L^2(\Gamma)$, $X = \text{Foc}\left(L^2(\Gamma)\right)$ *and* $f \in D^2_\infty(X)$, *we consider*

$$\begin{aligned} f^{(n)}(\gamma_1 \dots \gamma_n) &= E\left(D^{(n)}_{\gamma_1 \dots \gamma_n} f\right), \qquad \textit{then} \\ f^{(n)} &\in L^2_s(\Gamma^n) \qquad \textit{and} \\ f(x) &= \sum_{n=0}^{+\infty} \frac{1}{n!} \int f^{(n)} dx^{\otimes n} \end{aligned}$$

with convergence in D^2_∞.

Proof. As $f \in L^2(X)$, f has a chaos decomposition

4.1.1. $$f = \sum_{n=0}^{+\infty} u_n$$

where

$$u_n \in \mathcal{C}_n.$$

Denote by $\mathcal{L}$ the Ornstein-Uhlenbeck operator; then we have for all integers s,

$$(-\mathcal{L})^s f = \sum n^s u_n.$$

Therefore

4.1.2. $$\sum n^{2s} \|u_n\|^2_{L^2} = \|\mathcal{L}^s f\|^2_{L^2} < +\infty.$$

By the Krée-Meyer inequality

4.1.3. $$\|Dv\|^2_{L^2} \le p\|v\|^2_{L^2} \quad \text{if} \quad v \in \oplus_{r=1}^p \mathcal{C}_r.$$

By 4.1.2 and 4.1.3 the series 4.1.1. converges in D^2_∞. Therefore we can differentiate the series 4.1.1. term by term. We get

$$E(f) = u_0$$

$$E(Df) = E(Du_1) + \sum_{k>1} E(Du_k).$$

We have $Du_k \in \mathcal{C}_{k-1}$. Therefore Du_k is orthogonal to the constants. In the same way

$$E\left(D^{(2)} f\right) = E\left(D^{(2)} u_2\right) + \sum_{k>2} E\left(D^2 u_k\right).$$

All the terms of the series vanish again; we have therefore the general relation

$$E\left(D^{(k)} f\right) = E\left(D^{(k)} u_k\right).$$

Using Theorem 2.5.3 we can represent $u_k \in \mathcal{C}_k$ by an Itô-Wiener integral

$$u_k = \int g_k(\gamma_1 \dots \gamma_k) dx(\gamma_1) \dots dx(\gamma_k).$$

Using 2.5.1 we have

$$D^{(k)}u_k = k!g_{k'};$$

therefore

$$g_k = \frac{1}{k!}E\left(D^{(k)}f\right).$$

□

4.2. Corollary. *Given $\varphi \in D_1^2(X)$, denote by P_{C_1} the orthogonal projection in L^2 on the first chaos, then*

$$P_{C_1}\varphi = \int E\left(D_\tau\varphi\right)dx(\tau).$$

Proof. If $\varphi \in D^2_\infty$ this formula follows from 4.1.

Given $\psi \in D_1^2$ denote by P_t the Ornstein-Uhlenbeck semigroup and $\psi_t = P_t\psi$. Then

$$\psi_t \in D^2_\infty \qquad D\psi_t = e^{-t}P_tD\psi$$

$$P_{C_1}\psi_t = e^{-t}\int E\left[(P_tD\psi)_\tau\right]dx(\tau).$$

Now, when $t \to 0$, ψ_t converges to ψ in $L^2(X)$, $P_tD\psi$ converges to $P_td\psi$ and $E(P_tD\psi)$ converges to $E(\psi)$ in $L^2(\Gamma)$. The fact that r.h.s. is continuous for the $L^2(\Gamma)$ norm (see 2.5.3) finishes the proof. □

4.3. Corollary. *Consider $A \subset \Gamma$, then $L^2(\Gamma) = L^2(A)\oplus L^2(A^c)$ induces the product decomposition $X = X_1 \times X_2$ where $X_1 = \text{Seg}\left(L^2(A)\right)$, $X_2 = \text{Seg}\left(L^2(A^c)\right)$. Consider $F_1 \in L^2(X_1)$ and its chaos expansion $F_1 = \sum_q \int_{A^q} F_1^{(q)}dx_1^{\otimes q}$. Consider $f_2 \in L^2(X_2)$ belonging to the first chaos: $f_2(x_2) = \int_{A^c} f_2^{(1)}dx_2$. Consider $F(x) = F_1(x_1)f_2(x_2)$, then its chaos expansion is given by*

$$F^{(r)}\left(\gamma_1,\ldots,\gamma_r\right) = \sigma_r\left[F^{(r-1)}\left(\gamma_1,\ldots,\gamma_{r-1}\right)f_2^{(1)}(\gamma_2)\right]$$

where σ_r denotes the symmetrization operator in r variables.

Proof. By regularizing as in 4.2, it is sufficient to prove this formula under the assumption that $F_1 \in D_\infty(X_1)$. Then the Leibnitz formula gives

$$D^{(r)}F = f_2D^{(r)}F_1 + rDf_2D^{(r-1)}F_2.$$

Taking expectations on X, the first term vanishes and Taylor-Strook gives the result.

4.4. Proposition (Taylor formula with remainder term). *Given $f \in D_1^2$, denote*

$$s = E(f) + \int E(Df)dx,$$

then

$$E\left(|f-s|^2\right) \leq \int E\left[D_\tau f - E(D_\tau f)\right]^2\rho(d\tau).$$

Proof. Write $f_1 = f - s$, then f_1 satisfies

$$\begin{aligned} E(f_1) &= 0 \quad \text{and} \\ D_\tau f_1 &= D_\tau f - E(D_\tau f). \end{aligned}$$

This proposition will result from the following lemma.

4.4.1. Lemma. *Given $u \in D_1^2(X)$ such that $E(u) = 0$, then*

$$E(|u|^2) \leq \|Du\|^2.$$

Proof. We use a chaos decomposition $u = \sum_{n=1}^{+\infty} u_n$. Then

$$\|Du\|_{L^2}^2 = \sum \|Du_n\|_{L^2}^2 = \sum n\|u_n\|_{L^2} \geq \sum \|u_n\|_{L^2}^2 = \|u\|_{L^2}^2.$$

□

5. Skorokhod-Zakai-Nualart-Pardoux integral

5.1. Definition. Given a measure space $(\Gamma, \mathcal{B}; \rho)$ and given $B \in \mathcal{B}$, we have

$$L^2(\Gamma) = L^2(B) \oplus L^2(B^c).$$

The direct sum decomposition induces

$$\operatorname{Foc}\left(L^2(\Gamma)\right) = \operatorname{Foc}\left(L^2(B)\right) \times \operatorname{Foc}\left(L^2(B^c)\right).$$

We shall denote by E^{B^c} the natural projection of $L^2\left(\operatorname{Foc}\left(L^2(\Gamma)\right)\right) \rightarrow L^2\left(\operatorname{Foc}\left(L^2(B^c)\right)\right)$ (see 2.3.7).

Definition. Given a resolution $\{\mathcal{B}_n\}$ of $L^2(\Gamma)$, and given $F \in L^2(X \times \Gamma)$ (where $X = \left(\operatorname{Foc}\left(L^2(\Gamma)\right)\right)$, we define

5.1.1. $$I_n(F) = \sum_{\delta \in \Gamma_n} E^{\delta^c}\left[\left(E^{\mathcal{B}_n}(F)\right)(\delta)\right] w(\delta).$$

We recall that $w(\delta)$ denotes $\int \mathbf{1}_\delta dx$, and that if u is $\mathcal{B}_n$ measurable, then $u(\delta)$ denotes the constant value taken by u on the atom δ.

Definition. We shall say that F is Zakai-Nualart-Pardoux *integrable* if, for all resolutions $\{\mathcal{B}_n\}$ of $L^2(\Gamma)$, $I_n(F)$ *converges* in L^2 to an element of $\operatorname{Foc}\left(L^2(\Gamma)\right)$ independently of the choice of the resolution.

We denote this limit by

$$\int F(x, \gamma) dx(\gamma).$$

5.2. Theorem. *If $F \in D_1^2\left(X, L^2(\Gamma)\right)$ then F is Z.N.P. integrable and for every $\varphi \in D_1^2(X)$ we have*

$$E\left(\int F(x, \gamma)(D_\gamma \varphi)(x) d\rho(\gamma)\right) = E\left(\varphi \int F\, dx\right).$$

5.3. The Gaveau-Trauber duality

We can consider $F(x,\cdot)$ as a tangent vector to X. Therefore the map $x \to F(x,\cdot)$ defines a tangent vector field Z to X. We can then read Theorem 5.2 as an explicit construction of the divergence of this vector field:

$$\delta Z = \lim I_n.$$

The formula realizes a deep and unexpected relation between two apparently *different worlds*:

(i) The world of the *classical differential calculus* going back to Leibniz and developed in Chapter II in the spirit of Sobolev.
(ii) The world of the stochastic calculus discovered by K. Itô.

5.4.1. The case of a constant vector field

Assume that F does not depend on x

$$F(x,t) = f(t), \qquad \text{then}$$
$$E^{\delta^c} E^{B_n}(F) = E^{B_n}(f).$$

The sum 5.1.1 is the same as 2.4.1. We get the Wiener stochastic integral:

$$\int f(\gamma)dx(\gamma) = \lim I_n(x).$$

Given $\varphi \in D_1^2$, by 4.2

$$P_{\mathcal{C}_1}\varphi = \int E(D_\tau\varphi)dx(\tau),$$

and have

$$\int f(\gamma)dx(\gamma) \in \mathcal{C}_1;$$

therefore

$$E\left(\varphi\int f(\gamma)dx(\gamma)\right) = E\left(\left[\int E(D_\tau\varphi)dx(\tau)\right]\left[\int f(\tau)dx(\tau)\right]\right).$$

We now use 2.5.4 to get that the r.h.s. is equal to

$$E\left(\int (D_\tau\varphi)f(\tau)\rho(d\tau)\right);$$

and Theorem 5.2 is proved in this special case.

5.4.2. Orthogonality property

Given $A \in \mathcal{B}$ we have

$$\operatorname{Foc}\left(L^2(\Gamma)\right) = \operatorname{Foc}\left(L^2(A)\right) \times \operatorname{Foc}\left(L^2(A^c)\right).$$

Therefore if $\varphi \in \operatorname{Foc}\left(L^2(A^c)\right)$ we have $D_\gamma \varphi = 0$ for $\gamma \in A$, and in particular,

$$D_\gamma \left(E^{A^c}(\psi)\right) = 0 \quad \text{for} \quad \gamma \in A.$$

5.4.3. The divergence of a finite sum of indicator functions

We take

$$F(x, \tau) = \sum_k a_k(x) \mathbf{1}_{A_k}, \qquad A_k \in \mathcal{B}.$$

Using Chapter II, Proposition 1.3.3, we get

$$\delta\left(a_k \mathbf{1}_{A_k}\right) = a_k \delta\left(\mathbf{1}_{A_k}\right) - \int (D_\gamma a_k) \mathbf{1}_{A_k}(\gamma) \rho(d\gamma).$$

We now use 5.4.1

$$\delta\big(F(x, \cdot)\big) = \sum_k \left[a_k \int \mathbf{1}_{A_k}(\gamma) dx(\gamma) - \int D_\gamma a_k \mathbf{1}_{A_k}(\gamma) \rho(d\gamma)\right].$$

5.4.4. Approximation

We set

$$F_n = \sum_{\delta \in \Gamma_n} \left[E^{B_n}(F)\right](\delta) \mathbf{1}_\delta.$$

Then by II.5.1.

$$\|F_n - F\|_{D_1^2} \mapsto 0.$$

By the Watanabe Theorem II.6.6.2

$$\|\delta F_n - \delta F\|_{L^2} \to 0.$$

We use 5.4.3. to get

(i) $$\delta F_n = \sum_{\delta \in \Gamma_n} E^{B_n}(F)(\delta) w(\delta) - \int D_\gamma E^{B_n}(F) \mathbf{1}_\delta(\gamma) \rho(d\gamma).$$

6. Main lemma

$$\|\delta F_n - I_n\|_{L^2} \to 0.$$

Proof. The function F depends on two variables (x, λ). We denote by F_λ the partial function $x \to F(x, \lambda)$. Then

$$F_\lambda \in L^2(X).$$

The chaos expansion

$$F_\lambda = \sum_q (q!)^{-1/2} \int_{\Gamma^q} F_\lambda^{(q)} dx^{\otimes q},$$

leads to the following beautiful expression of E^{δ^c}

$$E^{\delta^c}(F_\lambda) = \sum_q (q!)^{-1/2} \int_{\Gamma^q} F_\lambda^{(q)} [\mathbf{1}_{\delta^c} dx]^{\otimes q}.$$

Furthermore

$$E\left(|F_\lambda|^2\right) = \sum_q \left\| F_\lambda^{(q)} \right\|^2_{L^2(\Gamma^q)};$$

and by 2.5.1

(i) $$D_{\gamma_0} F_\lambda = \sum_q (q!)^{-1/2} q \int_{\Gamma^q} F_\lambda^{(q)} \left(\gamma_0, \gamma_1 \dots \gamma_{q-1}\right) dx^{\otimes(q-1)}$$

which means that

(ii) $$\int E\left(D_\gamma F_\lambda\right)^2 \rho(d\gamma)\rho(d\lambda) = \sum q \, \| F_\cdot^{(q)} \|^2_{L^2(\Gamma^{q+1})} < +\infty.$$

As before, denote by E^{B_n} the conditional expectation in the variable λ. This operator commutes with the chaos expansion and we have

$$\begin{aligned} \left(E^{B_n} F_\cdot\right)_\lambda &= \sum_q (q!)^{-1/2} \int_{\Gamma^q} \left(E^{B_n} F_\cdot^{(q)}\right)_\lambda dx^{\otimes q} \\ E^{\delta^c}\left(E^{B_n} F_\cdot\right)_\lambda &= \sum_q (q!)^{-1/2} \int_{\Gamma^q} \left(E^{B_n} F_\cdot^{(q)}\right)_\lambda \mathbf{1}_{\delta^c}^{\otimes q} \, dx^{\otimes q}. \end{aligned}$$

Furthermore using 4.3, we have

$$\begin{aligned} &\left(\int \mathbf{1}_\delta dx\right) \int_{\Gamma^q} \left(E^{B_n} F_\cdot^{(q)}\right)_{\lambda_\delta} \mathbf{1}_{\delta^c}^{\otimes q} \, dx^{\otimes q} \\ &= \int_{\Gamma^{q+1}} \sigma \left[\left(E^{B_n} F_\cdot^{q}\right)_\lambda \mathbf{1}_\delta(\lambda) \mathbf{1}_{\delta^c}^{\otimes q}\right] dx^{\otimes(q+1)} \end{aligned}$$

where σ denotes the symmetrisation operator in the variables $\lambda, \gamma_1 \dots \gamma_q$.

The series

$$\begin{aligned} &\sum (q!)^{-1}(q+1)! \int \left\| \left[\left(E^{B_n} F_\cdot^{(q)}\right)_\cdot \mathbf{1}_{\delta^c}^{\otimes q}\right] \right\|^2_{L^2(\rho^{q+1})} \leq \\ &\sum (q+1) \left\| E^{B_n} \left(F_\cdot^{(q)}\right)_\cdot \right\|^2_{L^2(\Gamma^{q+1})} \end{aligned}$$

converges by (ii). Then

$$I_n = \sum_q (q!)^{-1/2} \int_{\Gamma^{q+1}}$$
$$\sum_{\delta \in \Gamma_n} \sigma \left[\left(E^{B_n} F^{(q)}_{.}\right)_{\gamma_1} \mathbf{1}_\delta(\gamma_1) \mathbf{1}_{\delta^c}(\gamma_2) \ldots \mathbf{1}_{\delta^c}(\gamma_{q+1}) \right] dx^{\otimes(q+1)}.$$

By dualizing (i), where F_λ is replaced by φ, we get the Skorokhod formula

$$\delta F_n = \sum_q (q!)^{-1/2} \int_{\Gamma^{q+1}} \sum_{\delta \in \Gamma_n} \sigma \left[(E^{B_n}(F^{(q)}_{.})_{\gamma_1} \mathbf{1}_\delta(\gamma_1)) \right] dx^{\otimes q+1}.$$

Therefore

$$E\left[(\delta F_n - I_n)^2\right] \leq$$
$$\sum_1 (q+1) \left\| \sum_{\delta \in \Gamma_n} \left(E^{B_n} F^{(q)}_{.}\right)_{\gamma_1} \mathbf{1}_\delta(\gamma_1) \left(\mathbf{1} - \mathbf{1}_{\delta^c}(\gamma_2) \ldots \mathbf{1}_{\delta^c}(\gamma_{q+1})\right) \right\|^2_{L^2(\Gamma^{q+1})}.$$

We remark that there exists $A^n_q \subset \Gamma^{q+1}$ such that

$$\mathbf{1}_{A^n_q} = \sum_{\delta \in \Gamma_n} \mathbf{1}_\delta(\gamma_1) \left(\mathbf{1} - \mathbf{1}_{\delta^c}(\gamma_2) \ldots \mathbf{1}_{\delta^c}(\gamma_{q+1})\right);$$

this means that in the r.h.s., at every point of Γ^{q+1} there is at most one term which is different from zero. We have

(iii) $$\rho(A^n_q) = \sum_{\delta \in \Gamma_n} \rho(\delta) \left(1 - (1 - \rho(\delta))^q\right) \leq \sup_{\delta \in \Gamma_n} \left(1 - (1 - \rho(\delta))^q\right) \to 0.$$

We deduce that

(iv) $$E\left((\delta F_n - I_n)^2\right) \leq \sum (q+1) \left\| \left(E^{B_n} F^{(q)}\right)_{.} \mathbf{1}_{A_q} \right\|^2_{L^2(\Gamma^{q+1})}.$$

Now

$$\left\| \left(E^{B_n} F^{(q)}\right) \mathbf{1}_{A_q} \right\|_{L^2(\Gamma^{q+1})} \leq \left\| F^{(q)} \right\|_{L^2(\Gamma^{q+1})}$$

which means that the series (iv) converges uniformly in n. Furthermore $\left(E^{B_n} F_{.}\right)_\gamma$ is, relative to n, uniformly integrable in $L^2(\Gamma)$: using (iii)

$$\lim_{n \to \infty} \left\| \left(E^{B_n} F^{(q)}\right) \mathbf{1}_{A_q} \right\|_{L^2(\Gamma^{q+1})} = 0$$

which proves the lemma. □

Corollary. *If $F \in \mathbb{D}^2_1\left(X; L^2(\Gamma)\right)$ then F is Zakai-Nualart-Pardoux integrable.*

Proof. The sum I_n converges in L^2 and its limit $\mathrm{div}(F)$ is independent of the choice of the resolution.

7. Energy identity

Theorem.

$$E\left([\text{div } F]^2\right) = \|F\|_{L^2}^2 + E\left(\int_{\Gamma^2} (D_\lambda F_\gamma)(D_\gamma F_\lambda)\rho(d\gamma)\rho(d\lambda)\right).$$

Proof. The Hilbert metric on H makes it possible to associate to the vector field Z the differential form

$$\langle \omega, h\rangle = (Z \mid h);$$

then

$$\text{div}(Z) = d^*(\omega).$$

We have the Shigekawa identity (II.6.7.6)

$$\|d^*\omega\|_{L^2}^2 + \|d\omega\|_{L^2}^2 = \|\omega\|_{L^2}^2 + \|D\omega\|_{L^2}^2.$$

We have

$$\begin{aligned} \|d\omega\|_{H\otimes H}^2 &= \frac{1}{2}\int_{\Gamma^2} (D_\lambda f_\gamma - D_\gamma F_\lambda)^2 \rho(d\gamma)\rho(d\lambda) \\ \|D\omega\|_{H\otimes H}^2 &= \int_{\Gamma^2} (D_\lambda F_\gamma)^2 \rho(d\gamma)\rho(d\lambda) \\ \|D\omega\|_{H\otimes H}^2 - \|d\omega\|_{H\otimes H}^2 &= \int_{\Gamma^2} (D_\lambda F_\gamma)(D_\gamma F_\lambda)\rho(d\gamma)\rho(d\lambda). \end{aligned}$$

Taking expectations we prove the theorem. □

8. Stroock Commutation lemma

Lemma. *Denote by $\mathcal{L}$ the Ornstein-Uhlenbeck operator, then for $Z \in \mathbb{D}_3^2(X; L^2(\Gamma))$ we have*

$$\mathcal{L}\left(\int F dx\right) = \int (\mathcal{L}F)dx - \int F dx.$$

Proof. Given the Gaveau-Trauber duality 5.3, the required identity is equivalent to the following divergence identity:

$$\mathcal{L}\delta = \delta\mathcal{L} + \delta.$$

We denote by □ the Rham-Hodge-Shigekawa operator on differential forms of degree one. Then we have:

$$\begin{aligned} \square_1 &= d^*d + d^*d \\ d^*\square_1 &= d^*dd^* + (d^*)^2 d. \end{aligned}$$

As $(d^*)^2 = 0$ we get

$$d^*\Box_1 = d^*dd^* = -\mathcal{L}d^*.$$

Furthermore, by the Shigekawa identity

$$\begin{aligned} \Box_1 &= -\mathcal{L}+1 \\ d^*(-\mathcal{L}+1) &= -\mathcal{L}d^* \\ d^*(\mathcal{L}-1) &= \mathcal{L}d^*. \end{aligned}$$

9. Stratonovich integral

9.1. Definition. We associate to $u \in \mathbb{D}_1^2\left(X; L^2(\Gamma)\right)$ the sequence

9.1.1.
$$s'_n(x) = \sum_{\delta\in\Gamma_n} E^{\mathcal{B}_n}(u)(\delta)w(\delta)$$

where Γ_n denotes the set atoms of $\mathcal{B}_n$ and where

$$w(\delta) = \int \mathbf{1}_\delta(\tau)dx(\tau).$$

We shall say that u is *Stratonovich integrable* if the following conditions (9.1.2 and 9.1.3) hold true:

9.1.2. $\lim s'_n(x)$ *exists*;

9.1.3. the limit does not depend upon the choice of the resolution. We denote

$$\lim s'_n(x) = \int_0^1 u(x,\tau)\ odx(\tau).$$

9.2. Definition of a trace. Given $u \in L^2(\Gamma^2)$ the integral $\int u(t,t)\rho(dt)$ may not exist. We shall say that the trace exists and belongs to L^p if for any resolution $\mathcal{B}_n$ of $L^2(\Gamma^2)$ and for every $\varphi \in L^q(\Gamma)$

$$\lim_{n\to\infty} \sum_{\delta\in\Gamma_n} \rho(\delta)\left[\int_{\delta^2} u(\gamma_1,\gamma_2)\ \frac{d\rho^{\otimes 2}}{\rho(\delta)^2}\right]\left(\int_\delta \varphi\ \frac{d\rho}{\rho(\delta)}\right)$$

exists and does not depend upon the resolution.

We shall denote this limit by

$$\int_\Gamma \left(\mathrm{Trace}_{\gamma_1=\gamma_2=\gamma}\left(u(\gamma_1,\gamma_2)\right)\right)\varphi(\gamma)\rho(d\gamma).$$

9.3. Theorem. *Given $u \in \mathbb{D}_1^2\left(X; L^2(\Gamma)\right)$, assume that*

$$\mathrm{Trace}\big(D_{.}u(x,\cdot)\big) \quad \textit{exists in} \quad L^p,$$

then u is Stratonovich integrable and

$$\int u(x,\gamma)\ o dx(\gamma) = \int u(x,\gamma)dx(\gamma) + \int \mathrm{Trace}_\gamma\big(D_{.}u(x,\cdot)\big)\rho(d\gamma).$$

Proof. We choose a resolution $\mathcal{B}_n$ of $L^2(\Gamma)$. We denote

$$u_n = E^{\mathcal{B}_n}\big(u(x,\cdot)\big),$$

then

$$u_n \in \mathbb{D}_1^2\left(X; L^2(\Gamma)\right)$$
$$\|u - u_n\|_{\mathbb{D}_1^2} \to 0,$$

which implies

$$\delta(u_n) \to \delta(u).$$

We will compute $\delta(u_n)$ as in 5.4.3:

$$\delta u_n = \sum_{\delta\in\Gamma_n}\left(u_n(\delta)x(\delta) - \int_\delta D_\gamma u_n \rho(d\gamma)\right).$$

We have

$$u_n(\delta) = \frac{1}{\rho(\delta)}\int_\delta u(x,\xi)\rho(d\xi),$$

then

$$\begin{aligned}(\delta u_n)(x) &= \sum_{\delta\in\Gamma_n}\left[\left(\frac{1}{\rho(\delta)}\int_\delta u(x,\gamma)\rho(d\gamma)\right)x(\delta)\right] \\ &\quad - \sum_{\delta\in\Gamma_n}\int_{\delta^2}(D_\tau u)(x,\xi)\frac{\rho(d\xi)\rho(d\tau)}{\rho(\delta)}\ ;\end{aligned}$$

this gives the identity

$$s_n'(x) = (\delta u_n)(x) + \sum_{\delta\in\Gamma_n}\int_{\delta^2}(D_\tau u)(x,\xi)\frac{\rho(d\xi)\rho(d\tau)}{\rho(\delta)}.$$

Letting $n \to +\infty$ we get the theorem. □

Chapter VII

Itô's Theory of Stochastic Integration

Contents: The probability space of Brownian motion and its filtration – Energy identity for stochastic integral of an adapted process – Itô's stochastic integral of an adapted process – Chaos expansion in terms of iterated Itô stochastic integrals – Itô representation of a martingale by a stochastic integral – Clark-Bismut-Ocone representation of a martingale in $\mathbb{D}_1^p$ – Itô calculus on semi-martingales – Covariance under C^∞-maps of the Stratonovich representation of semi-martingales – Change of variables formula – Appendix: Estimates for Brownian martingales.

Information and probability are the two main tools in a rational decision procedure. Information *increases* in time and can sometimes make the probabilistic point of view ultimately obsolete. Itô's concept of *filtration* is the mathematical formulation of these basic facts.

We define the white noise space as Fock space of $L^2([0, 1])$. Multiplication by $\mathbf{1}_{[0,t]}$ defines on $L^2([0, 1])$ a continuous increasing family of projection operators; this family will induce the Itô filtration on the Fock space.

In Chapter V we needed a $\mathbb{D}_1^2$ hypothesis in order to be able to define the ZNP stochastic integral. In the case of *adapted integrands*, a remarkable collapsing makes it possible to dominate the L^2-norm of the ZNP stochastic integral in terms of the L^2-norm of the integrand alone. This will be our starting point from which to introduce Itô's integrals of adapted integrands.

The notion of filtration makes it possible to write the Wiener-Itô expansions in chaos as a series of *iterated* stochastic integrals. This fact will imply the Itô representation of every L^2-martingale by a stochastic integral.

The discussion of the basic Itô martingale representation will occupy most of the remainder of the chapter. The problem discussed will be that of the effective *determination* of the integrand.

(i) In the case of a $\mathbb{D}_1^2$ martingale, the integrand will be given by a *global formula* involving the conditional expectation of the gradient of its final value. This formula of Clark-Bismut-Ocone establishes another deep connection between the usual differential calculus of Chapter II and the stochastic differential calculus.

(ii) *Itô's calculus* makes it possible to calculate the integrand by an *infinitesimal dynamical analysis* of the martingale.

Because of the requirements of *stochastic differential geometry* to be developed in Chapter VIII, our approach to *Itô's calculus* will be circuitous. We shall first develop a theory of *Stratonovich integration*, with an emphasis on the following two facts:

(i) the Stratonovich integral behaves under a change of variables like the following familiar object of differential geometry: *the integral of a differential form of degree one along a smooth path.*
(ii) For regular integrands, the Itô and Stratonovich integrals differ by a correcting term involving only a Lebesgue integral.

An appendix at the end of this chapter will make some applications of the Itô calculus.

1. The probability space of Brownian motion and its filtration

1.1. The probability space of Brownian motion

We consider $L^2([0,1])$ relative to Lebesgue measure, and we set

$$W = \operatorname{Foc}\left(L^2([0,1])\right).$$

We shall define an injective map Φ of W into the space of continuous *functions* on $[0, 1]$ by writing

$$x_w(t) = w([0,t]).$$

We shall call the *universal Wiener space* X the space of $\{x_w(\cdot)\}$. Then

$$X \quad \textit{is isomorphic to} \quad \operatorname{Foc}\left(L^2([0,1])\right).$$

We shall abbreviate x_w by x.

1.2. The Cameron-Martin space

The derivative on W was taken in Chapter V relative to $L^2([0,1])$. We want to transfer this fact by a unitary isomorphism. We associate to $u \in L^2([0,1])$ the function $v = \Phi(u)$ defined by

$$v(t) = \int_0^t u(\xi)d\xi.$$

We define the *Cameron-Martin space* by

$$H = \left\{v; \quad v \text{ continuous}, \quad v(0) = 0, \quad \dot{v} \text{ exists a.e. and } \dot{v} \in L^2\right\}.$$

Then H is a Hilbert space with the norm

$$\|v\|_H^2 = \int_0^1 |\dot{v}(\tau)|^2 d\tau.$$

The map $\Phi : u \to v$ is a Hilbert space isometry of $L^2([0,1])$ onto H. Our definitions agree with those of Chapter I.8.7.

By this isometry we can transfer the results of the preceding chapter. For instance, derivatives of functions defined on X will be taken along H. We shall also denote by $\dot{H}$ the space of derivatives of functions in H. Of course, $\dot{H} = L^2([0,1])$; nevertheless it will be convenient to use both these notations.

1.3. Filtration on X

1.3.1. We have on $L^2([0,t])$ the increasing family of projection operators

$$P_t f = \mathbf{1}_{[0,t]} f.$$

The range of P_t is $L^2([0,t])$. We have

$$L^2([0,1]) = L^2([0,t]) \oplus L^2([t,1]).$$

This direct sum decomposition induces on Fock space the following product decomposition

$$\mathrm{Foc}\left(L^2([0,1])\right) = \mathrm{Foc}\left(L^2([0,t])\right) \times \mathrm{Foc}\left(L^2([t,1])\right).$$

We shall denote by $\tilde{\mathcal{N}}_t$ the σ-field generated by the random variables which depend only on the first factor of the right hand side.

Then Φ maps $\tilde{\mathcal{N}}_t$ into a σ-field $\mathcal{N}_t$ on X. Obviously, $\mathcal{N}_t$ is an increasing family of σ-fields; $\mathcal{N}_0$ is the trivial σ-field; $\mathcal{N}_1$ is the σ-field of all measurable events on X.

Given $F \in L^2(X; \dot{H})$ we shall say that F is *adapted* if for almost all t

$$F_t \quad \text{is } \mathcal{N}_t\textit{-measurable}.$$

1.3.2. Proposition (Continuity of the filtration). *Given $f \in L^2$ we denote $E^{\mathcal{N}_t}(f) = M_t$, then $t \to M_t$ is a continuous map of $[0,1]$ into L^2.*

Proof.

$$E\left(\|x(t) - x(t')\|^2\right) = |t - t'|.$$

Therefore $x(t)$ converges towards $x(t_0)$ in probability. It is the same for every polynomial in the $x(\xi)$ variable, which proves the proposition if f has a finite number of terms in its chaos expansion; the general case follows by passage to the limit. □

1.3.3. Paul Lévy construction of the Brownian motion

We choose the resolution of $L^2([0,1])$ in the sense of Chapter VI, 3.1, constituted by the sequence of σ-fields $\mathcal{B}_n$ having for atoms the dyadic intervals $]k2^{-n},(k+1)2^{-n}]$.

The probability space constructed by the scheme VI.3.1 is based on the determination of Gaussian variables G_δ, $\delta \in \Gamma_n$ entangled by the relations

$$G_{\delta'} + G_{\delta''} = G_\delta \quad \text{where} \quad \delta = \delta' \cup \delta'', \ \delta \in \Gamma_n.$$

We solve this system by introducing $\Delta_\delta = G_{\delta'} - G_{\delta''}$. Then Δ_δ is a Gaussian variable independent of the G_δ, $\delta \in \Gamma_n$ which satisfies $E(\Delta_\delta^2) = 2^{-n}$: these conditions determine uniquely the random variables $\cup_n \{\Delta_\delta;\ \delta \in \Gamma_n\}$ and the probability space associated to $\operatorname{Foc}\left(L^2([0,1])\right)$.

Paul Lévy geometrized this approach by constructing a sequence of continuous paths φ_n determined recursively by the conditions:

$$\varphi_n \quad \text{is linear on each dyadic interval} \quad \left[k2^{-n},(k+1)2^{-n}\right]$$

$$\begin{aligned} \varphi_n(k'2^{-s}) &= \varphi_s(k'2^{-s}), \quad \text{if} \quad s \le n \\ \varphi_n(k2^{-n}) &= \varphi_{n-1}(k2^{-n}) + \frac{1}{2}\Delta_\delta, \\ & \quad k \text{ being an arbitrary odd integer;} \quad \varphi_0 = 0. \end{aligned}$$

Then, when $n \to \infty$, the paths φ_n converge uniformly towards the Brownian path.

2. Energy identity for stochastic integrals

2.1. Lemma. *Given $F(x,t) \in \mathbb{D}_1^2(X;\dot{H})$, F is adapted if and only if*

$$D_\lambda F(x,t) = 0 \qquad \text{if} \quad \lambda > t.$$

Proof. We fix t_0 and set

$$\varphi(x) = F(x,t_0),$$

then $\varphi(x)$ is $\mathcal{N}_t$-measurable which means

$$\varphi(x) = u(y)$$

where $x = (y,z)$ is associated to the product

$$\operatorname{Foc}\left(L^2([0,1])\right) = \operatorname{Foc}\left(L^2([0,t])\right) \times \operatorname{Foc}\left(L^2([t,1])\right);$$

therefore

$$(\partial_z\varphi)(x) = (\partial_z u)(y) = 0.$$

Conversely, using Chapter VI.4.3, we get that the vanishing of the derivatives implies that the partial function $F_{x_1}(.,t)$ is constant where the splitting $x = (x_1, x_2)$ is induced by the product decomposition of the Fock spaces considered above).

2.2. Theorem. *Given F adapted, $F \in \mathbb{D}_1^2(X; \dot{H})$, consider the ZNP integral,*

$$I(x) = \int_0^1 F(x,t)dx(t).$$

We have then

$$E\left([I(x)]^2\right) = E\left(\int_0^1 [F(x,t)]^2 dt\right).$$

Proof. We use Chapter V.7

$$E\left((I(x))^2\right) - E\left(\int_0^1 [F(x,t)]^2 dt\right) = E\left[\int_{[0,1]^2} D_\lambda F(x,t) D_t F(x,\lambda) dt\, d\lambda\right].$$

Outside a set of measure zero we have $\lambda < t$ or $\lambda > t$. Then by Lemma 2.1,

$$D_t F(x,\lambda) = 0 \quad \text{in the first case and}$$
$$D_\lambda F(x,t) = 0 \quad \text{in the second case.}$$

□

3. Itô's stochastic integral

Theorem. *Given any $F \in L^2([0,1], \dot{H})$, its Itô stochastic integral exists and is uniquely characterized by the following properties*

3.1. $$E\left(\int_0^1 F(x,t)d_I x(t)\right) = 0.$$

3.2. $$E\left(\left[\int_0^1 F(x,t)d_I x(t)\right]^2\right) = E\left(\int_0^1 [F(x,t)]^2 dt\right) \quad \text{(Energy identity).}$$

3.3. $\int_0^t F(x,\xi)d_I x(\xi)$ *is an $\mathcal{N}_t$ martingale.*

3.4. $$\int_0^1 F(x,t)d_I x(t) = \sum \alpha_s\left(x(t_{s+1}) - x(t_s)\right) \quad \textit{for every} \quad f \in \mathcal{E}_a.$$

(the definition of $\mathcal{E}_a$ is given below).

Proof. A function $F \in L^2(X; \dot{H})$ is called *elementary* if there exist a finite number of deterministic values of time $0 < t_1 < t_2 < t_q < 1$ such that $F(x,t)$ is constant on each of the intervals $\left[t_s, t_{s+1}\right[$:

$$F(x,t) = \sum \alpha_s(x) \mathbf{1}_{[t_s,t_{s+1}[};$$

an elementary function F is *adapted* if and only if

$$\alpha_s(x) \quad \textit{is } \mathcal{N}_{t_s}\textit{-measurable.}$$

We denote by $\mathcal{E}_a$ the vector space of *adapted elementary functions*. Now given $g \in \mathcal{E}_a$, its integral is defined by 3.4

$$\int_0^1 F(d_I x) = \sum_s \alpha_s(x)\left[x(t_{s+1}) - x(t_s)\right].$$

This definition is independent of the choice of the partition. Now 3.1, 3.2, 3.3, are straightforward for $F \in \mathcal{E}_a$.

Given $F \in L^2(X; H)$, where F is adapted, we can find $F_n \in \mathcal{E}_a$ such that

$$E\left(\int [F(x,t) - F_n(x,t)]^2\, dt\right) \to 0;$$

3.2 then implies that

$$\int_0^1 F_n d_I x \quad \textit{converges} \text{ in } \quad L^2(X).$$

3.5. Corollary. *Given $F, G \in L^2(X; \dot{H})$, adapted, define*

$$M(s) = \int_0^s F(x,t) d_I x(t) \quad \textit{and} \quad N(s) = \int_0^t G(x,t) d_I x(t);$$

then

$$E\big(M(s)N(s)\big) = E\left(\int_0^s F(x,t)G(x,t)dt\right).$$

Proof. We have

$$E\big(M(s) + \lambda N(s)\big)^2 = E\left(\int_0^s [F(x,t) + \lambda G(x,t)]^2 dt\right).$$

□

4. Itô's integral as extension of the ZNP integral

4.1. Theorem. *Let $F \in \mathbb{D}_1^2(X; \dot{H})$. Assume that F is adapted. Then*

$$\int F(x,t)dx(t) = \int F(x,t)(d_I x)(t).$$

Proof. According to Chapter V, in order to define the left hand side, we must choose a resolution $\mathcal{B}_1 \subset \mathcal{B}_2 \subset \ldots \subset \mathcal{B}_n \subset \ldots$ of $L^2([0,1])$. To avoid complication of notation, we shall use the resolution associated to dyadic intervals. The atoms of $\mathcal{B}_n$ will then be the intervals $\left[k2^{-n}, (k+q)2^{-n}\right[$.

A function F will be called *dyadic elementary* at the level n_0 if $F(x,\cdot)$ is $\mathcal{B}_{n_0}$ measurable.

4.2. Lemma. *Given F dyadic elementary at the level n_0, then for all $n > n_0$*

$$E^{\delta_c}\left(E^{B_n}F\right) = F(x, k2^{-n})$$

where

$$\delta = \left]k2^{-n},\ (k+1)2^{-n}\right[.$$

Proof.

$$E^{\mathcal{B}_n}F = \int_{k2^{-n}}^{(k+1)2^{-n}} F(x,\xi)2^n d\xi = F(x, k2^{-n}).$$

As $\{k2^{-n}\} \in \delta^c$ we get by continuity

$$E^{\delta^c}\left(F(x, k2^{-n})\right) = F\left(x, k2^{-n}\right).$$

□

4.3. *Proof of the theorem for the dyadic elementary functions*

Given F dyadic elementary, then by 4.2 for $n > n_0$

$$\int E^{\delta^c}\left(E^{B_n}F\right)dx = \sum_k F(x, k2^{-n})\left[x\left((k+1)2^{-n}\right) - x(k2^{-n}))\right].$$

The right hand side is the Itô sum when we put together all the intervals contained in $\left[j2^{-n_0}, (j+1)2^{-n_0}\right]$. The left hand side converges towards the ZNP integral. □

4.4. *Proof in the general case*

Given m, we define for $\lambda \in [0, 1]$

$$\lambda^- = 2^{-m} \quad [\text{ integral part of } (2^m\lambda)]$$

$$F_m(x,\lambda) = \int_0^1 F(x, \lambda^- - 2^{-m}s)ds$$

(where $\lambda^- - 2^{-m}s$ is negative, we take $F(x, 0)$ as the value at zero of the integrated function).

Then F_m is dyadic elementary. Furthermore,

$$D_t F_m(x,\lambda) = \int_0^1 \left(D_\tau F(x, \lambda^- - 2^{-m}s)\right)ds \quad \text{and} \quad F_m \in \mathbb{D}_1^2(X; \dot{H}).$$

Therefore it is possible to define the ZNP integral

$$I(F_m) = \int F_m(x,t)dx(t).$$

As

$$\|F_m - F\|_{L^2(X;\dot{H})} \to 0,$$

where F_m is adapted, we deduce from 2.2 that $I(F_m) \to I(F)$; as the same is true for the Itô integral, the theorem follows. □

4.5. Convention of notation

In the following, we shall not use the ZNP integral any more. For this reason, there is no ambiguity in writing the Itô integral using the classical notation

$$\int F(x,t)dx(t)$$

for F adapted and omitting the I in the differential $d_I x$.

5. Chaos expansion in terms of iterated Itô integrals

On $L^2([0,1])$, as on any Fock space, we have the chaos expansion introduced in Chapter V. For every $u \in L^2(X)$ we have:

$$u(x) = \sum_n \frac{1}{n!} \int_{[0,1]} U_n\,(t_n, \ldots, t_n)\, dx^{\otimes n}\,(t_1 \ldots t_n)$$

where U_n is a symmetric function. We denote by Δ_n the simplex $0 < t_n < t_{n-1} \ldots < t_1 < 1$. Then we have for the Lebesgue integral of a symmetric function u

$$\frac{1}{n!} \int_{[0,1]^n} u d\tau^{\otimes n} = \int_{\Delta_n} u d\tau^{\otimes n}.$$

We want to extend this formula to Wiener-Itô stochastic integrals.

5.1. Theorem. *Given an L^2-symmetric function U of n real variables and given $0 < t_n < t_{n-1} < t_1 < 1$, we consider the sequence of functions defined on X by the following recursion formula:*

$$\begin{aligned} V_1\,(t_1 \ldots t_{n-1}, x) &= \int_0^{t_{n-1}} U\,(t_1, \ldots t_{n-1}, \tau_n)\, dx(\tau_n) \\ V_2\,(t_1, \ldots t_{n-2}, x) &= \int_0^{t_{n-2}} V_1\,(t_1, \ldots \tau_{n-1}, x)\, dx(\tau_{n-1}) \\ V_n(x) &= \int_0^1 V_{n-1}(\tau, x) dx(\tau_1). \end{aligned}$$

5.1.1. *Then all the stochastic integrals are well defined and*

5.1.2.
$$\frac{1}{n!} \int_{[0,1]^n} U dx^{\otimes n} = V_n.$$

Proof. We fix $t_1 \ldots t_{n-1}$; then by the Fubini theorem, $U(t_{n-1}, \cdot)$ is a.e. in $L^2([0,1])$ and V_1 is defined as a stochastic integral.

For the next step, we have to integrate a function V_1 which now depends upon x. But as $t < t_{n-1}$ is the first integration we have that for t_1, t_2, t_{n-2} *fixed* that V_1 is $\mathcal{N}_{t_{n-1}}$ measurable. Therefore the stochastic integral in $dx(\tau_{n-1})$ is well defined.

By iterated application of the energy identity we get

$$E\left([V_n(x)]^2\right) = \int_{\Delta_n} [U(t_1 \dots t_n)]^2 \, dt_1 \dots dt_n.$$

Using this equality it is sufficient to prove 5.1.2 for $U \in L^2_{s,s}$. In fact we shall extend 5.1.2 by continuity to the closure of $L^2_{s,s}$ which is L^2.

Given $U \in K^2_{s,s}$, then there exists a dyadic partition $\mathcal{B}_m$ of [0, 1] such that U is measurable for $B_m^{\otimes n}$. Furthermore U is equal to zero on the atoms of $B_m^{\otimes n}$ which intersect the diagonal. Then

(i) $$\frac{1}{n!}\int U dx^{\otimes n} = \frac{1}{n!}\sum_{\delta_1 \dots \delta_n} U(\delta_1 \dots \delta_n) x(\delta_1) \dots x(\delta_p).$$

The sum in the right hand side is finite. The order on [0, 1] induces a natural order on the atoms of $\mathcal{B}_n$. We can therefore limit ourselves to take in the summation

$$\delta_n < \delta_{n-1} \dots < \delta_2 < \delta_1,$$

then the right hand side of (i) is the Itô iterated integral.

□

5.2. Theorem (Itô's representation of a Brownian martingale). *Let M be an L^2-martingale on X relative to the filtration $\mathcal{N}_t$, and such that $M_0 = 0$. Then there exists $F \in L^2(X; \dot{H})$ such that*

$$M(x,t) = \int_0^t F(x,\tau) dx(\tau).$$

Proof. Denote by φ the final value of M. Then

$$M_t = E^{\mathcal{N}_t}(\varphi).$$

As $\varphi \in L^2$, using VI.2.5.3, it can be expanded in a series of multiple Wiener-Itô integrals which will be transformed by 5.1.2 into iterated stochastic integrals. Then

$$\varphi = E(\varphi) + \sum_{n=1}^{+\infty} V_n.$$

As $M_0 = 0$ we have $E(\varphi) = 0$. Furthermore

$$V_n = \int_0^1 dx(\tau_1) \int_0^{\tau_1} dx(\tau_2) \int_0^{\tau_2} \dots \int_0^{\tau_n} U_n(\tau_1, \tau_2, \tau_n) dx(\tau_n).$$

Denote

$$A_n(x,\tau_1) = \int_0^{\tau_1} dx(\tau_2) \int_0^{\tau_2} \dots \int_0^{\tau_n} U_n(\tau_1, \dots \tau_n) dx(\tau_n).$$

We get

$$V_n(x) = \int_0^1 A_n(x, \tau_1) dx(\tau_1).$$

Define

$$B_q = \sum_{n \le q} A_n.$$

Then if $q < q'$

$$E\left(\int \left[B_{q'} - B_q\right]^2 d\tau\right) = \left(\left[\int_0^1 \left(\sum_{q<n\le q'} A_n\right) dx\right]^2\right)$$
$$= E\left(\left(\sum_{q<n\le q'} V_n\right)^2\right).$$

As the series $\sum V_n$ converges to φ, we deduce that $\{B_q\}$ is a Cauchy sequence in $L^2(X \times [0, 1])$. Denote its limit by F. Using again the energy identity we get

$$\varphi = \int_0^1 F\, dx.$$

□

5.3. Remark. The representation given by Theorem 5.2 is unique: the difference between two different representations leads to a vanishing Itô stochastic integral:

$$\int_0^1 a\, dx = 0.$$

Using the energy identity we obtain $E(\int a^2\, dt) = 0$ which implies $a(x, t) = 0$.

6. Clark-Bismut-Ocone formula

Under an additional smoothness hypothesis on φ we want to compute the integrand F in Theorem 5.2. We need the following preliminary lemma.

6.1. Lemma. *Given $F \in L^2(X \times [0, 1])$, then*

$$q(x, t) = \left(E^{\mathcal{N}_t} F\right)(x, t)$$

is well defined almost everywhere in (t, x) and

$$\|q\|_{L^2(X\times[0,1])} \le \|F\|_{L^2(X\times[0,1])}.$$

Proof. Denote by $F_t(\cdot)$ the partial function $F_t(\cdot) = F(\cdot, t)$ and consider

$$A = \left\{t \in [0,1]; \qquad F_t \in L^2(X)\right\},$$

then, by the Fubini theorem, $\lambda(A^c) = 0$.

Fix $t_0 \in A$. Then define

$$u(s) = E^{\mathcal{N}_s}\left(F_{t_0}\right);$$

we then know by 1.2.3. that u is a *continuous* map of $[0, 1]$ into $L^2(X)$. This means that $u(t_0)$ is a well defined L^2-valued function. We denote

$$q(x, t_0) = u(t_0).$$

Furthermore, for all s,

$$\|u(s)\|_{L^2} \leq \|F_{t_0}\|_{L^2}$$

which implies

$$E\left([q(x, t_0)]^2\right) \leq E\left([F(x, t_0)]^2\right);$$

by integrating in t_0 we get the inequality. □

6.2. Theorem (Clark-Bismut-Ocone). *Let $\varphi \in \mathbb{D}_1^2(X)$. Assume*

$$E(\varphi) = 0.$$

Denote

$$F(x, t) = (D_t\varphi)(x).$$

Denote by q the function associated to F by Lemma 6.1. Then

$$\varphi(x) = \int_0^1 q(x, \tau)dx(\tau)$$

or, in short,

$$\varphi(x) = \int_0^1 E^{\mathcal{N}_t}\left(D_t(x)\right)dx(t).$$

Proof. For a smooth vector field $Z \in \mathbb{D}_1^2(X; \dot{H})$ we have

$$E(\delta Z\ \varphi) = E\left(\int_0^1 (D_\tau\varphi)(x)Z(x, \tau)d\tau\right).$$

Assume that Z is *adapted*. Then by VI.5.2 and by 4.1

$$E\left(\left[\int_0^1 Z(x, \tau)dx(\tau)\right]\varphi(x)\right) = E\left(\int_0^1 (D_\tau\varphi)(x)Z(x, \tau)d\tau\right).$$

By Theorem 5.2,

$$\varphi(x) = \int_0^1 A(x, \tau)dx(\tau),$$

or

$$E\left(\left[\int_0^1 Z(x,\tau)dx(\tau)\right]\left[\int_0^1 A(x,\tau)dx(\tau)\right]\right) = E\left(\int_0^1 Z(x,\tau)A(x,\tau)d\tau\right).$$

Finally

$$E\left(\int_0^1 Z(x,\tau)\left[A(x,\tau) - D_\tau\varphi(x)\right]d\tau\right) = 0$$

for all $Z \in \mathbb{D}_1^2(X; \dot{H})$ where Z is *adapted*.

We reduce this expression; introducing

$$q(a,t) = E^{\mathcal{N}_t}\left[(D_\tau\varphi)(x)\right],$$

then we get

(i) $$E\left(\int_0^1 Z(x,\tau)\left[A(x,\tau) - q(x,\tau)\right]d\tau\right) = 0.$$

Denote

$$\tilde{Z}(x,\tau) = A(x,\tau) - q(x,\tau).$$

Then we apply the Ornstein-Uhlenbeck semigroup $e^{\varepsilon\mathcal{L}}$ to $\tilde{Z}(\cdot,t)$. We get $\tilde{Z}_\varepsilon(\cdot,t)$ which is smooth and *adapted* (according to the following commutation lemma). We can therefore substitute Z by $\tilde{Z}_\varepsilon$ in (i). We get

$$\left(\tilde{Z}_\varepsilon \mid \tilde{Z}\right)_{L^2(X\times[0,1])} = 0.$$

As $\tilde{Z}_\varepsilon$ converges towards $\tilde{Z}$ when $\varepsilon \to 0$, we get $\tilde{Z} = 0$. □

6.3. Lemma. (Commutation of Ornstein-Uhlenbeck semigroup with the Itô filtration). *Given $\varphi \in L^2(X)$, then*

$$E^{\mathcal{N}_{t_0}}\left(e^{\varepsilon\mathcal{L}}\varphi\right) = e^{\varepsilon\mathcal{L}}\left(E^{\mathcal{N}_{t_0}}\varphi\right).$$

Proof. We shall use the Wiener-Itô representation of the chaos:

$$\varphi = \sum \frac{1}{n!}\int_{[0,1]^n} U_n dx^{\otimes n}.$$

The multiplication by $\mathbf{1}_{[0,t]}$ defines a projection operator $P_{t_0} : L^2([0,1) \to L^2([0,t]);$ P_{t_0} extends to $P_{t_0}^{\otimes 2} : L^2([0,1]) \otimes L^2([0,1]) = L^2([0,1]^2)$ and so on. Furthermore

$$P_{t_0}^{\otimes 2}U_2(t_1,t_2) = U_2(t_1,t_2)\mathbf{1}_{[0,t_0]}(t_1)\mathbf{1}_{[0,t_0]}(t_2)$$

and so on. Therefore

(i) $$E^{\mathcal{N}_{t_0}}\varphi = \sum_n \frac{1}{n!}\int_{[0,1]^n} U_n \mathbf{1}_{[0,t_0]}^{\otimes n} dx^{\otimes n}.$$

By the chaos diagonalization of $\mathcal{L}$ we have

(ii) $$e^{\varepsilon\mathcal{L}}\varphi = \sum_n \frac{1}{n!} e^{-\varepsilon n} \int_{[0,1]^n} U_n \, dx^{\otimes n}.$$

Repeated use of (i) and (ii) proves the lemma. □

6.4. Theorem (Taylor formula with remainder). *Consider $\varphi \in D_2^2$; then*

$$\varphi - E(\varphi) - \int_0^1 E(D_t\varphi)\, dx(t) =$$
$$\iint_{0<t_1<t_2<1} \left[E^{\mathcal{N}_{t_1}}\left(D^2_{t_1,t_2}\varphi\right)\right] dx(t_1)dx(t_2)$$

where the integral of the r.h.s. is an iterated stochastic integral, the integration being performed at first with respect to t_1 and afterwards with respect to t_2.

Proof. The CBO formula 6.2 plays the role of the following identity of first year calculus: $f(1) - f(0) = \int_0^1 f'$; by applying this identity to f' the calculus teacher obtains the Taylor formula at order 2, with remainder term expressed as an integral. We proceed along the same lines.

For fixed t_2, the function $D_{t_2}\varphi$ can be expressed through the CBO formula as

$$D_{t_2}\varphi - E\left(D_{\tau_2}\varphi\right) = \int_0^1 E^{\mathcal{N}_{\tau_1}}\left(D^2_{\tau_1,\tau_2}\varphi\right) dx(\tau_1).$$

We will apply the CBO formula to φ; we need to compute $E^{\mathcal{N}_{t_2}}\left(D_{\tau_2}(\varphi)\right)$, an expression which is equal to the previous integral taken on $[0, t_2]$; we get the desired formula, the necessary domination is obtained by considering $D_*\varphi \in D_1^2(X; \dot{H})$.

7. Itô's calculus on semi-martingales

7.1. Definition. Given an L^2 adapted process $S(x, \cdot)$, we shall say that S in an L^2-semi-martingale if there exists an L^2-adapted process $p \in L^2(X; \dot{H})$ such that

7.1.1. $$M(x,t) = S(x,\tau) - \int_0^1 p(x,\tau)d\tau$$

is an L^2-martingale.

Using Theorem 5.2, there exists an adapted process $a \in L^2(X; \dot{H})$ such that

7.1.2. $$S(x,t) = S(0) + \int_0^t a(x,\tau)dx(\tau) + \int_0^t p(x,\tau)d\tau.$$

We shall assume furthermore that

7.1.3. there exist versions of a and p such that a.s. the partial functions a_x and p_x are continuous.

7.2. Theorem. (Stability of semi-martingales under a change of variable). *Given a function ψ with bounded first and second derivatives and given a semi-martingale $S(x,t)$, then*

$$S_1(x,t) := \psi\big(S(x,t)\big)$$

is a semi-martingale and

$$S_1(x,t) = \psi\big(S(0)\big) + \int_0^t \tilde{a}(x,\tau)dx(\tau) + \int_0^t \tilde{p}(x,\tau)d\tau$$

where $\tilde{a}$ and $\tilde{p}$ are computed by the Itô rule:

7.2.1.
$$\begin{cases} \tilde{a}(x,t) = \psi'\big(S(x,t)\big)a(x,t) \\ \tilde{p}(x,t) = \psi'\big(S(x,t)\big)p(x,t) + \frac{1}{2}\psi''\big(S(x,t)\big)a^2(x,t) \end{cases}.$$

Proof. Define

$$Q(x,t) = \psi\big(S(x,t)\big) - \int_0^t \tilde{a}(x,\tau)dx(\tau) - \int_0^t \tilde{p}(x,\tau)d\tau.$$

Using 7.1.3 we have

$$\begin{aligned} &E^{\mathcal{N}_t}\big(\big[Q(x,t+\varepsilon) - Q(x,t)\big]\big) = \\ &\qquad E^{\mathcal{N}_t}\big\{\psi\big(S(x,t+\varepsilon)\big) - \psi\big(S(x,t)\big)\big\} - \varepsilon\tilde{p}(x,t) + o(\varepsilon). \end{aligned}$$

$$\psi(S_{t+\varepsilon}) = \psi\left(S_t + \int_t^{t+\varepsilon}(adx + pd\tau)\right).$$

We expand ψ by the Taylor formula:

$$\begin{aligned} E^{\mathcal{N}_t}(Q_{t+\varepsilon} - Q_t) &= \psi'(S_t)E^{\mathcal{N}_t}\left(\int_t^{t+\varepsilon} adx\right) + \psi'(S_t)E^{\mathcal{N}_t}\left(\int_t^{t+\varepsilon} pd\tau\right) \\ &\quad + \psi''(S_t)E^{\mathcal{N}_t}\left(\left(\int_t^{t+\varepsilon} adx\right)^2\right) - \varepsilon\tilde{p} + o(\varepsilon) \\ E^{\mathcal{N}_t}(Q_{t+\varepsilon} - Q_t) &= o(\varepsilon) \end{aligned}$$

which implies that Q_t is a martingale.

$$\begin{aligned} &E^{\mathcal{N}_t}\left(\left[Q_{t+\varepsilon} - Q_t\right]^2\right) = \\ &\qquad E\left(\left[\psi'(S_t)\int_t^{t+\varepsilon} adx - \int_t^{t+\varepsilon}\tilde{a}dx\right]^2\right) + o(\varepsilon) = o(\varepsilon). \end{aligned}$$

Therefore $E\left([Q(t) - Q(0)]^2\right) = 0$ which implies that Q is constant: $Q = \psi\big(S(0)\big)$. □

7.3. Theorem. *Consider the probability space X^n of Brownian motion on $\mathbb{R}^n$. Denote by S an R^d-valued semi-martingale defined on X^n by the formula*

7.3.1. $$S(x,t) = S(0) + \int_0^t A(x,\tau) \cdot dx(\tau) + \int_0^t P(x,\tau) d\tau$$

where

$$P(x,t) \in \mathbb{R}^d, \qquad A(x,t) \in \mathcal{L}(R^n; R^d).$$

Denote by ψ a bounded C^2 function defined on $\mathbb{R}^d$ and denote $s(x,t) = \psi\big(S(x,t)\big)$. Then

$$s(x,t) = s(0) + \int_0^t a(x,\tau) \cdot dx(\tau) + \int_0^t p(x,\tau) d\tau$$

where

$$\begin{aligned} a_t &= \psi'(S_t) \cdot A_t \\ p_t &= \psi'(S_t) P_t + \frac{1}{2} \sum_{\alpha=1}^{n} \psi''(S_t)(Ae_\alpha, Ae_\alpha) \end{aligned}$$

and where

$$dx = \sum e_\alpha dx^\alpha \qquad (a_\alpha \quad \text{being an orthonormal basis of} \quad \mathbb{R}^n).$$

Proof. Similar to the proof of 7.2.

Dynamical meaning of Itô calculus

In ordinary differential calculus, knowledge of the speed determines the trajectory if its initial position is known.

For an $\mathbb{R}^d$-valued semi-martingale S, we introduce its *speed in the mean*

$$\lim_{\varepsilon \to 0} \varepsilon^{-1} E^{\mathcal{N}_t}\big(S(t+\varepsilon) - S(t)\big) = p$$

and its infinitesimal variance

$$\lim_{\varepsilon \to 0} \varepsilon^{-1} E^{\mathcal{N}_t}\left(\big(S^i(t+\varepsilon) - S^i(t)\big)\big(S^j(t+\varepsilon) - S^j(t)\big)\right) = a_{i,j}.$$

Then the Itô calculus shows that these two invariants *determine* the law of the *stochastic process*. Furthermore

$$a_{i,j} = \sum_k A_k^i \, A_k^i.$$

7.4. Proposition. (Stratonovich's representation of semi-martingales). *Consider the martingale S defined in 7.3.1. Assume $D_\tau A(x, \lambda)$ exists in $L^2(X; \dot{H}^{\otimes 2})$ and*

7.4.1. $$\text{Trace}_{\tau_1=\tau_2}\left(D_{\tau_1} A(x, \tau_2)\right) \quad \textit{exists in } L^2.$$

Then we have

7.4.2. $$S(x,t) = S(0) + \int_0^t A \cdot odx + \int_0^t \tilde{P}\, d\tau$$

where the stochastic integral is in the Stratonovich sense and where

7.4.3. $$\tilde{P}_t^i(x) = P_t^i(x) - \frac{1}{2}\sum_{\alpha=1}^{n} \text{Trace}_{\tau_1=\tau_2=t}\left(D_{\tau_1}^{(\alpha)} A_\alpha(x, \tau_2)\right).$$

We have denoted by $D^{(\alpha)}$ the partial derivative on $X^{(\alpha)}$, where $X^n = \prod_{\alpha\in[1,n]} X^{(\alpha)}$.

7.4.4. Remark. Under the assumption 7.4.1 we can transform an Itô representation into a Stratonovich representation and *vice-versa* obtaining P_t from 7.4.3 in terms of $\tilde{P}_t$.

Proof. We write 7.3.1 for its i-th component:

$$S_t^i = S_t^i(0) + \sum_{\alpha=1}^{n} \int_0 A_\alpha^i\, dx^\alpha + \int_0^t P^i\, dt.$$

The product decomposition

$$X^n = X^{(\alpha)} \times \prod_{\beta\neq\alpha} A^{(\beta)} \qquad \beta \in [1,n]$$

defines a structure on X^n of $X^{(\alpha)}$ partial Wiener space on which we apply VI.9.3.

$$\int_0^t A_\alpha^i\, odx^\alpha = \int_0^t A_\alpha^i\, dx^\alpha + \int_0^t \text{Trace}_\tau\left(D_\cdot^{(\alpha)} A_\alpha^i(\cdot)\right) d\tau.$$

Then as A_α^i is adapted, the ZNP integral in the right hand side is an Itô integral. Summing over α we get 7.4.3. □

7.5. Theorem. *Consider a semi-martingale*

7.5.1. $$S_t = S_0 + \int_0^t A \cdot odx + \int_0^t Pdt$$

where A satisfies 7.4.1. Given $\psi : R^d \to R$, ψ of class C^2 with bounded derivatives, denote

$$s_t = \psi(S_t).$$

Then s_t has the Stratonovich representation:

7.5.2. $$s_t = s_0 + \int_0^t \big(\psi'(S_\tau)A_\tau\big)\ odx + \int_0^t \psi'(S_\tau)P_\tau d\tau.$$

Proof. We translate 7.5.2 into the Itô representation. Then we get

$$\hat{P}_\tau = \psi'(S_\tau)P_\tau + \mathrm{Trace}_{\tau_1=\tau_2=\tau}\left(D_{\tau_1}\psi'(S_{\tau_2})A_{\tau_2}\right).$$

We have, by adaptedness,

$$D_\lambda\big(\psi'(S(x,\tau))A(x,\tau)\big) = 0 \qquad \text{if} \quad \lambda > \tau.$$

In order to compute $D_\lambda S$ for $\lambda < \tau$, we use the Itô representation

$$S_t = S_0 + \int_0^t A\ dx + \int_0^t \hat{P}\ d\tau.$$

Then

$$D_\lambda^{(\alpha)} S_t = \int_\lambda^t D_\lambda^{(\alpha)} A\ dx + A_\alpha(\lambda) + \int_\lambda^t D_\lambda^{(\alpha)}\hat{P}\ d\tau$$

where A_α denotes the α-column of the matrix A; furthermore

$$\begin{aligned} &Tr_{\tau_1=\tau_2}\left(D_{\tau_1}^{(\alpha)}\psi'\big(S(x,\tau_2)\big)A_\alpha(x,\tau_2)\right) = \\ &\quad \mathrm{Trace}_{\tau_1=\tau_2}\left[\psi'\left(S(x,\tau_2)\right) D_{\tau_1}^{(\alpha)} A_\alpha(x,\tau_2)\right] \\ &\quad + \mathrm{Trace}_{\tau_1=\tau_2}\left[\psi''\left(S(x,\tau_2)\right) D_{\tau_1}^{(\alpha)}\left(S(x,\tau_2)\right)\right] A_\alpha(x,\tau_2). \end{aligned}$$

We have

$$\begin{aligned} D_\lambda^{(\alpha)} A_t &= A_\alpha(\lambda) \quad \text{if} \quad \lambda = t-0 \\ D_\lambda^{(\alpha)} A_t &= 0 \quad \text{if} \quad \lambda > t. \end{aligned}$$

The resolution $\mathcal{B}_n$ is defined by the partition of [0, 1] into the family of intervals δ, $\delta \in \Gamma_n$. Then we have

$$\mathrm{Trace}\ \left(D_\cdot^{(\alpha)} S_\cdot\right) = \lim_{n\to\infty}\left[\sum_{\delta\in\Gamma_n} \mathbf{1}_\delta(\tau) \int_{\delta^2} D_{\tau_1}^{(\alpha)} S_{\tau_2}\ \frac{d\tau_1 d\tau_2}{[\rho(\delta)]^2}\right].$$

We write $\delta^2 = \delta_+^2 \cup \delta_-^2$ where δ_+^2, δ_-^2 are disjoint and

$$\delta_+^2 = \left\{(\tau_1,\tau_2) \in \delta^2, \tau_1 < \tau_2\right\}.$$

Then we get

$$\mathrm{Trace}_{\tau_1=\tau_2}\left(D_{\tau_1}^\alpha\left(S(x,\tau_2)\right)\right) = \frac{1}{2}A_\alpha.$$

Then we get

$$\begin{aligned}\hat{P}_t &= \sum_\alpha \psi'(S_t)\left[P_t + \sum_\alpha \operatorname{Trace}_{\tau_1=\tau=t} D_\tau^{(\alpha)} A_\alpha(x,\tau_2)\right] \\ &\quad + \frac{1}{2}\sum_\alpha \psi''(S_t)(A_\alpha, A_\alpha)\end{aligned}$$

and we are back to 7.2. □

7.6. Stochastic differentials

By convention we write 7.3.1. in the form

7.6.1. $$dS = \sum_\alpha A_\alpha dx^\alpha + \hat{P} d\tau$$

where A_α, P are vector valued processes in $\mathbb{R}^d$.

We shall say that 7.6.1 is *an Itô stochastic differential.* In the same way we can write 7.4.2 as

7.6.2. $$dS = \sum A_\alpha \, o dx^\alpha + P d\tau.$$

We shall say that 7.6.2 is a *Stratonovich stochastic differential.* Given a C^2 map from $\psi : \mathbb{R}^d \to \mathbb{R}^s$, we denote $s = \psi(S)$. We then have

$$ds = \sum_\alpha a_\alpha \, o dx^\alpha + p d\tau$$

where

$$\begin{cases} a_\alpha &= \psi'(S) A_\alpha \\ p &= \psi'(S) P \end{cases}$$

and where ψ' denotes the Jacobian matrix of ψ.

Therefore we get the following important theorem:

7.6.3. Theorem. *Stratonovich stochastic differentials behave under a smooth change of variable like ordinary differentials.*

A supplementary derivation is therefore hidden in the Stratonovich representation; this will be an advantage for formal computations in stochastic differential geometry. On the other hand this is a permanent inconvenience for getting estimates which can only be obtained by Itô's representation.

8. Change of variables formula

Our approach, following Airault [4], is based on the ideas of Cruzeiro's Lemma II.1.2.4 for divergences; to be quick, we proceed under unnecessarily strong hypotheses; results under weaker hypotheses can be found in the monographs Nualart [9], Üstünel [10] and in the original papers Ramer [1], Kusuoka [1], [7], Üstünel-Zakai [4], [5], [6], [7], [8], Zakai-Zeitouni [1], Buckdahn [1].

8.1. Carleman determinant

Given a Hilbert space $\mathcal{H}$ and an operator B on $\mathcal{H}$ we denote by $\|B\|_\infty$ the operator norm; we denote by $\mathcal{I}_p$ the ideal of compact operators B such that

$$\|B\|_p = \left[\sum |\lambda_k|^{p/2}\right]^{1/p} < \infty$$

where λ_k are the eigenvalues of B^*B ; then $\mathcal{I}_2$ is the class of Hilbert-Schmidt operators and $\mathcal{I}_1 \subset \mathcal{I}_2$; for $B \in \mathcal{I}_1$ we define $\operatorname{tr}(B) = \sum (Be_i|e_i)$ where e_i is an orthonormal basis of $\mathcal{H}$: this expression is independent of the choice of the basis and $|\operatorname{tr}(B)| \leq \|B\|_1$.

For $B \in \mathcal{I}_1$, $\|B\|_\infty < 1$ (where $\| \quad \|_\infty$ is the usual operator norm) we define

(8.1.1) $$\det_1(I+B) = \exp\left(\sum_{k=1}^{k=\infty} \frac{(-1)^{k+1}}{k} \operatorname{tr}(B^k)\right)$$

(8.1.2) $$\det_2(I+B) = e^{-\operatorname{tr} B} \det_1(I+B) = \exp\left(\sum_{k=2}^{k=\infty} \frac{(-1)^{k+1}}{k} \operatorname{tr}(B^k)\right).$$

Theorem. *Consider a kernel $Q(u,v) \in L^2([0,1]^2)$, assume that $\|Q\|_{L^2} < 1$, then*

(8.1.3) *the series (8.1.2) converges and defines* $\det_2(I+B)$.

We associate to $\sigma^n = (\sigma_1^n, \ldots, \sigma_n^n) \in [0,1]^n$ the $n \times n$ matrix $Q(\sigma^n)$ given by $Q(\sigma)_i^j = (1-\delta_i^j) Q(\sigma_i, \sigma_j)$ where δ_i^j denotes the Kronecker symbol. Then

(8.1.4) $$\det_2(I+Q) = 1 + \sum_{n=2}^{\infty} \frac{1}{n!} \int_{[0,1]^n} \det\left(Q(\sigma^n)\right) d\sigma^n.$$

Proof. For (8.1.3), we remark that $\|Q\|_\infty \leq \|Q\|_{L^2}$ and that for $k \geq 2$ we have $|\operatorname{tr} Q^k| \leq \|Q_\infty\|^{k-2} \|Q\|_{L^2}^2$. □

For (8.1.4) see Carleman [1] or Simon [1] p.108. □

8.2. Theorem (Ramer-Kusuoka formula). *Denote by X the probability space of one-dimensional Brownian motion. Given a function $K(x,s) = K_s$ defined on $X \times [0,1]$, and given a parameter $\lambda \in [0,1]$, consider the change of variable*

$$T^\lambda(x)_t = x_t + \lambda \int_0^t K_s(x)\, ds.$$

We assume that $\forall \lambda \in [0,1]$ the map $T^\lambda : X \to X$ has an inverse A^λ which preserves the class of the Wiener measure μ; denote by G^λ the Radon-Nikodym derivative of $(A^\lambda)_*\mu$ relative to μ. Set

$$G_1^\lambda(x) = \exp\left(-\lambda \int_0^1 K_s\,dx_s - \frac{\lambda^2}{2}\int_0^1 |K_s|^2\,ds\right),$$
$$G_2^\lambda(x) = \det_2 (1 + \lambda D_* K_*),$$

where the stochastic integral considered is a ZNP integral. Assume that

$$\sup_{\tau,s,x} |D_\tau K_s(x)| < 1, \quad \textit{and that} \quad D_{s-0}K_s,\ D_{s+0}K_s \quad \textit{exist}.$$

Then $G^\lambda = G_1^\lambda G_2^\lambda$ *and we have the change of variables formula*

(8.2.1) $$\int_X f\left(T^\lambda(x)\right) G^\lambda(x)\,\mu(dx) = \int_X f(y)\,\mu(dy).$$

Proof. Write $T^\lambda(x) = y$, $A^\lambda(y) = x$. Introduce $g^\lambda(x) = f\left(T^\lambda(x)\right)$; then $(A^\lambda)^* g^\lambda = f$ and by the definition of the direct image of a measure we obtain $E^Y(f) = E^X(g^\lambda G^\lambda)$, which identity gives (8.2.1). □

We have

(8.2.2) $$A^\lambda(y)_t = y_t - \lambda \int_0^t K_s\left(A^\lambda(y)\right)\,ds.$$

Differentiating relative to λ we obtain

$$\frac{d}{d\lambda} A^\lambda(y)|_\tau = -\int_0^\tau K_s\left(A^\lambda(y)\right)\,ds.$$

For any test function φ defined on X we have

$$E^X\left(\varphi(x)\,\frac{dG^\lambda}{d\lambda}(x)\right) = \frac{d}{d\lambda} E^Y\left(\varphi(A^\lambda(y)\right)$$
$$= -E^Y\left[\int_0^1 D_\tau^Y\left[\varphi\left(A^\lambda(y)\right)\right] K_\tau\left(A^\lambda(y)\right)\,d\tau\right].$$

By using the Gaveau-Trauber duality on Y we get that this last expression is equal to

$$-E^Y\left(\varphi\left(A^\lambda(y)\right)\Phi^\lambda(y)\right),$$

where

(8.2.3) $$\Phi^\lambda(y) = \int_0^1 K_\tau\left(A^\lambda(y)\right)\,dy(\tau).$$

We now make the change of variables $y = T^\lambda(x)$ to get

$$E^X\left(\varphi(x)\,\frac{\partial G^\lambda}{\partial\lambda}(x)\right) = -E^X\left(G^\lambda(x)\varphi(x)\Phi^\lambda(T^\lambda(x))\right)$$

which implies Cruzeiro's expression

(8.2.4) $$G^{\lambda_0}(x) = \exp\left(-\int_0^{\lambda_0} \Phi^\lambda\left(T^\lambda(x)\right)\,d\lambda\right).$$

The computation of the change of variables in the integral Φ^λ will be done by the following lemma:

8.2.5. Lemma (Change of variables for ZNP integrals). *Given* $u_* \in D_1^p(Y; L^2)$ *such that* $D_{s-0}^Y u_s$ *and* $D_{s+0}^Y u_s$ *exist, consider its ZNP integral*

$$\psi(y) = \int_0^1 u_s(y)\, dy(s);$$

then

$$\psi(T(x)) = \int_0^1 u_s(T(x))\, dx(s) + \int_0^1 u_s(T(x)) K_s(x)\, ds + C \quad \textit{where}$$

$$C = \int_0^1 \left[D_s^X (u_s(T(x))) - (D_s^Y u_s)(Tx) \right] ds.$$

Proof. The change of variables formula for Stratonovich integrals is proved via their approximation by Riemann sums: we obtain the same identity, but with $C = 0$. Therefore the origin of C lies in the stochastic contraction which appears when we pass from Stratonovich to ZNP integrals.

Introducing $D_s^Y u_s = \frac{1}{2}(D_{s-0} u_s + D_{s+0} u_s)$, we have

$$\begin{aligned} \int_0^1 u_s(y)\, dy(s) &= \int_0^1 u_s(y)\, o dy(s) - \int_0^1 D_s^Y u_s\, ds \\ \int_0^1 u_s(Tx)\, dx(s) &= \int_0^1 u_s(Tx)\, o dx(s) - \int_0^1 \left[D_s^X u_s(Tx) \right] ds. \end{aligned}$$

The difference between the two stochastic contractions gives C. □

Proof of 8.2.1

Applying (8.2.5) to Φ^λ defined in (8.2.3), and using (8.2.4) we get

$$\begin{aligned} \log G_2^{\lambda_0}(x) &= -\lambda_0 \int_0^1 D_s^X K_s\, ds + C_2^{\lambda_0} \quad \text{where} \\ C_2^{\lambda_0} &= \int_0^{\lambda_0} \int_0^1 \left[D_s^Y \left(K_s \circ A_s^\lambda \right) \right] \circ T_s^\lambda\, ds\, d\lambda. \end{aligned} \tag{8.2.6}$$

Differentiating (8.2.2) relative to y, we obtain for every test function $\theta(x)$

$$D_s^Y(\theta \circ A^\lambda) = (D_s^X \theta) \circ A^\lambda - \lambda \int_0^1 \left((D_\alpha^X \theta) \circ A^\lambda \right) D_s^Y (K_\alpha \circ A^\lambda)\, d\alpha.$$

Taking $\theta(x) = K_\sigma(x)$ in this identity we get

$$D_s^Y(K_\sigma \circ A^\lambda) = (D_s^X K_\sigma) \circ A^\lambda - \lambda \int_0^1 \left((D_\alpha^X K_\sigma) \circ A^\lambda \right) D_s^Y (K_\alpha \circ A^\lambda)\, d\alpha,$$

and upon introducing the kernel $B_{s,\sigma}^\lambda(x) = D_s^Y(K_\sigma \circ A^\lambda) \circ T^\lambda$ and making the change of variable $y \to T^\lambda(x)$ we get

$$B^{\lambda}_{s,\sigma} = D_s K_\sigma - \lambda \int_0^1 D_\alpha K_\sigma \; B^{\lambda}_{s,\alpha} \, d\alpha. \tag{8.2.7}$$

Denote by $\mathcal{B}^\lambda$, $\mathcal{K}$ the operators corresponding to the kernels B, K; then the relation (8.2.7) can be writen as $\mathcal{B}^\lambda = \mathcal{K} - \lambda \mathcal{B}^\lambda \mathcal{K}$; inverting by geometric series we get

$$\mathcal{B}^\lambda = \sum_{n=0}^{\infty} (-1)^n \lambda^n \mathcal{K}^{n+1}.$$

Taking the trace of this identity and using (8.1.1) we get

$$\operatorname{tr}(\mathcal{B}^\lambda) = \frac{d}{d\lambda} \log \left[\det_1 (I + \lambda D_* K_*) \right];$$

combining this identity with (8.2.6) we get (8.2.1). □

8.3. Theorem (Girsanov formula). *Under the assumptions of Theorem 8.2 and the further assumption that K_* is an adapted process, then*

$$\det_2(1 + D_* K_*) = 1$$

and the change of variables formula becomes

$$E^X \left(\exp\left(-\lambda \int_0^1 K_s \, dx(s) - \frac{\lambda^2}{2} \int_0^1 |K_s|^2 \, ds \right) f(T^\lambda(x)) \right) = E^Y(f(y)),$$

where the stochastic integral is an Itô integral.

Proof. We want to show that all the determinants appearing in the series (8.1.4) are equal to zero; as a permutation of the indices will not change determinants we can restrict ourself to the case where σ^n is an increasing sequence; the adaptedness implies that $Q(u, v) = D_u K_v = 0$ for $u > v$; therefore the matrix $Q(\sigma^n)$ is a matrix for which all of those subdiagonal elements vanish; as the diagonal elements also vanish, we obtain that $\det Q(\sigma^n) = 0$. □

Remarks. When K_s does not depend on x, we recover the Cameron-Martin formula; therefore Girsanov can appear as a generalization of Cameron-Martin.

Our approach proceeds from the general to the particular by considering Girsanov's formula as a special case of the Ramer-Kusuoka formula. The classical approach to the Girsanov formula proceeds directly by Itô calculus; it works under much weaker hypotheses: see the monographs Nualart [9], Ikeda-Watanabe [6], Malliavin [9], Zakai-Zeitouni [1].

9. Appendix: Estimates of Brownian martingales

The purpose of this appendix is to illustrate the power of the Itô calculus by providing proofs of some classical inequalities used in some other parts of this book; in this way we make an effort to provide a self-contained book.

We call *Brownian martingales* an L^p $(p > 1)$ martingales $M(t)$ defined on the probability space of the Brownian motion, relative to its natural filtration. This martingale is completely defined by its final value. When $p \geq 2$, using Theorem 5.2, we represent f by a stochastic integral:

$$M(x,t) = M(x,0) + \int_0^t a(x,t)dx(t).$$

We normalize the situation by assuming that $M(0) = 0$. We denote by $M^* = \sup |M|$ the maximal function. We introduce the *increasing processes*

$$\beta(x,t) = \int_0^t a^2(x,\tau)d\tau$$

and the *area function*

$$A(x) = [\beta(x,1)]^{\frac{1}{2}}.$$

9.1. Exponential martingales

Theorem. *We have for all real λ and for all stopping times T*

$$E\left(\exp\left(\lambda M(x,T) - \frac{\lambda^2}{2}\beta(x,T)\right)\right) \leq 1.$$

Proof. Consider the semi-martingale

$$S_\lambda(t) = \lambda M(t) - \frac{\lambda^2}{2}\int_0^t a^2\, d\tau.$$

Denote by T_n the first time where $|S(*)|$ takes the value n. On the interval $[-n,n]$, the function $\exp(*)$ is bounded, and so are its first two derivatives; we can apply 7.2 to obtain that $\exp(S_z(t \wedge T_n))$ is a martingale; as this martingale takes the value 1 for $t = 0$, we get that for every stopping time T

$$E\left(\exp\left(S_\lambda(T \wedge T_n)\right)\right) = 1.$$

Then by Fatou's lemma

$$E\left(\exp\left(S_\lambda(T)\right)\right) \leq \liminf_{n\to\infty} E\left(\exp\left(S_\lambda(T \wedge T_n)\right)\right) \leq 1.$$

□

9.2. Üstünel's lemma. *Given a Hilbert space H, consider its associated Gaussian probability space* $\mathrm{Seg}(H)$; *consider a function* $f \in D_1^2$ *such that* $\|Df(*)\|_H$ *is bounded by* γ, *then*

$$E\left(\exp(cf^2)\right) < \infty \qquad \forall c < \frac{\gamma^{-2}}{2}.$$

Proof. We choose a basis e_k of H; by sending $\varepsilon_k \to \sqrt{2}\,\frac{\sin k\pi t}{k\pi}$ we construct a surjective isometry of H onto $H^1([0,1], R)$, the Cameron-Martin space of the Brownian motion. The unitary equivalence induces a probability space isomorphism between $\mathrm{Seg}(H)$ and X the probability space of the Brownian motion. Therefore, without any loss of generality, it is sufficient to prove the lemma on X. We can then use the Clark-Bismut-Ocone formula:

$$f(x) = E(f) + \int_0^1 E^{\mathcal{N}_t}(D_t f)dx(t).$$

Normalizing by $E(f) = 0$, we get that f is a Brownian martingale; using the exponential martingale inequality:

$$E\left(\exp\left(\lambda f - \frac{\lambda^2}{2}\gamma\right)\right) \le 1;$$

by Tchebyschev's inequality

$$\mathrm{Prob}(|f| > r) \le 2\exp\left[\inf_\lambda\left(\frac{\lambda^2}{2}\gamma - \lambda r\right)\right] = 2\exp\left(-\frac{r^2}{2\gamma^2}\right).$$

□

9.3. Integrability of Gross radonifying norms

Theorem. *Consider an abstract Wiener space B, then there exists a constant $c > 0$, such that*

$$E\left(\exp\left(c\|b\|_B^2\right)\right) < \infty.$$

Proof. Denote by H the Cameron-Martin space of B; the inclusion $H \subset B$ implies that there exists a constant γ such that $\|h\|_B \le \gamma |h|_H$; denote $\varphi(b) = \|b\|_B$; then the triangle inequality gives

$$|\varphi(b+h) - \varphi(b)| < \gamma \|h\|_H.$$

Consider $\varphi_\varepsilon = P_\varepsilon \varphi$ where P_t is the Ornstein-Uhlenbeck semigroup. Then φ_ε satisfies the hypothesis of (9.2); as φ_ε converges in probability to φ, Fatou's lemma leads to 9.3.

□

9.4. Theorem (Burkholder inequalities). *Given $1 < p < \infty$, any continuous L^p martingale M defined on X with respect to the filtration $\mathcal{N}_t$, and such that $M_0 = 0$, can be represented by the Itô stochastic integral*

(i) $$M_1 = \int_0^1 a_k\, dx^k.$$

Denoting $M^* = sup_t|M_t|$ and $|||a|||_p = \left[E\left(\left[\int_0^1 \sum_k a_k^2\, dt\right]^{p/2}\right)\right]^{1/p}$, we have

(ii) $$\|M^*\|_{L^p} \le p(p-1)|||a|||_p \quad \textit{for} \quad 2 \le p < \infty,$$

(iii) $$|||a|||_{p'} \le p(p-1)\|M_1\|_{L^{p'}}, \quad \textit{where} \quad 1 < p' \le 2,$$

where p is the conjugate exponent of p'.

Proof. We denote by T_R the first time that $|M_*|$ reaches the value R or if this value is not reached we take $T_R = 1$; we denote by $M_t^R = M_{t\wedge R}$ the stopped martingale; as M^R is an L^2 martingale, Theorem 5.2 provides its representation as a stochastic integral. Passing to the limit it will therefore be sufficient to prove (ii), (iii) for M^R; without loss of generality we can therefore restrict ourselves to bounded martingales.

As $p \ge 2$, the function $\xi \to |\xi|^p$ is C^2 and its second derivative is $p(p-1)|\xi|^{p-2}$; applying the Itô formula 7.3, we get for any stopping time T

$$E\left(|M_T|^p\right) = p(p-1)E\left(\int_0^T |M_t|^{p-2}\, \|a_t\|^2\, dt\right).$$

Given a positive constant c, the transformation $M \to cM$ multiplies the norms by c; this operation does not affect the inequalities; so using this multiplication we can suppose that $\|M^*\|_{L^p} > \delta^{-1}$, where δ is a fixed parameter, $0 < \delta < 1$. To each $k \in N$, the set of strictly positive integers, we associate T_k, the stopping time where $|M_t|$ reaches for the first time the value $(1+\delta)^{k-1}$; if this value is not reached for $t \in [0, 1]$ then we take $T_k = 1$; using the Markov property at T_{k-1} we get

(iv) $$\begin{aligned} E\left(|M_{T_k}|^p - |M_{T_{k-1}}|^p\right) &= p(p-1)E\left(\int_{T_{k-1}}^{T_k} |M_t|^{p-2}\, \|a_t\|^2\, dt\right) \\ &\le p(p-1)E\left(J_k[M^*]^{p-2}\right) \end{aligned}$$

$$\text{where} \quad J_k = \int_{T_{k-1}}^{T_k} \|a_s\|^2\, ds.$$

We set $j(x) = \sup\{k;\ T_k(x) < 1\}$ and when $T_k = 1$, $\forall k \in N$ we take $j(x) = 0$; then $M^* \le (1+\delta)^j$, and

(v) $$(M^*)^p \leq 1_{j=0} + (1+\delta)^p \sum_{k=1}^{k=j} \left[|M_{T_k}|^p - |M_{T_{k-1}}|^p \right],$$

where $M_{T_0} = 0$.

Using (iv) and (v) we obtain

$$E([M^*]^p) \leq 1 + (1+\delta)^p p(p-1) E\left([M^*]^{p-2} \sum_k J_k \right).$$

We use the inequalities $1 \leq \delta \|M^*\|_{L^p}$ and $\sum J_k \leq \int_0^1 |a|^2$ together with Hölder's inequality with the exponents $\frac{p}{p-2}$, and $\frac{p}{2}$ to get

$$\|M^*\|_{L^p}^p \leq \frac{(1+\delta)^p}{1-\delta} \, p(p-1) \|M^*\|_{L^p}^{\frac{p-2}{p}} \left(E\left(\left[\int_0^1 |a_t|^2 \, dt \right]^{\frac{p}{2}} \right) \right)^{\frac{2}{p}}.$$

We divide both sides by $\|M^*\|_{L^p}^p$, and let $\delta \to 0$ to get the inequality (ii). □

As $M^* \geq |M_1|$ we obtain from (ii)

(vi) $$\|M_1\|_{L^p} \leq p(p-1) |||a|||_p.$$

The norm $|||*|||_p$ is the norm of the Banach space $L^p\left(X;\ L^2([0,1])\right)$ which has as its dual $L^{p'}\left(X;\ L^2([0,1])\right)$. Given two martingales $M,\ \tilde{M}$, the energy identity for L^2 stochastic integrals gives the duality coupling

$$E(M_1 \tilde{M}_1) = E\left(\int_0^1 a_t . \tilde{a}_t \, dt \right)$$

which reduces (iii) to (vi). □

Remark. A complete discussion of Burkholder type inequalities can be found in the book of Revuz-Yor [1].

Notes and comments on Part III

The Itô stochastic integral of an *adapted* processes is the cornerstone of the Itô calculus and its subsequent developments: from this point of view, our procedure of delaying the discussion of this central case until the end of this Part might seem strange; nevertheless it fits our general programme of going from the general to the particular; furthermore, the presentation of anticipative integrals in a structure free from any ordering has some conceptual advantage. A direct exposition of Itô's theory can be found in McKean [1].

Multiple Itô stochastic integrals differ from multiple Wiener stochastic integrals by contraction terms which produce the orthogonality between integrals of different degree. Another basic fact which also appears in Itô [3] is the product formula which describes the structure algebra of multiple stochastic integrals in a spirit close to Segal [2]. The statement that all the theory of Gaussian spaces is hidden in this product formula is true, and at the same time paradoxical in the sense that it constitutes a claim for an algebraic approach to probability theory that is much more radical than Segal's point of view (see De Concini-Malliavin-Malliavin [1] for another approach in the same spirit).

The Stroock-Taylor formula is taken from Stroock [8].

The notation $D_\tau f$ for the derivative on the white noise space has been introduced in Nualart-Zakai [1]: this notation conveys the fruitful image of "partial derivative" on the "product space" $\mathbb{R}^{[0,1]}$.

The anticipative stochastic integral appears for the first time in Ramer [1]; it was defined relative to some basis of the Cameron-Martin space; this point of view was systematically developed by Ogawa [1], [2], [3]. Skorokhod [2] used the canonical basis of the second quantization constituted by multiple stochastic integrals. The approach of Nualart-Pardoux-Zakai obtains the integral by an effective *renormalized limit*, which is canonically constructed on the white noise space. A basic reference for anticipative stochastic calculus is the book of Nualart [9].

The duality formula between gradient and anticipative integrals was discovered in Gaveau-Trauber [2]; this duality is in some sense characteristic of Gaussian spaces; it does not hold true on path spaces on a Riemannian manifold (see Part V and Cattiaux-Roëlly-Zessin [1]). The energy identity for anticipative integrals appears as a consequence of the energy identity for harmonic differential forms proved in Shigekawa [4]. All these facts establish strong links between differential calculus and stochastic calculus (see Malliavin [17], Üstünel [1], [2], [3]).

On the path space of a Riemannian manifold, the energy identity for Skorokhod-type integrals is obtained in Cruzeiro-Fang [1].

The filtered probability space corresponds to path space. A well-known fact in algebraic topology is that path spaces are contractible; the probabilistic counterpart of this contractibility is the L^2-continuous resolution of the identity on the probability space, furnished by the family of conditional expectations. On a contractible manifold, the Poincaré lemma says that every differential form of degree 1 is the differential of a function that can be computed by a well-defined integral of the differential form in radial coordinates. This key principle is at the root of the Clark-Bismut-Ocone formula (Clark [1], Haussmann [1], Bismut [2], [11], Ocone [1], Ocone-Karatzas [1]). This formula also provides an unexpected link between Sobolev space differential calculus and Itô calculus. A Riemannian version of this formula appeared in Airault-Malliavin [4].

Cruzeiro [1] shows that infinitesimal quasi-invariance can imply integrated quasi-invariance; inspired by Zakai-Zeitouni [1] and Airault [4], we have used this approach to prove the Ramer [1] Kusuoka [2] formula of change of variables. This formula has been greatly extended by Üstünel-Zakai [4], [5], [7], [8], [9], [10], in the framework of Getzler degree.

Part IV

Stochastic Differential Equations

The theory of SDEs has raised the probability space of Brownian motion to the status of a *universal probability space* on which the semigroup generated by any elliptic operator can be realized; this realization can be done in a canonical form for elliptic operators which are written as a sum of squares of Lie derivatives along *driving vector fields*.

In the Stratonovich formalism, the theory of stochastic differential equations behaves under changes of variables like the theory of ordinary differential equations. We will go further than this formal analogy by enforcing the theory of SDE as a limiting case of the theory of ODE: this is the substance of the *limit theorem*.

The approximating ODEs are constructed from the *control map* which is generated by the driving vector fields. The structure of the control map depends heavily upon the Lie algebra $\mathcal{A}$ generated by the driving vector fields: when $\mathcal{A}$ is commutative, the solution of the SDE is locally equal to a Brownian motion; the non-commutativity of $\mathcal{A}$ induces a *hypoellipticity* phenomenon which constitutes one of the main difficulties in proving the limit theorem.

A by-product of the limit theorem is a *principle of transfer* which PROVES automatically results for SDE from classical results on ODE.

This transfer principle makes it possible to differentiate the solution of an SDE relative to its initial value. This *stochastic calculus of variations* has the consequence that solutions of SDEs are functionals on the Wiener space X which belong to $D_\infty(X)$: the results of Chapter III on the regularity of laws become applicable to fundamental solutions of parabolic equations. So Stochastic Analysis provides an alternative approach to what is known in classical PDE theory as parabolic regularity; this approach goes further in degenerate cases or in infinite dimension.

Chapter VIII

From Ordinary Differential Equations to Stochastic Flow: The Transfer Principle

Contents: Stratonovich SDE – Intrinsic stochastic integral – Itô SDE, uniqueness of the Cauchy problem – The Stroock-Varadhan piecewise linear approximation – Algebraic analysis on the group of C^∞-diffeomorphisms: the exponential map, the adjoint action – Reduced variation of the Stroock-Varadhan approximation sequence of a smooth SDE, its a priori bound; Bootstrap and proof of the limit theorem – Cauchy problem for SDE with Lipschitz coefficients – Critique of our approach.

Prologue

The classical Wiener space X has been defined as the probability space of Brownian motion; all Gaussian probability spaces are (non-canonically) isomorphic to this classical Wiener space.

We denote by Δ the Laplacian defined on $\mathbb{R}^d$ by

$$\Delta = \frac{1}{2} \sum_{k=1}^{d} \frac{\partial^2}{\partial(\xi^k)^2}.$$

By the Itô calculus along the $\mathbb{R}^d$-valued Brownian path x, we have that $\forall f \in C_b^2$

(i) $$M_f(t) = f(x(t)) - \int_0^t [\Delta f]\big(x(s)\big)\, ds \quad \textit{is a martingale.}$$

Stroock-Varadhan summarize (i) in the following concept: *the Brownian motion is the solution of the martingale problem associated to the Laplacian.*

Consider now a general elliptic operator:

$$\Gamma = \frac{1}{2} \sum_{i,j} a_{i,j}(\xi)\, \frac{\partial^2}{\partial \xi^i \partial \xi^j} + \sum_i c^i(\xi)\, \frac{\partial}{\partial \xi^i},$$

$a_{*,*}$ being a non-negative symmetric matrix.

A solution of the *martingale problem associated to* Γ is given by a Borel measure ν_Γ on the path space $P(\mathbb{R}^d)$, (the space of continuous maps $[0,1] \to \mathbb{R}^d$) such that, $\forall f \in C_b^2(\mathbb{R}^d)$,

$$M_f(t) = f\big(y(t)\big) - \int_0^t [\Gamma f]\big(y(\tau)\big)\, d\tau$$

is a martingale relative to the natural filtration of $P(\mathbb{R}^d)$ and relative to the measure ν_Γ.

Denote by $p_t(\xi_0, \xi)$ the fundamental solution of the heat equation:

$$\frac{\partial}{\partial t} p_t(\xi_0,\xi) = \Gamma_{\xi_0} p_t(\xi_0,\xi); \qquad \lim_{t\to 0} \int \varphi(\xi) p_t(\xi_0,\xi)\, d\xi = \varphi(\xi_0).$$

The integral of a cylinder function $f(y) = F\,(y(t_1), \ldots, y(t_r))$ relative to the solution ν_Γ of the martingale problem can be expressed in terms of the fundamental solution p_t by the formula:

(ii)
$$\int_{P_0(R)} f\, d\nu_\Gamma = \int_{\mathbb{R}^{rd}} F(\xi_1, \xi_2, \ldots, \xi_r) p_{t_1}(0,\xi_1) p_{t_2-t_1}(\xi_1,\xi_2) \ldots p_{t_r - t_{r-1}}(\xi_{r-1},\xi_r)\, d\xi_1 d\xi_2 \ldots d\xi_r$$

The formula (ii) provides an approach to the construction of the measure ν_Γ.

In this framework, we have essentially as many different probability spaces $\big(P(\mathbb{R}^d), \nu_\Gamma\big)$ as we have elliptic operators having distinct second order terms $a_{*,*}$. The theory done in the first seven chapters of this book has to be done again when we change our basic elliptic operator. For each new elliptic operator Γ we would have to investigate the Cameron-Martin space for ν_Γ, the possibility of getting a Krée-Meyer type inequality, the construction of stochastic integrals and so on: a frightening prospect!

The Itô theory of *stochastic differential equations* (SDE) realizes a great unification: the solution of an SDE produces the *Itô map* which is an isomorphism of probability spaces carrying the Wiener measure μ into the measure ν_Γ: the probability space of Brownian motion thus acquires the status of *a universal object* amongst probability spaces associated to elliptic operators.

Assuming that Γ has the form

(iii)
$$\Gamma = \frac{1}{2}\sum_{k=1}^{\ell} \mathcal{L}_{A_k}^2 + \mathcal{L}_{A_0},$$

then the Itô map $\mathcal{I}$ is canonical: it is constructed by solving the following *stochastic differential equation* (in the Stratonovich sense):

(iv)
$$\begin{cases} dm_x = A_0 d\tau + \sum_{k>0} A_k o dx^k \\ m_x(0) = m_0 \end{cases}.$$

Given a *smooth* curve $\tau \to h(\tau)$ on $\mathbb{R}^\ell$, the *control* map $\mathcal{C}$ is defined as the solution of the ordinary differential equation (or ODE)

(v)
$$\begin{cases} \dot{m}_h = A_0 + \sum_{k=1}^n A_k \dot{h}^k \\ m_h(0) = m_0 \end{cases}.$$

Following a schema of Stroock-Varadhan, the Brownian motion x can be approximated by a sequence x_n of Lipschitz curves. The basic *limit theorem* of this chapter will assert that the *Itô map is the limit of the control map*

(vi)
$$\big(\mathcal{J}(x)\big) = \lim_{n\to+\infty} [\mathcal{C}(x_n)].$$

A feature of our approach is that the *existence theorem* for the solution of (iv) could be obtained as a by-product of the limit theorem.

The limit theorem makes it possible to *transfer* properties from ODE to SDE. This *transference method* will provide simultaneously *statements* of stochastic propositions *and their proofs*. The transference method applied to the corpus of *Infinitesimal Differential Geometry* (as it happens in *local chart*) will give rise to *Infinitesimal Stochastic Differential Geometry.*

1. Stochastic differential equations: their Stroock-Varadhan approximation

1.1 Definition (Intrinsic stochastic integration). We denote by M a finite dimensional C^∞-smooth manifold of dimension d. We suppose we are given $(\ell+1)$ C^∞-smooth vector fields $A_0, A_1 \dots A_\ell$ on M. We consider the (formal) stochastic differential equation written as

1.1.
$$dm_x(t) = A_0\,(m_x(t))\,dt + \sum_{k=1}^{\ell} A_k\,(m_x(t))\;\; o dx^k(t).$$

In the equation, $x = (x^1 \dots x^\ell) \in X$, where X denotes the Wiener space on R^ℓ. As the stochastic differentials have been defined by an integration procedure, we have to give a meaning to 1.1 in terms of a Stratonovich stochastic integration. We assume that $m_x(t)$ is adapted to the filtration $\mathcal{N}^t$. This implies that $m_x(0)$ is deterministic.

We define a chart φ^α of M as a diffeomorphism of an open set O^α of M to an open set Ω_α in $\mathbb{R}^d$. Given φ^α, a vector field A can be mapped to a vector field A^α on $\Omega_{\alpha'}$ by the formula

$$A^\alpha(\xi) = (\varphi^\alpha)'(m)\big(A(m)\big) \quad \text{where} \quad \xi = \varphi(m).$$

We shall denote

$$A^\alpha = \varphi_*^\alpha A.$$

We shall say that A^α is the *direct image* of A under φ^α.

Now the meaning of 1.1 will be given by the following definition.

1.2. Definition. Given $t_0 \in [0, 1[$, given a chart φ^α containing $m_x(t_0)$, then there exists a stopping time T^α, ($T^\alpha > t_0$ a.s.), such that for $t \geq t_0$,

$$m_x^\alpha(t \wedge T) - m_x^\alpha(t_0) = \int_{t_0}^{t\wedge T} A_0^\alpha\left(m_x^\alpha(\tau)\right) dt + \sum_{k=1}^{\ell} \int_{t_0}^{t\wedge T} A_k^\alpha\left(m_x^\alpha(\tau)\right) \; o dx^k(\tau).$$

This definition is justified by the following

1.3. Proposition. *The Definition 1.2 is independent of the choice of the chart. Therefore the stochastic integrals considered are of intrinsic nature.*

Proof. This means that given another chart φ^β containing $m_x(t_0)$, we have

(i) $$\theta^{\beta,\alpha}\left(m_x^\alpha(\tau)\right) = m_x^\beta(\tau) \qquad \tau \in \left[t_0, T^\alpha \wedge T^\beta\right[$$

where $\theta^{\beta,\alpha} = \varphi^\beta \circ (\varphi^\alpha)^{-1}$ is the *change of chart map*. Then

$$(\theta^{\beta,\alpha})_* A_k^\alpha = (\varphi^\beta)_*[(\varphi^\alpha)^{-1}]_*(\varphi^\alpha)_* A_k = (\varphi^\beta)_* A_k = A_k^\beta.$$

Therefore the stochastic differential of

$$v_x(\tau) = \theta^{\beta,\alpha}\left(m_x(\tau)\right)$$

is given by VII.7.6.3

(ii) $$dv_x^\beta(\tau) = A_0^\beta\left(v_x(\tau)\right) d\tau + A_k^\beta\left(v_x(\tau)\right) \; o dx^k(\tau).$$

Therefore $v_x(\tau)$ and $m_x^\beta(\tau)$ will satisfy the same stochastic differential equation. Furthermore, $v_x(t_0) = m_x^\beta(t_0)$. We get (i) given *the uniqueness of the Cauchy problem* for S.D.E. with smooth coefficients. This uniqueness will be proved below in Section 1.5.

1.4. Itô's stochastic differential equation on $\mathbb{R}^d$

We consider on $\mathbb{R}^d$ the stochastic differential equation:

1.4.1. $$d\xi_x = A_0(\xi_x)dt + \sum_{k=1}^{\ell} A_k(\xi_k)dx^k, \qquad \xi_x(0) = 0,$$

where the stochastic differential is taken in the *Itô sense*. We consider also

1.4.2. $$d\xi_x = A_0'(\xi_x)dt + \sum_{k=1}^{\ell} A_k(\xi_x)\ o dx^k, \qquad \xi_x(0) = 0,$$

where the stochastic differential is taken in the Stratonovich sense.

1.4.3. Proposition. *Assume the vector fields A_k are C^1; then the equations 1.4.1 and 1.4.2 are equivalent if*

$$A_0' = A_0 + \frac{1}{2}\sum_{s,k} \frac{\partial A_k}{\partial \xi^\alpha} A_k^s$$

(where A^s denotes the components of A).

Proof. VII.7.4. □

1.5. Proposition (Uniquenss for the Cauchy problem). *Given $(\ell + 1)$ C^2 fields A_k on an open set Ω of R^d, given $\xi \in \Omega$, we consider two solutions $\xi^{(1)}$, $\xi^{(2)}$ of*

$$\xi_x^{(i)}(t) = \xi_0 + \int_0^{t_i} A_0'\left(\xi_x^{(i)}(\tau)\right) d\tau + \sum_{k=1}^{t_i} A_k\left(\xi_x^{(i)}(\tau)\right)\ o dx^k(\tau).$$

Then there exists a stopping time T ($T > 0$ a.s.) such that

$$\xi_x^{(1)}(t \wedge T) = \xi_x^{(2)}(t \wedge T).$$

Proof. We denote

$$r_x(t) = \|\xi_x^{(1)}(t) - \xi_x^{(2)}(t)\|^2.$$

Then by Itô calculus we have

$$\begin{aligned} \frac{1}{2} d' r_x(t) &= \left(\xi_x^1(t) - \xi_x^2(t)\right) \cdot \Big[\left(A_0^1(\xi_x^1) - A_0^1(\xi_x^2)\right) d\tau \\ &\quad + \sum_k \left(A_k(\xi_x^1) - A_k(\xi_x^2)\right) o dx^k \Big]. \end{aligned}$$

On some neighbourhood O of ξ_0 we can find a constant K such that

$$\left|A_k(\xi) - A_k(\xi')\right| \le \frac{1}{2} K |\xi - \xi'|, \quad \left|A_0'(\xi) - A_0'(\xi')\right| < \frac{1}{2} K |\xi - \xi'|.$$

We transform this Stratonovich differential into an Itô differential and we get:

$$\begin{aligned} dr_x &= (\xi_x^1 - \xi_x^2) \cdot \left(A_k(\xi_x^1) - A_k(\xi_x^2)\right) dx^k + a_0 d\tau \quad \text{where} \\ a_0 &= \sum_k \left\|A_k(\xi^1) - A_k(\xi^2)\right\|^2 + \left(A_0(\xi^1) - A_0(\xi^2)\right) \cdot (\xi^1 - \xi^2) \end{aligned}$$

where A_0 is computed from A_0' by 1.4.3.

Denoting by T the entrance time in the complement of O we have:

$$r_x(t_\wedge T) = \int_0^{t_\wedge T} \sum a_k dx^k + \int_0^{t_\wedge T} a_0 d\tau$$

where $|a_k| < K\ r$.

Then using energy identity and the Schwarz inequality

$$\begin{aligned} E\left(\left|\int_0^{t_\wedge T} \sum a_k dx^k\right|^2\right) &\leq nK^2 E\left(\int_0^{t_\wedge T} r_x^2(\tau)d\tau\right) \\ E\left(\left[\int_0^{t_\wedge T} a_0\, d\tau\right]^2\right) &\leq tK^2 E\left(\int_0^{t_\wedge T} r_x^2(\tau)d\tau\right). \end{aligned}$$

We denote $\psi(t) = E\left(r_x^2(t_\wedge T)\right)$ and we obtain finally

$$\psi(t) \leq \left(2nK^2 + tK^2\right)\int_0^t \psi(\tau)d\tau$$

which implies $\psi = 0$. □

1.6. The Stroock-Varadhan approximation

To an element $x \in X$ we associate the sequence x_n of *continuous, piecewise linear paths*, defined by the following properties:

$$\begin{cases} x_n(k2^{-n}) = x(k2^{-n}) \\ x_n(\tau) \quad \text{is linear on} \quad \left[k2^{-n}, (k+1)2^{-n}\right] \end{cases}.$$

This transparent geometric definition has the following analytic formulation. We denote

$$\begin{aligned} t^- &= 2^{-n} \quad (\text{integral part of} \ \ 2^n t) \\ t^+ &= t^- + 2^{-n}, \quad \text{then} \\ x_n(t) &= x(t^-) + 2^n(t - t^-)\left[x(t^+) - x(t^-)\right]. \end{aligned}$$

The derivative $\dot{x}_n$ of x_n exists at all points distinct from $k2^{-n}$. At those points right and left derivatives exist. Going back to the notation of 1.1 we consider the system

(i) $$\begin{cases} \dot{m}_{x_n} &= A_0 + \sum_k A_k\, \dot{x}_n^k \\ m_{x_n}(0) &= m_0 \end{cases}.$$

This differential equation is well defined at the points distinct from $k2^{-n}$. At those points we will impose continuity of m_{x_n}. This implies that m_{x_n} is defined without ambiguity.

We introduce the *control map* $U^{x_n}_{\tau\leftarrow 0}(m_0)$ which is the solution of (i) at the time τ. We want to prove that $U^{x_n}_{\tau\leftarrow 0}$ *converges in the group* G of diffeomorphisms.

Our proof will have to deal with two kinds of difficulties:

(I) Algebraic complications coming from the fact that the vector fields A_k do not commute.

(II) Convergence questions.

We shall first present in Section 2 the methodology used to overcome (I).

2. Algebraic analysis on the group of diffeomorphisms

We shall denote by M a C^∞ manifold and by Diff(M) the group of C^∞-diffeomorphisms of M. The reader can suppose without inconvenience that $M = \mathbb{R}^d$.

2.1. Algebraic properties of Diff(M)

We shall often abbreviate Diff(M) by G. Given $g \in G$ we denote by $g(m_0)$ the image under g of the point $m_0 \in M$. We shall denote by $g'(m_0)$ the Jacobian matrix of g.

2.1.1. Inverse image of a function

Given $f \in C^\infty(M)$ we denote

$$\left((g)^* f\right)(m) = f\left(g(m)\right);$$

then $g^* f$ is called the *inverse image* of f by g. We have:

$$(g_1 g_2)^* = g_2^* g_1^*.$$

2.1.2. Direct image of a vector field

Given a smooth vector field A on M, we define its *direct image* $g_* A$ as

$$(g_* A)_{m_1} = g'(m_0) A_{m_0} \quad \text{where} \quad m_0 = g^{-1}(m_1).$$

Then

$$(g_1 g_2)_* = (g_1)_* (g_2)_*.$$

Furthermore, given a vector field A and a function f we write $(\mathcal{L}_A f)(m_0)$ for the directional derivative of f at m_0 in the direction A_{m_0}. Then $f \mapsto \mathcal{L}_A f$ defines an operator on $C^\infty(M)$; we have

$$\begin{aligned} g^*\left(\mathcal{L}_{g_* A} f\right) &= \mathcal{L}_A(g^* f) \quad \text{or} \\ \mathcal{L}_{g_* A} &= (g^*)^{-1} \mathcal{L}_A g^*. \end{aligned}$$

2.1.3. Exponential of a regular vector field

Given a smooth vector field Z on M we associate to Z the differential equation

$$\text{(i)} \qquad \frac{d\varphi}{d\tau} = Z_{\varphi_\tau}.$$

We shall say that Z is *regular* if all the maximal solutions of (i) are defined for all values of τ.

In the case $M = \mathbb{R}^d$, the vector fields satisfying a global Lipschitz condition $\|A(\xi) - A(\xi')\| \leq c\|\xi - \xi'\|$ are regular.

Given m_0 we denote by $\varphi_\tau(m_0)$ the solution of (i) satisfying $\varphi_\tau(0) = m_0$. Then φ_t is a diffeomorphism. We shall denote

$$\begin{aligned} \varphi_\tau &= \exp(\tau Z) \quad \text{and} \\ \varphi_\tau(m_0) &= \exp(\tau Z)(m_0). \end{aligned}$$

We have $\exp(\tau Z) \in \mathrm{Diff}(M)$ and

$$\exp\big((\tau' + \tau)Z\big) = \exp(\tau' Z)\exp(\tau Z).$$

2.1.4. Adjoint action on Diff(*M*)

We denote by $\mathrm{Ad}(g)$ the action on G defined by

$$\mathrm{Ad}(g) : g_0 \mapsto g\ g_0 g^{-1}.$$

Proposition. *We have:*

$$Ad(g)\big(\exp(\tau A)\big) = \exp\big(\tau(g_* A)\big)$$

$$\left\{\left(\frac{d}{d\varepsilon}\big(\exp(\varepsilon B)\big)\right)_* A\right\}_{\varepsilon=0} = [A, B]$$

where the bracket of vector fields A and B is defined by

$$\mathcal{L}_{[A,B]} f = \mathcal{L}_A \mathcal{L}_B f - \mathcal{L}_A \mathcal{L}_B f.$$

Proof. We have

$$\mathrm{Ad}(g)\exp(\tau A) = g\ \exp(\tau A) g^{-1}$$

$$\begin{aligned} \frac{d}{d\tau}\big(\mathrm{Ad}(g)\exp(\tau A)\big)_{m_0} &= g\left(\frac{d}{d\tau}\exp(\tau A)\right)_{m_1} \quad \text{where } m_1 = g^{-1}(m_0) \\ &= g'(m_1)A_{m_1} = (g_* A)m_0. \end{aligned}$$

□

$$\begin{aligned}\frac{\partial}{\partial\varepsilon}\exp(\varepsilon B)_*A &= \frac{\partial^2}{\partial\varepsilon\partial\tau}\left[\mathrm{Ad}\big(\exp(\varepsilon B)\exp(\tau A)\big)\right] \\ &= \frac{\partial^2}{\partial\varepsilon\partial\tau}\big(\exp(\varepsilon B)\exp(\tau A)\exp(-\varepsilon B)\big).\end{aligned}$$

Apply this identity to a smooth test function. We get

$$\begin{aligned}\langle\frac{\partial}{\partial\varepsilon}\big(\exp(\varepsilon B)\big)_*A, df\rangle &= \frac{\partial^2}{\partial\varepsilon\partial\tau}\big[\exp(\varepsilon B)\exp(\tau A)\exp(-\varepsilon B)\big]^* f \\ &= \frac{\partial^2}{\partial\varepsilon\partial\tau}\left[\big(\exp(-\varepsilon B)\big)^*\big(\exp(\tau A)\big)^*\big(\exp(\varepsilon B)\big)^* f\right] \\ &= -\mathcal{L}_B\mathcal{L}_A f + \mathcal{L}_A\mathcal{L}_B f.\end{aligned}$$

□

2.1.5. Theorem. *Given two vector fields A, Z on M we have*

$$\begin{aligned}\left[\frac{d}{d\varepsilon}\exp(\varepsilon Z + A)\right]_{\varepsilon=0} &= \exp(A)\left(\int_0^1 \mathrm{Ad}\big(\exp(-\tau A)\big)Z\,d\tau\right) \\ \text{(i)}\qquad &= \left(\int_0^1 \mathrm{Ad}\big(\exp((1-\tau)A)\big)Z\,d\tau\right)\exp(A);\end{aligned}$$

(ii) $$\frac{d}{d\tau}Z_\tau = ad(A)Z_\tau \text{ where } Z_\tau = \mathrm{Ad}\big(\exp(\tau A)\big)Z,\quad ad(A)B = [A, B].$$

Proof. The proof will be based on the variation of an ODE; it is parallel to the theory of the variation of SDE made in Chapter IX.

Introduce an auxiliary parameter $\tau \in [0, 1]$; define m_τ^ε by the ODE

(iii) $$\frac{d}{d\tau}m_\tau^\varepsilon = (\varepsilon Z + A)(m_\tau), \qquad m^\varepsilon(0) = m_0.$$

Define the *reduced variation* Q_τ by

$$Q_\tau = \exp(-\tau A)\frac{\partial m^\varepsilon}{\partial\varepsilon}\bigg|_{\varepsilon=0};$$

differentiate (iii) with respect to ε and interchange the order of the differentiation relative to ε and τ, we get the following ODE determining the reduced variation:

$$\frac{d}{d\tau}Q_\tau = \mathrm{Ad}\big(\exp(-\tau A)\big)\left(Z(m_\tau^0)\right).$$

□

The identity (ii) is based on the equation

(iv) $$\mathrm{Ad}(g_2 g_1) = \mathrm{Ad}(g_2)\mathrm{Ad}(g_1)$$

which is applied to $g_1 = \exp(\tau A)$, $g_2 = \exp(\varepsilon A)$: we apply this identity to Z, differentiate with respect to ε and use 2.1.4. □

2.1.6. Corollary. *Assume that* $[A, B] = 0$*; then*

(i)
$$\mathrm{Ad}\big(\exp(A)\big)B = B$$

(ii)
$$\exp(A + B) = \exp(A)\exp(B) = \exp(B)\exp(A).$$

Proof. Introduce the parameter τ and consider the vector field

$$\varphi(\tau) = \mathrm{Ad}\big(\exp(\tau A)\big)B,$$

then using 2.1.5(iv) and 2.1.4 we have

$$\frac{d}{d\tau}\varphi(\tau) = \mathrm{Ad}\big(\exp(\tau A)\big)[A, B] = 0;$$

therefore $\varphi(\tau) = \varphi(0) = B$. □

To prove the second identity we use 2.1.5(i)

$$\frac{d}{d\tau}\exp(\tau B + A) = \mathrm{Ad}\big(\exp(A)\big)B\exp(\tau B + A) = B\exp(\tau B + A);$$

the r.h.s. is the ODE characterizing $\exp(\tau B)$; by uniqueness of the solution of the Cauchy problem for this ODE we obtain (ii). □

2.1.7. Calculus in local coordinates

In the case of an open subset of $\mathbb{R}^d$ a vector field A is given by its components A^*; then the derivation operator can be written

$$\mathcal{L}_A f = \sum_s A^s \frac{\partial f}{\partial \xi^s};$$

the components of the bracket are given by

$$[A, B]^s = \sum_r \left(A^r \frac{\partial B^s}{\partial \xi^r} - B^r \frac{\partial A^s}{\partial \xi^r}\right).$$

A diffeomorphism is given by d scalar valued functions $g^1(\xi_1), \ldots, g^d(\xi_d)$; the Jacobian matrix is the following $d \times d$ matrix $g' = \left(\left(\frac{\partial g^i}{\partial \xi^j}\right)\right)$. The direct image $B = g_* A$ of the vector field A is given by

$$B^i(\xi) = \sum \frac{\partial g^i}{\partial \xi^j}(\xi')A^j(\xi'), \quad \text{where} \quad \xi = g(\xi').$$

3. Algebraic integration of certain control ODE

The purpose of this paragraph is to give an explicit formula to solve an ODE driven by a smooth control y when the vector fields A_k satisfy certain algebraic relations. It will not be used in any other part of this book: therefore it can be skipped in a first reading. Nevertheless, the material covered is an elementary illustration of bracket phenomena which are at the root of the main lemma used in the proof of the limit theorem in Section 5.

Using the limit theorem proved later in Section 5, and replacing the smooth control y by a Brownian motion x we shall get an explicit solution of the corresponding Stratonovich SDE. These formulae have been directly obtained in the context of SDE by Doss, Stroock-Taniguchi, Gaveau. The propositions of this paragraph will be stated in terms of SDE, and proved in the ODE context.

3.1. Proposition (Doss, Stroock-Taniguchi). *Assume that the vector fields A_k are regular; assume that $[A_k, A_l] = 0$, $\forall k, l$, then the solution to the Cauchy problem for the following Stratonovich SDE*

$$dm_x = \sum_{k=1}^{n} A_k \, odx^k; \qquad m_x(0) = m_0$$

is given by

$$\Gamma_x(t)(m_0) \quad \text{where} \quad \Gamma_x(t) = \exp\left(\sum_{1\le k\le n} x^k(t)A_k\right).$$

Proof. The solution of the Cauchy problem for the control ODE is given by $m(t) = \Gamma_y(t)(m_0)$ where

$$\Gamma_y(t) = \exp\left(\sum_{1\le k\le n} y^k(t)A_k\right);$$

in fact using 2.1.6 (ii)

$$\Gamma_y(t+\varepsilon)m_0 = \exp\left(\sum_{1\le k\le n} \left(y^k(t+\varepsilon) - y^k(t)\right) A_k\right)(m(t));$$

we differentiate in ε and taking $\varepsilon = 0$ we get:

$$\frac{d}{dt}m(t) = \sum_k \dot{y}_n^k A_k.$$

□

3.2. The Gaveau diffusion

We consider a non-commutative case which is the closest to commutativity: the *nilpotent case of step 2*. More precisely, denoting $C_{i,j} = [A_i, A_j]$, we assume that the following commutation relations hold true:

$$(3.2.1) \qquad [C_{i,j}, C_{k,l}] = 0, \qquad [C_{i,j}, A_k] = 0 \quad \text{where} \quad C_{i,j} = [A_i, A_j].$$

We denote by $so(n)$ the space of real antisymmetric $n \times n$ matrices. On $R^n \times so(n)$ we define the vector fields

$$(3.2.2) \qquad B_k(\zeta) = \partial_k + \frac{1}{2}\left[\sum_{s<k} u_s \partial_{s,k} - \sum_{s>k} u_s \partial_{k,s}\right] \quad \text{where } \zeta = (u_., u_.^{\cdot})$$

where ∂_k denotes the coordinate vector field on $\mathbb{R}^n$ and $\partial_{i,j}$, $i < j$, the coordinate vector field on $so(n)$ identified with its upper diagonal elements.

A direct computation shows that the B_k satisfy (3.2.1).

Theorem (Gaveau). *The solution of the Cauchy problem for the Stratonovich SDE*

$$d\zeta_x = \sum_k B_k \, o dx^k, \qquad \zeta(t) \in \mathbb{R}^n \times so(n)$$

can be written as $\zeta_x(t) = \Theta_x(t)(\zeta_0)$ *where*

$$(3.2.3.) \qquad \begin{aligned} \Theta_x(t) &= \Delta_x(t)\Gamma_x(t), \\ \Gamma_x(t) &= \exp\left(\sum_{1\le k\le n} x^k(t) B_k\right), \\ \Delta_x(t) &= \exp\left(\sum_{i<j} I_x^{i,j}(t)\partial_{i,j}\right), \end{aligned}$$

and where the $I_x^{i,j}$ *are given by the following Itô stochastic integrals*

$$(3.2.4) \qquad \begin{aligned} I_x^{i,j}(t) = \frac{-1}{2}\int_0^t & \left(x^i(s) - x^i(0)\right) dx^j(s) \\ & - \left(x^j(s) - x^j(0)\right) dx^i(s), \end{aligned}$$

the integral $I_x^{i,j}$ *coincides with the same expression integrated in the Stratonovich sense.*

Remark. The Brownian motion is generally normalized by the condition $x(0) = 0$; we have written the expression for $I_x^{i,j}$ free of normalization in order to be Markovian.

Proof. We define on $\mathbb{R}^n \times so(n)$ the associative product $*$

$$\zeta_1 * \zeta_2 = (\xi_1, \gamma_1) * (\xi_2, \gamma_2) = (\xi_1 + \xi_2, \gamma_1 + \gamma_2 + \xi_1 \wedge \xi_2) \tag{3.2.5}$$

where

$$[\xi_1 \wedge \xi_2]^{i,j} = \left(\frac{1}{2}\xi_1^i \xi_2^j - \xi_1^j \xi_2^i\right).$$

Multiplication by $(0,0)$ is the identity operator; the identity $\zeta * (-\zeta) = (0,0)$ implies that $-\zeta$ is the inverse of ζ for the $*$ product: the $*$ product defines a Lie group structure on $G = R^n \times so(n)$. Using (3.2.5)

$$\zeta * \exp(\varepsilon \partial_k) = \varepsilon B_k(\zeta) + o(\varepsilon)$$

As the l.h.s. is invariant under left multiplication we see that B_k is a left invariant vector field on G. The Lie algebra of G is therefore generated by the vectors B_k; it is the most general Lie algebra satisfying the commutation relations (3.2.1).

We denote $B = \sum c^k B_k$, $ad(B)Z = [B, Z]$; by Taylor expansion and $[ad(B)]^2 = 0$ we get

$$\mathrm{Ad}\big(\exp(\tau B)\big) = \exp\big(\tau ad(B)\big) = I + \tau ad(B),$$

which implies

$$\int_0^1 \mathrm{Ad}\big(\exp(1-\tau)B\big)\, d\tau = I + \frac{1}{2} ad(B). \tag{3.2.6}$$

We proceed now as in 3.1 in replacing the Brownian motion x by a smooth control y. Using 2.1.5(i) and (3.2.6) we get

$$\begin{aligned} \dot{\Gamma}_y(t) &= \left(I + \frac{1}{2}\sum y^j(t) ad(B_j)\right)\left(\sum_k \dot{y}^k B_k\right)\Gamma_y(t) \\ &= \sum_k \dot{y}^k B_k + \frac{1}{2}\sum_{j<k}\left(y^j \dot{y}^k - y^k \dot{y}^j\right)\partial_{j,k}. \end{aligned}$$

We remark that

$$\begin{aligned} \frac{d}{dt} I_y^{j,k}(t) &= \frac{-1}{2}\left(y^j(t)\dot{y}^k(t) - y^k(t)\dot{y}^j(t)\right), \quad \text{therefore} \\ \frac{d}{dt}\left(\Delta_y(t)\Gamma_y(t)\right) &= \left(\sum_k \dot{y}^k(t) B_k\right)\Delta_y(t)\Gamma_y(t). \end{aligned}$$

□

3.3. An example of the limit theorem

In the special case of the Gaveau diffusion, we shall prove directly the following version of the limit theorem:

Proposition. *Consider the control map $U^{y}_{t\leftarrow 0}$ associated to the vector fields B_k defined in (3.2.2), then almost surely $U^{x_n}_{t\leftarrow 0}(0,0)$ converges uniformly in $t \in [0,1]$.*

Proof. Using the representation of the control map given in 3.2, the statement is equivalent to

$$\text{(i)} \qquad \lim_{n\to\infty} \int_0^t \left(x_n^1 \dot{x}_n^2 - x_n^2 \dot{x}_n^1\right) d\tau \quad \text{exists uniformly in} \quad t \in [0,1].$$

We consider the finite σ-field $\mathcal{A}_n$ defined by its atoms which are the dyadic intervals of length 2^{-n}. Then the orthogonal projection $L^2([0,1]) \to L^2(\mathcal{A}_n)$ induces the conditional expectation $E^{\mathcal{A}_n}$ which is a map from the Wiener space X to a finite-dimensional Gaussian space. Consider the Itô stochastic integral

$$I_\infty(x) = \int_0^1 x^1\, dx^2 - x^2\, dx^1.$$

Then $I_\infty \in L^2(X)$; therefore $J_n = E^{\mathcal{A}_n}(I_\infty)$ is an L^2 martingale which converges towards I.

We have $E^{\mathcal{A}_n}(x^k) = x_n^k$, for $k = 1, 2$. Furthermore x^1, x^2 are independent; it is the same for x_n^1, x_n^2. Therefore

$$\begin{aligned} E\left(F_1(x^1)F_2(x^2)f_1(x_n^1)f_2(x_n^2)\right) &= E(F_1 f_1)E(F_2 f_2) \\ &= E\left(E^{\mathcal{A}_n}(F_1)f_1\right) E\left(E^{\mathcal{A}_n}(F_2)f_2\right); \end{aligned}$$

this identity implies that

$$E^{\mathcal{A}_n}(F_1 F_2) = E^{\mathcal{A}_n}(F_1)E^{\mathcal{A}_n}(F_2).$$

This relation implies that $I_n = J_n$. □

Remark. The Lyons theory of ODE driven by rough signals provides a conceptual machinery which reduces the proof of the limit Theorem 5.2 to the convergence (i).

4. Analysis on a torus

We denote by $\mathbb{T}^d$ the quotient of $\mathbb{R}^d$ by the lattice Z^d. Continuous functions on $\mathbb{T}^d$ can be identified with continuous periodic functions on $\mathbb{R}^d$. We introduce the Fourier coefficients

$$\hat{f}(n) = \int_{\mathbb{T}^n} f(\xi) e^{-2i\pi\xi\cdot n} d\xi \quad \text{with} \quad n \in Z^d.$$

Then

$$\int_{\mathbb{T}^d} |f(\xi)|^2 d\xi = \sum_{n\in Z^d} |\hat{f}(n)|^2.$$

To each positive integer s we associate the following Sobolev space on $\mathbb{T}^d$

(i) $$\mathcal{H}_s = \left\{ f \in L^2(\mathbb{L}^d);\ \|f\|^2_{\mathcal{H}_s} = \sum (|n|^2+1)^s |\hat{f}(n)|^2 < +\infty \right\}.$$

We denote by C_k the space of k-times continuously differentiable functions with norm

(ii) $$\|f\|_{C_k} = \sup \left| \frac{\partial^i f}{\partial \xi^{i_1} \partial \xi^{i_d}} (\xi) \right| \quad \text{with } i = i_1 + \ldots + i_d,\ i \in [0,k],\ \xi \in \mathbb{T}^d.$$

We have now the Sobolev embedding theorem

(iii) *For every s, k such that $s > k + \frac{1}{2}d$ there exists $\gamma_{k,s}$* such that

$$\|f\|_{C_k} \leq \gamma_{k,s} \|f\|_{\mathcal{H}_s}.$$

(iv) C^∞-convergence. We say that φ_n is C^∞-convergent to φ_∞ if $\|\varphi_n - \varphi_\infty\|_{C_k} \to 0$ for all k.

By (ii) C^∞-convergence of φ_n to φ_∞ is equivalent to

$$\|\varphi_n - \varphi_\infty\|_{\mathcal{H}_s} \to 0 \quad \text{for all} \quad s.$$

(v) A *C^∞-vector field on* $\mathbb{T}^d$ is a C^∞ map of $\mathbb{T}^d$ to $\mathbb{R}^d$.

(vi) A *system of coordinates* on $\mathbb{T}^d$ is a set $u^1 \ldots u^m$ of C^∞-functions such that for all $\xi \in \mathbb{T}^d$ the matrix $\left(\frac{\partial u^i}{\partial \xi^k} \right)(\xi)$ is of rank d, for instance the system of functions

$$\begin{aligned} u^i &= \cos(\xi^i) \qquad i \in [1,d] \\ u^i(\xi) &= \sin(\xi^{i-d}) \qquad i \in [d+1, 2d]. \end{aligned}$$

(vii) We denote by G the group of C^∞-diffeomorphisms of $\mathbb{T}^d$. Given $\{g_n\} \in G$, then $\{g_n\}$ C^∞ converges towards g_∞ if for all $i \in [1,m]$ and all k.

$g_n^* u^i$ and $(g_n^{-1})^* u^i$ are C_k-convergent to $g_\infty^* u^i$ and $(g_\infty^{-1})^* u^i$.

4.1. Theorem. *Given $\{g_n\} \in G$, assume that*

(i) $(g_n^{-1})^* u^i$ *converges in* $\mathcal{H}_s, i \in [1,m], s \in \mathbb{N}$ *and assume that*

(ii) $|\det(g_n')| \leq M$ *where M is a constant.*

Then g_n converges in C^∞ to $g_\infty \in G$.

Proof. Set

$$\begin{aligned} v^i &= \lim (g_n^{-1})^* u^i \\ v^i(\xi) &= \lim u^i \left(g_n^{-1}(\xi) \right). \end{aligned}$$

By the implicit function theorem we deduce that

$$g_n^{-1}(\xi) \text{ converges to } \gamma(\xi) \text{ and that}$$
$$v^i(\xi) = u^i(\gamma(\xi)).$$

Again by the implicit function theorem γ is a C^∞ map of $\mathbb{T}^d$ into $\mathbb{T}^d$, we have

$$\left[\det\left(\gamma'(\xi)\right)\right] = \lim\left[\left(\det\left(g_n'(\xi)\right)\right)^{-1}\right];$$

then (ii) implies that the Jacobian matrix $\gamma'(\xi)$ is invertible. Therefore

γ *is a local C^∞-diffeomorphism.*

Furthermore we have the matrix identity

$$g_n'(\xi) = \det\left[\left(g_n'(\xi)\right)^{-1}\right] M_n(\xi)$$

where M_n is the matrix of cofactors of $(g_n^{-1})'(\xi)$. This identity implies that $g_n'(\xi)$ converges uniformly.

We fix ξ_0 and denote $g_n(\xi_0) = \eta_n$. We extract a subsequence σ such that $\{\eta_n\}$, $n \in \sigma$ converges and we denote its limit by η_∞. As g_n^{-1} converges uniformly to γ we deduce that

$$\gamma(\eta_\infty) = \lim_{n \in \sigma} g_n^{-1}(\eta_n) = \xi_0,$$

which means that γ is surjective.

Finally, the identity

$$\int \det(g_n^{-1}) = \text{ volume } \mathbb{T}^d$$

implies, upon passing to the limit, the same identity for γ; this establishes its bijectivity. □

5. Limit theorem on the torus

5.1. Lévy's construction of Brownian motion

We resume the construction done in VII.1.3.3. Denote by S_n the set $\{k2^{-n}\}$ where k is an integer, define for $\lambda \in S_n$

$$\varphi_\lambda(\tau) = \mathbf{1}_{[\lambda, \lambda^+ - 2^{-n-1}]} - \mathbf{1}_{[\lambda^+ - 2^{-n-1}, \lambda^+]}.$$

Define

$$\Gamma_\lambda = \int \varphi_\lambda(\tau) dx(\tau),$$

then $\{\Gamma_m\}$ are independent Gaussian variables. They are also independent of the σ-field generated by x_n. Furthermore

$$\Gamma_\lambda = \frac{1}{2}\left[x(\lambda) + x(\lambda^+)\right] - x(\lambda + 2^{-n-1}).$$

We shall denote

$$\begin{aligned}\sigma_\lambda &= x(\lambda^+) - x(\lambda), \\ A(\sigma_\lambda) &= 2^{-n}A_0 + \sum_{k=1}^{\ell} \sigma_\lambda^k A_k.\end{aligned}$$

5.2. Limit theorem. *Consider* $(\ell+1)$ C^∞*-vector fields* $A_0, A_1 \ldots A_\ell$ *on* $\mathbb{T}^d$*. Define for* $\lambda_0 \in S_n$ *a diffeomorphism of* $\mathbb{T}^d$ *by*

$$g_{x_n}(\lambda_0) = \prod_{\lambda_0 > \lambda \geq 0} \exp\left(A(\sigma_\lambda)\right).$$

when the order on factors is such that $\lambda = 0$ *appears on the extreme right and so on.*

Define

$$\begin{aligned} g_{x_n}(\tau) &= \exp\left(A\left[2^n(\tau - \tau^-)(\sigma_{\tau^-} - 2\Gamma_{\tau^-})\right]\right) g_{x_n}(\tau^-) \\ &\qquad \text{where } \tau \in \left[\tau^-, \frac{\tau^+ + \tau^-}{2}\right]. \end{aligned}$$

(The case $\tau \in \left[\frac{\tau^+ + \tau^-}{2}, \tau^+\right]$ *is defined in a similar way).*

Then almost surely $g_{x_n}(\tau)$ *converges in* C^∞ *in* G*, uniformly in* $\tau \in [0, 1]$*. This convergence also holds in* $L^p(X; \mathcal{H}_s)$ *for all* (p, s)*.*

Denote

$$\xi_x(\tau) = \lim g_{x_n}(\tau)(\xi_0),$$

then ξ_x *satisfies the following Stratonovich SDE*

$$d\xi_x = A_0(\xi_x)d\tau + A_j(\xi_x)\, o dx^j.$$

5.3. Algebraic machinery of the proof

The reduced variation δ_n is defined by

$$\delta_n(t) = g_n^{-1}(t) g_{n+1}(t).$$

The concept of reduced variation is transparent: the reduced variation is the *virtual variation* of the *initial point* which would produce the *observed variation at time* t.

5.3.1. Lemma.

$$\delta_n(\lambda_0^+) = \prod_{\lambda_0 \geq \lambda} \Phi\left(A^\lambda(\sigma_\lambda), A^\lambda(\Gamma_\lambda)\right)$$

where

$$A_k^\lambda = \left(g_n^{-1}(\lambda)\right)_* A_k,$$

where

$$\Phi(z, Z) = \exp(-z) \exp\left(\frac{1}{2}z + Z\right) \exp\left(\frac{1}{2}z - Z\right),$$

and where λ_0, λ *denote elements of* S_n *and where*

$$A^\lambda(\sigma_\lambda) = 2^{-n} A_0^\lambda + \sum_k \sigma_\lambda^k A_k^\lambda, \quad A^\lambda(\Gamma_\lambda) = 2^{-n-1} A_0^\lambda + \sum \Gamma_\lambda^k A_k^\lambda.$$

Furthermore for $\tau \notin S_n$, *we have*

$$\begin{aligned} \delta_n(\tau) &= \operatorname{Ad}\left(g_n^{-1}(\lambda)\right)\left[\exp\left(-2^n(\tau - \lambda)A(\sigma_\lambda)\right)\right. \\ &\quad \left.\exp\left(2^n(\tau-\lambda)A(\sigma_\lambda - 2\Gamma_\lambda)\right)\right]\delta_n(\lambda) \end{aligned}$$

if $\tau \in \left[\lambda, \frac{\lambda + \lambda^+}{2}\right]$, *and a similar formula holds if* $\tau \in \left[\frac{\lambda+\lambda^+}{2}, \lambda^+\right]$.

Proof. We remark that

$$\Phi\left(A^\lambda(\sigma_\lambda), A^\lambda(\Gamma_\lambda)\right) = \operatorname{Ad}\left(g_n^{-1}(\lambda)\right) \Phi\left(A(\sigma_\lambda), A(\Gamma_\lambda)\right).$$

We introduce

$$\gamma_\lambda = \delta_n(\lambda^+)[\delta_n(\lambda)]^{-1},$$

then

$$\delta_n(\lambda_0) = \prod_{\lambda_0 > \lambda \geq 0} \gamma_\lambda$$

where

$$\begin{aligned} \gamma_\lambda &= g_n^{-1}(\lambda^+) g_{n+1}(\lambda^+) \left[g_{n+1}(\lambda)\right]^{-1} g_n(\lambda) \\ &= \operatorname{Ad}\left(g_n^{-1}(\lambda)\right)\left[\left(g_n(\gamma^+) g_n^{-1}(\lambda)\right)^{-1} g_{n+1}(\lambda^+)\left(g_{n+1}(\lambda)\right)^{-1}\right]. \end{aligned}$$

We use now the differential equation defining g_n and g_{n+1}:

$$\begin{aligned}
g_n(\lambda^+) &= \exp\left(A(\sigma_\lambda)\right) g_n(\lambda) \\
g_{n+1}\left(\frac{\lambda+\lambda^+}{2}\right) &= \exp\left(\frac{1}{2}A(\sigma_\lambda) - A(\Gamma_\lambda)\right) g_{n+1}(\lambda) \\
g_{n+1}(\lambda^+) &= \exp\left(\frac{1}{2}A(\sigma_\lambda) + A(\Gamma_\lambda)\right) g_{n+1}\left(\frac{\lambda+\lambda^+}{2}\right) \\
g_{n+1}(\lambda^+) &= \exp\left(\frac{1}{2}A(\sigma_\lambda) + A(\Gamma_\lambda)\right) \exp\left(\frac{1}{2}A(\sigma_\lambda) - A(\Gamma_\lambda)\right) \\
&\quad g_{n+1}(\lambda).
\end{aligned}$$

By multiplying by γ_λ we get the desired expression for $\delta_n(\lambda_0)$.

Consider now the case $\tau \in \left]\lambda, \frac{\lambda+\lambda^+}{2}\right]$; then in the interval $[\tau, \lambda]$, g_n and g_{n+1} satisfy linear differential equations which lead to the relations

$$\begin{aligned}
g_n(\tau) &= \exp\left(2^n(\tau-\lambda)A(\sigma_\lambda)\right) g_n(\lambda) \\
g_{n+1}(\tau) &= \exp\left(2^{n+1}(\tau-\lambda)A\left(\frac{1}{2}\sigma_\lambda - \Gamma_\lambda\right)\right) g_{n+1}(\lambda) \\
\delta_n(\tau) &= g_n^{-1}(\lambda)\exp\left(-2^n(\tau-\lambda)A(\sigma_\lambda)\right) \\
&\quad \exp\left(2^n(t-\lambda)A(\sigma_\lambda - 2\Gamma_\lambda)\right) g_{n+1}(\lambda) \\
&= \mathrm{Ad}\left(g_n^{-1}(\lambda)\right)\left[\exp\left(-2^n(\tau-\lambda)A(\sigma_\lambda)\right)\right. \\
&\quad \left.\exp\left(2^n(\tau-\lambda)A(\sigma_\lambda - 2\Gamma_\lambda)\right)\right]\delta_n(\lambda).
\end{aligned}$$

The last case is treated in a similar way. □

5.4. Theorem (An a priori bound). *For $u \in \mathcal{H}^s(T^d)$, we have*

$$P\left(\sup_\tau \|(g_n^{-1}(\tau))^* u\|_{\mathcal{H}_s} \geq a\right) \leq \frac{c_s'}{a^2}\|u\|_{\mathcal{H}^s}$$

where c_s' is a constant depending only on s.

Proof. It will be of importance to emphasize the action of G on $\mathcal{H}_s$. We shall denote

$$\rho_s(g)f = (g^{-1})^* f \quad \text{for} \quad f \in \mathcal{H}_s.$$

We have the fundamental fact that ρ_s is a *representation* of G which means that

$$\rho_s(g_1 g_2) = \rho_s(g_1)\rho_s(g_2).$$

We will denote by F_λ the σ-field generated by the $x(\lambda')$ for $\lambda' \leq \lambda$.

5.4.1. Lemma. *There exists a constant c_s such that*

$$E^{F_\lambda}\left[\rho_s^*\left(\exp((A(\sigma_\lambda)))\right)\rho_s\left(\exp\left(A(\sigma_\lambda)\right)\right)\right] \leq \left(1 + 2^{-n}c_s\right) I$$

where I denotes the identity operator and where the inequality is taken in the sense of the natural order for the Hermitian operators on $\mathcal{H}_s$.

Proof. We shall treat first the case $s = 0$. Then given $f \in L^2$ we have to evaluate

$$\left(\left(E^{F_\lambda}(\rho_0^*\rho_0) - I\right) f | f\right) = E^{F_\lambda}\|\rho_0 f\|_{L^2}^2 - \|f\|_{L^2}^2.$$

We abbreviate $A(\sigma_\lambda)$ as A_σ

$$\|\rho_0 f\|^2 = \int_M f^2\left(\exp(-A_\sigma)(m')\right) dm',$$

we denote by δA_k the divergence of A_k for the volume measure of $\mathbb{T}^d$, then by Cruzeiro's Lemma II.1.2.4

$$= \int_M f^2(m)\left(\exp\left[\left[\int_0^1 \delta A_\sigma(\exp(\xi A(\sigma))m)d\xi\right]\right)dm$$

$$E^{F_\lambda}\|\rho_0 f\|^2 - \|f\|^2 = \int_M f^2 q_0 dm \quad \text{where}$$

$$q_0(m) = E^{F_\lambda}\left[\exp\left[\left[\int_0^1 \delta A_\sigma(\exp(\xi A(\sigma))m)d\xi\right] - 1\right]\right].$$

5.4.2. Sublemma. *Given $\eta \in \mathbb{R}^\ell$, denote*

$$u(\eta, m) = \exp(A(\eta))m$$

then u is infinitely differentiable in (η, m); its derivatives are bounded uniformly in (η, m) when $\|\eta\| < c$.

Proof. Calculus of variations on the ordinary differential equation $\dot{m} = A_m(\eta)$ as developed in Chapter IX. □

5.4.3. Sublemma. *Denoting*

$$\tilde{q}_0 = 2^{-n}\left(\delta A_0 + \frac{1}{2}\sum_{j\geq 1}\left[(\delta A_j)^2 + \mathcal{L}_{A_j}(\delta A_j) + \mathcal{L}(\delta A_j)\right]\right)(m)$$

then we have uniformly in m

$$q_0 = \tilde{q}_0 + O\left(\left(2^{-n}\right)^{3/2}\right).$$

Proof.

$$u(\xi\sigma, m) = m + \xi A_m(\sigma) + O(\sigma^2), \quad \xi \in [0, 1].$$

$$J = \int_0^1 (\delta A)(\sigma)\left(u(\xi\sigma, m)\right) = (\delta A)(\sigma) + \frac{1}{2}(A(\sigma)|\nabla\delta A(\sigma)) + O(\sigma^{3/2})$$

$$\exp(J) = 1 + (\delta A)(\sigma) + \frac{1}{2}(A(\sigma)|\nabla\delta A(\sigma)) + \frac{1}{2}(\delta A(\sigma))^2 + O(\sigma^{3/2}).$$

Then

$$\begin{aligned} E^{F_\lambda}(\exp(J)-1) &= 2^{-n}\sum_{k=1}^{\ell}\mathcal{L}(\delta A_k)+2^{-n}\delta A_0 \\ &\quad +\frac{1}{2}E^{F_\lambda}\left[(\delta A(\sigma))^2+(A(\sigma)|\nabla\delta A(\sigma))\right]+O(2^{-3n/2}). \end{aligned}$$

□

5.4.4. *Proof of Lemma 5.4.1*

The case $s=0$ results from 5.4.3. For $s=1$ we use the same methodology. We have to compute

$$E^{F_\lambda}\left(\|\rho_1 f\|^2_{\mathcal{H}_1}-\|f\|^2_{\mathcal{H}_1}\right) \quad \text{where}$$

$$\|f\|^2_{\mathcal{H}_1}=\|f\|^2_{L^2}+\|\nabla f\|^2_{L^2}$$

$$\nabla\left(\rho_1(g)f\right)=\nabla\left[f\left(g^{-1}(m)\right)\right]=(\nabla f)\left(g^{-1}(m)\right)(g^{-1})'(m).$$

We consider the following $d\times d$ matrices:

$$M_j=\partial A_j, \quad \text{and} \quad M_\sigma=\sum M_j\sigma^j, \quad \text{then}$$

$$(g^{-1})'(m)=\exp\left(*\int_{[0,1]}M_\sigma(u(\xi\sigma,m))d\xi\right):=Q$$

$$\|\nabla\left[f(g^{-1}(m))\right]\|^2=\left(\nabla f|QQ^*\nabla f\right)(g^{-1}(m))$$

$$\int\|\nabla f(g^{-1}(m))\|^2dm=\int_M\left(\nabla f\mid QQ^*\nabla f\right)\exp\left(\int_0^1\delta A\right)dm.$$

We have now to expand to Q. We use the Picard series expansion of the resolvant

$$\begin{aligned} Q(m) &= I+\int_0^1 M_\sigma\big(u(\xi\sigma,m)\big)d\xi \\ &\quad +\int_0^1 M_\sigma\left(u(\xi_1\sigma,m)\right)d\xi_1\left[\int_0^{\xi_1}M_\sigma\left(u(\xi_2\sigma,m)\right)d\xi_2\right]+O(\sigma^3) \end{aligned}$$

$$Q(m)=I+M_\sigma(m)+\frac{1}{2}\mathcal{L}_{A(\sigma)}M_\sigma(m)+\frac{1}{2}M_\sigma^2(m)+O(\sigma^3)$$

$$\begin{aligned} QQ^* &= I+M_\sigma+M_\sigma^* \\ &\quad +\frac{1}{2}\left[M_\sigma^2+(M_\sigma^*)^2+\mathcal{L}_{A(\sigma)}\big(M(\sigma)+M^*(\sigma)\big)\right]+O(\sigma^3). \end{aligned}$$

Denote

$$E^{F_\lambda}\left(QQ^*\exp\left(\left(\int_0^1 \delta A_\sigma\big(u(\xi\sigma)d\xi\big)\right)\right)\right) = I + q_1 = I + \tilde{q}_1 + O(2^{-3/2n})$$

where

$$\tilde{q}_1 = \tilde{q}_0 I + \frac{1}{2}\sum_{j\ge1}\left[M_j^2 + M_j^{*2} + \mathcal{L}_{A_j}(M_j + M_j^*)\right] + \sum_{j\ge1}\delta A_j(M_j + M_j^*).$$

By the construction of q_1 we have

$$E^{F_\lambda}\left(\|\rho_1\big(\exp(A(\sigma))\big)f\|^2_{\mathcal{H}_1}\right) - \|f\|^2_{\mathcal{H}_1} = \int |f|^2 q_0\,dm + \int (q_1\nabla f \mid \nabla f)\,dm.$$

This relation proves 5.4.1 for $s = 1$.

The same methodology will give the result for $s > 1$. □

5.4.5. Remark. An easy computation gives

(i) $$E^{F_\lambda}\left(\rho_s(g_x(\lambda^+))\right) = L_n\rho_s(g_x(\lambda))$$

where L_n is an operator which will tend for large n to the operator $\mathcal{L}_{A_0} + \frac{1}{2}\sum_{j=1}^{\ell}\mathcal{L}^2_{A_j}$. Therefore an inequality $\|L_n\varphi\|_{\mathcal{H}_s} < c\|\varphi\|_{\mathcal{H}_s}$, holding uniformly in n, is impossible, for it would imply that one could dominate the L^2-norm of the Laplacian of φ by the L^2-norm of φ!

The case of $E^{F_\lambda}\left(\rho_s^*(g_x(\lambda^+))\rho_s(g_x(\lambda^+))\right)$ that we have treated is *fundamentally different*. An heuristic explanation of this difference can be founded in the fact that the first order differential operator $\sqrt{-1}\mathcal{L}_A$ is, *up to low order terms* (in the sense of Calderon's pseudo-differential operators), self-adjoint in L^2. The equality $\mathcal{L}_A^* = -\mathcal{L}_A$ would imply the unitarity of $\rho(g_n(\lambda))$. As $\mathcal{L}_A^* = -\mathcal{L}_A + \delta A$, we are working on a perturbation of the unitary case. In fact, if $\delta A_k = 0$, $k = 0, \dots, \ell$, and $s = 0$, we are in a unitary situation.

5.4.6. *Proof of Theorem 5.4*

Denote

$$z_\lambda = \rho_s\left(g_{x_n}(\lambda)\right)u$$

and

$$M(\lambda) = \|z_\lambda\|^2_{\mathcal{H}_s},$$

then

$$\begin{aligned} E^{F_\lambda}\big(M(\lambda^+)\big) &= \left(E^{F_\lambda}\left(\rho_s^*(\exp(A(\sigma_\lambda))\rho_s(A(\sigma_\lambda))z_\lambda \mid z_\lambda\right)\right) \\ &\le \left(I + 2^{-n}c_s\right)\|z_\lambda\|^2 \le M(\lambda)\exp(2^{-n}c_s). \end{aligned}$$

Then $\tilde{M}(\lambda) \doteq M(\lambda)\exp(-c_s\lambda)$ is a supermartingale.

We finish the proof by applying Doob's maximal inequality. □

5.5. Proposition. *Almost surely we have for all $\varepsilon > 0$*

$$\sup_{n,\lambda} \left\| \det(g'_{x_n}(\lambda)) \right\|_{L^\infty(T^d)} = O(n^\varepsilon).$$

Proof. Introduce $q_k = \text{trace}\ (\partial_. A_k^.)$. Then we get

$$\log\left(\det\left(g'_{x_n}(\lambda_0)\right)\right) = \int_0^{\lambda_0} \sum_k \dot{\sigma}_\lambda^k q_k\left(g_{x_n}(\tau)\right) d\tau.$$

This expression is equal, up to a bounded term, to the martingale

$$M(\lambda_0) = \sum_{\lambda<\lambda^+} \sigma_{\lambda^+}^k\ q_k(g_x(\lambda)).$$

We introduce $R = \sum_k \|q_k\|_{L^\infty}^2$, then

$$\begin{aligned} E\left(\exp\left(c \sup_\lambda M(\lambda) - \frac{c^2}{2}\, r\right)\right) &< 1 \quad \text{or} \\ P\left(\sup_\lambda M(\lambda) > \gamma\right) &< \exp\left(-\frac{1}{2}\frac{\gamma^2}{R}\right) \\ P\left(\sup_\lambda\left[\exp(M(\lambda))\right] > a\right) &\leq \exp\left(-\frac{1}{2}\frac{(\log a)^2}{R}\right). \end{aligned}$$

Taking $a = n^\varepsilon$ we get the result. □

5.6. Proposition. *For all integers k there exists q_k such that, almost surely*

$$\sup_\lambda \left\| \left(g_{x_n}^{-1}(\lambda)\right)_* A_j \right\|_{C^k} = O(n^{q_k}) \qquad j \in [0, \ell].$$

Proof. By definition we have

$$\left(g_{x_n}^{-1}(\lambda) A\right)_m = \left(g_{x_n}^{-1}(\lambda)\right)'(m_1) A_{m_1}$$

where $m_1 = g_{x_n}(\lambda)_m$.

Now by 5.4, taking $a = n^{1/2+\varepsilon}$

$$\sup_\lambda \left\| \left(g_n^{-1}(\lambda)\right)^* u^i \right\|_{C^k} = O(n^{1/2+\varepsilon}) \quad \text{a.s..}$$

Using 5.5 we deduce

$$\sup_\lambda \left\| (g_n(\lambda))^* u^i \right\|_{C^k} = O(n^{1/2+\varepsilon}) \quad \text{a.s..}$$

The derivatives of $\left(g_{x_n}^{-1}(\lambda)\right) A_j$ can be computed in terms of products of derivatives of $(g^{-1})^* u^i$ or $(g)^* u^i$, so 5.6. follows. □

5.7. Proposition. *Given a smooth vector field A there exists a constant c_s such that*

$$\| \exp(\tau A) f - f \|_{\mathcal{H}_s} \le c_s |\tau| \; \| f \|_{\mathcal{H}_{s+1}}.$$

Proof. We shall first prove the statement for $s = 0$.

Denote by δA the divergence of A relative to Lebesgue measure

$$\delta A = -\sum_{i=1}^{d} \partial_i A^i.$$

By our assumption, $\delta A \in L^\infty(M)$. Denote $\varphi_t(m) = exp(\tau A)(m)$. We apply Cruzeiro's lemma and we get

$$(\varphi_t) * dm = \exp\left(- \int_0^t (\delta A)\,(\varphi_{-t}(m_0))\, dt \right) dm_0 \doteq k_t(m_0) dm_0.$$

From this formula we get immediately

$$\|k_t\|_{L^\infty} \le \exp\left(t \|\delta A\|_{L^\infty}\right),$$
$$\|k_t - 1\|_{L^\infty} = 0(t) \qquad t \to 0, \quad \text{and}$$
$$\|\varphi_t^* f\|_{L^2}^2 \le \exp\left(\frac{t}{2}\, \|\delta A\|_{L^\infty} \right) \|f\|_{L^2}^2$$

□

$$f\,(\varphi_t(m)) - f(m) = t \int_0^1 (\langle df, A\rangle\, (\varphi_{t\tau}(m))\, d\tau$$

$$\int_M |f\,(\varphi_t(m)) - f(m)|^2\, dm \le t^2 \int_M \int_0^1 |\langle df, A\rangle(\xi)|^2 k_\tau(\xi) d\xi$$

$$\|\varphi_t^* f - f\|_{L^2} \le t \|f\|_{\mathcal{H}_1} \|A\|_{L^\infty} \exp\left(t \|\delta A\|_{L^\infty}\right).$$

The proposition is established for $s = 0$. Now for $s = 1$ we have to express the Jacobian of the flow φ_t from the resolvant equation:

$$\frac{dJ}{di} = \partial_. A' J, \qquad J(0) = \text{Identity}.$$

As $\|\partial_. A\| \le c$ we deduce that

$$\|J(m,t) - I\| \le \exp(ct) \qquad t > 0$$

and the proof follows the same lines that for $s = 0$. The case $s = 2$ is obtained differentiating twice the equation $\dot{m} = A_m$ and so on. □

5.8. *Proof of the existence of the limit*

We denote

$$\begin{cases} \Phi(z, Z) = \exp(-z) \exp\left(\frac{1}{2} z + Z\right) \exp\left(\frac{1}{2}\right) \\ u_\lambda = \rho_s\left(g_n^{-1}(\lambda) g_{n+1}(\lambda)\right) u \end{cases}$$

then u_λ satisfies the recurrence system

$$u_{\lambda^+} = \rho_s \left(\Phi\left(A^\lambda(\sigma_\lambda), A^\lambda(\Gamma_\lambda)\right)\right) u_\lambda.$$

5.8.1. Main lemma. *When z, $Z \to 0$ for the topology C^∞, we have for all k:*

(i) $$\Phi(z, Z) = \exp\left(\frac{1}{2}\,[Z, z]\right)\exp(r)$$

$$\text{where}\quad \|r\|_{C^k} \leq c\|z\|^3_{C^{k+1}} + c\|Z\|^3_{C^{k+1}}$$

Consider the discrete analogue of (3.2.4):

$$I_\lambda^{i,j} = \frac{1}{2}\left(\sigma_\lambda^i \Gamma_\lambda^j - \sigma_\lambda^j \Gamma_\lambda^i\right).$$

Then, $\forall \beta < \frac{3}{2}$, we have almost surely

(ii) $$\Phi\left(A^\lambda(\sigma_\lambda), A^\lambda(\Gamma_\lambda)\right) = \exp\left(\sum_{i<j} I_\lambda^{i,j}[A_j^\lambda, A_i^\lambda]\right)\exp\left(r_1(\lambda)\right)$$

$$\text{where}\quad \sup_\lambda \|r_1(\lambda)\|_{C^k} = O(2^{-\beta n}).$$

Proof. We shall first sketch a formal approach to (i) using formal Taylor series, modulo terms of order at least 3:

$$\exp(-z) \equiv I - z + \frac{1}{2}z^2,$$

$$\exp\left(\frac{z}{2} + Z\right) \equiv I + \frac{z}{2} + Z + \frac{1}{2}\left(\frac{z}{2} + Z\right)^2,$$

$$\exp\left(\frac{z}{2} - Z\right) \equiv I + \frac{z}{2} - Z + \frac{1}{2}\left(\frac{z}{2} - Z\right)^2.$$

By multiplying these identities we find:

$$\begin{aligned}\Phi(z, Z) &= \exp(-z)\exp\left(\frac{z}{2} + Z\right)\exp\left(\frac{z}{2} - Z\right)\\ &\equiv I + \frac{1}{2}(Zz - zZ) \equiv \exp\left(\frac{1}{2}\,[Z, z]\right).\end{aligned}$$

This formal analytic approach does not provide immediately the desired estimate for the remainder term.

We shall proceed in the differential geometry context; we introduce two auxiliary parameters $\tau_1, \tau_2 \in [0, 1]$, we fix $\xi_0 \in T^d$ and we consider the function

$$\varphi(\tau_1, \tau_2) = \Phi(\tau_1 z, \tau_2 Z)(\xi_0);$$

then $\varphi(0, \tau_2) = \varphi(\tau_1, 0) = \xi_0$; therefore

$$\varphi(1, 1) - \xi_0 = \int_0^1 \int_0^1 \frac{\partial^2 \varphi}{\partial \tau_1 \partial \tau_2}\, d\tau_1 d\tau_2.$$

The derivatives on the right could be computed using 2.1.5; we avoid these lengthy computations by making the following remarks:

a) the second derivative $\partial^2\varphi$ depends continuously in the C^k norm on the C^{k+1} norm of the data;
b) the formal calculus proves that

$$\left.\frac{\partial^2\varphi}{\partial\tau_1\partial\tau_2}\right|_{\tau_1=\tau_2=0} = [Z, z].$$

These remarks imply (i). We obtain (ii) using (i) and the following upper bounds which, for n large enough, are true almost surely

$$|\sigma_\lambda| \leq 2^{-n\gamma}, \qquad |\Gamma_\lambda| \leq 2^{-\gamma},$$

where γ is fixed , $\gamma < \frac{1}{2}$. □

5.8.2. Conclusion

We shall bootstrap using the exponential decay 5.8.1 and the polynomial estimates obtained in 5.6.

Denote by $\mathcal{G}_\lambda$ the filtration generated by Γ_λ, with $\lambda' < \lambda$ and by the $x_n(\cdot)$. Then

$$E^{\mathcal{G}_\lambda}(u_{\lambda^+}) = E^{\Gamma_\lambda}(\rho_s(\varphi(A(\sigma_\lambda), A(\Gamma_\lambda)))u_\lambda.$$

(i) $$\left\| E^{\Gamma_\lambda}(\rho_s(\Phi(A^\lambda(\sigma_\lambda), A^\lambda(\Gamma_\lambda))u_\lambda) - u_\lambda \right\|_{\mathcal{H}_s} \leq c_n\|\sigma_\lambda\| \, \|u_\lambda\|_{\mathcal{H}_s}$$

with $c_n = c\, n^q 2^{-n}$.

The evaluation of

$$\sup_\lambda \|u_\lambda\|_{\mathcal{H}_s}$$

will be made using the same argument as in 5.4.1. Denote $M(\lambda) = \|u_\lambda\|^2_{\mathcal{H}_s}$, then

$$E^{\mathcal{G}_\lambda}\left(M(\lambda^+)\right) \leq (1 + 2^{-n}c)M(\lambda)$$

where c is chosen such that

$$E^{\mathcal{G}_\lambda}(\rho_s^*(\Phi(A(\sigma_\lambda), A(\Gamma_\lambda))\rho_s(\Phi(A(\sigma_\lambda), A(\Gamma_\lambda))) \leq (1 + 2^{-n}c) \quad \text{(identity)}$$

then (i) implies that

$$\sup \|u_\lambda - u_0\| \leq c_n \ \sup \|\sigma^\lambda\| = O(2^{\frac{-n}{2+\varepsilon}}).$$

We have $u_0 = u$. Furthermore

$$\sup_\lambda \|\rho_s(g_{n+1}(\lambda)u - \rho_s(g_n(\lambda))u\|_{\mathcal{H}^s} \leq \sup_\lambda \|\rho_s(g_n(\lambda))\|_{\mathrm{End}(\mathcal{H}^s)} \sup_\lambda \|u_\lambda - u\|_{\mathcal{H}^s}.$$

As, almost surely by 5.4, the norm of ρ_s grows at most polynomially, we deduce that almost surely, for all dyadic rational μ, $(g_n(\mu))_* u$ converges in $\mathcal{H}^s$. Convergence for $\forall\tau \in [0, 1]$ follows from the almost sure estimate

$\|\rho_s(g_n^{-1}(\tau^-)g_n(\tau)\|_{\mathrm{End}(\mathcal{H}^s)} = O(2^{-n/3})$, where τ^- is the largest integer multiple of 2^{-n} to the left to τ. Furthermore this convergence is uniform in τ.

Applying this result to the vector fields $-A_k$ we find that almost surely $\left(g_n^{-1}(t)\right)_* u$ converges in $\mathcal{H}^s$. We deduce a uniform bound on the Jacobian of $g_n^{-1}(t)$. We can apply now 4.2 to see that $g_\infty(t) = \lim g_n(t)$ is a diffeomorphism of T^d.

By considering the preceding estimates

$$\sup_n E\left(\|g_{x_n}(t)\|^p_{\mathcal{H}^s}\right) < \infty \qquad \forall p < \infty;$$

these inequalities imply that $\left\{\|g_{x_n}(t)\|^{p_0}_{\mathcal{H}^s}\right\}$ is uniformly integrable: almost-sure convergence implies convergence in L^{p_0}, $\forall p_0 < \infty$ that is convergence in $L^{\infty-0}(X; \mathcal{H}^s)$.

5.9. Identification of the limit

We write the solution of the differential equation associated to x_n in the following integral form

$$\xi_{x_n}(\tau) = \xi_0 + \int_0^\tau A_0\left(\xi_{x_n}(\lambda)\right) d\lambda + \int_0^\tau A_j\left(\xi_{x_n}(\lambda)\right) \dot{x}_n^j(\lambda) d\lambda.$$

The first integral converges to

$$\int_0^\tau A_0\left(\xi_x(\lambda)\right) d\lambda.$$

The second integral will be written as $I_1 + I_2 + I_3$ where

$$\begin{aligned}
I_1(\tau) &= \int_0^{\tau^+} A_j\left(\xi_{x_n}(\lambda^-)\right) dx_n^j(\lambda) \\
I_2(\tau) &= \int_0^{\tau^+} \left[A_j\left(\xi_{x_n}(\lambda)\right) - A_j\left(\xi_{x_n}(\lambda^-)\right)\right] dx_n^j(\lambda) \\
I_3(\tau) &= \int_\tau^{\tau^+} \left[A_j\left(\xi_{x_n}(\lambda)\right)\right] dx_n^j(\lambda).
\end{aligned}$$

The last integral is $O\left(x_n(\tau^+) - x_n(\tau)\right)$ and tends to zero uniformly.

We denote by $\psi_n(\tau)$ the following adapted processes

$$\psi_{n,j}(\lambda) = \sum_{\lambda \in S_n} A_n\left(\xi_{x_n}(\lambda^-)\right) \mathbf{1}_{[\lambda^-, \lambda^+[}$$

then

$$E\left(\int_0^1 \|\psi_{n,j}(\lambda) - A_j(\xi_x(\lambda))\|^2 d\lambda\right) \to 0.$$

This relation implies by the energy identity for Itô's stochastic integral that

$$I_1(\tau) = \int_0^{\tau^+} \psi_{n,j}(\lambda) dx_n^j(\lambda)$$

converges in $L^2(X)$ to

$$\int_0^{\tau} A_j\,(\xi_x(\lambda))\, dx^j(\lambda).$$

The Stratonovich SDE appears in Itô's form as

$$d\xi_x = A_0 d\tau + \sum_j A_j dx^j + \frac{1}{2} \sum_j \partial_. A_j A_j^. d\tau.$$

Therefore the following lemma will end the proof:

Lemma.

$$\lim I_2 = \frac{1}{2} \int_0^{\tau} \sum_{k=1}^{\ell} \left(\partial_k A_j\right) A_j^k.$$

Proof.

$$A_j\left(\xi_{x_n}(\lambda)\right) - A_j\left(\xi_{x_n}(\lambda^-)\right) = \partial_k A_j\left(\xi_{x_n}(\lambda^-)\right)\left(\xi_{x_n}^k(\lambda) - \xi_x^k(\lambda^-)\right) + r_1$$

$$\xi_{x_n}^k(\lambda) - \xi_{x_n}^k(\lambda^-) = \sum_j A_j\left(\xi_{x_n}(\lambda^-)\right)\left(x_n^j(\lambda) - x_n^j(\lambda^-)\right) + r_2,$$

where almost surely $r_i = O(2^{-n+\varepsilon})$, $i = 1, 2$.

$$\int_{\lambda^-}^{\lambda^+} A_j\left(\xi_x(\lambda^-)\right)\left(x_n^j(\lambda - x_n^j(\lambda^-)\right) dx^k(\lambda) = A_j\left(\xi_{x_n}(\lambda^-)\right) I(x, \lambda^+, \lambda^-)$$

where

$$I(x, \lambda^+, \lambda^-) = \int_0^{\lambda} \left(x^j(\lambda) - x^j(\lambda^-)\right)\left(x^k(\lambda^+) - x^k(\lambda^-)\right) \frac{d\lambda}{\lambda^+ - \lambda^-}.$$

We have

$$\begin{aligned} E\left(I(x, \lambda^+, \lambda^-)\right) &= 0 \quad \text{if} \quad k \neq j \\ &= \frac{1}{2}(\lambda^+ - \lambda^-) \quad \text{if} \quad k = j \\ E\left(\left[I(x, \lambda^+, \lambda^-) - E\left(I(x, \lambda^+, \lambda^-)\right)\right]^2\right) &= \frac{1}{2}(\lambda^+ - \lambda^-)^2 \end{aligned}$$

which proves the lemma. □

6. Extension of the limit theorem

We shall extend the result of the previous section to $\mathbb{R}^d$ and to an arbitrary compact manifold.

6.1. Theorem. *Consider, on $\mathbb{R}^d$, $(\ell+1)$ C^∞- vector fields with bounded derivatives of all orders. Consider the flow $U_{x_{n,\tau}}(\xi_0)$ on $\mathbb{R}^d$ defined by*

$$\left|\begin{array}{lll} d\xi_{x_n} & = & (\xi_{x_n})d\tau + \sum_{s=1}^{\ell} A_s(\xi_{x_n})dx_n^s \\ \xi_n(0) & = & \xi_0 \end{array}\right. .$$

Then a.s. $U_{x_{x,.}}(\cdot)$ converges uniformly (as do all its derivatives in ξ) on every compact subset of $[0, 1] \times \mathbb{R}^d$.

Denote

$$\xi_x(t) = \lim_n U_{x_{n,t}}(\xi_0),$$

then $\xi_n(\tau)$ satisfies the following Stratonovich SDE

$$d\xi_x = A_0(\xi_x)d\tau + A_j(\xi_x)odx^j.$$

Proof. We denote by R a fixed constant; we fix φ a smooth function of one variable such that $\varphi(t) = 1$ if $|t| < \frac{1}{2}$, $\varphi(t) = 0$ if $|t| \geq 1$. We denote $\Phi(\xi) = \prod_{i=1} \varphi\left(\frac{\xi_i}{M}\right)$. Then ΦA_j defines a periodic smooth vector of period $2R$. Denote by $\gamma_R : \mathbb{R}^d \to \mathbb{T}^d$ the map which associates to ξ the class of $\frac{\xi}{2R}$ modulo Z^d. Then γ_M restricted to $|\xi| < \frac{R}{2}$ is a diffeomorphism $\tilde{\gamma}_M$. Therefore the vector field $(\gamma_M)_* \Phi A_j = A_j^R$ is a smooth C^∞ vector field on $\mathbb{T}^d$. Furthermore

$$\|A_j^R\|_{C^k(\mathbb{T}^d)} \leq \frac{d_k}{M}.$$

Let K be a given compact set, then for M large enough

$$\text{distance } (\tilde{\gamma}(\xi), 0) < \frac{1}{4} \quad \text{for} \quad \xi \in K.$$

By the limit theorem on the torus, there exists an integer k and a function ψ with $\psi(t) \to 0$ when $t \to 0$, such that

$$P\left(\sup_{n,\eta,\tau}\left[\text{distance}\left(U^Z_{x_{n,\tau}}(\tilde{\eta}), \tilde{\eta}\right) > \frac{1}{4}\right]\right) \leq \psi(\|Z\|_{c^k}).$$

Therefore we can find a sequence R_s such that

$$P\left(\sup_{n,\eta,\tau}\left[\text{distance}\left(U^{A^{R_s}}_{x_{n,\tau}}(\tilde{\eta}), \tilde{\eta}\right)\right] > \frac{1}{4}\right) \leq n^{-2}.$$

Then almost surely for n and s large enough

$$\gamma\left(U_{x_{n,\tau}}(\eta)\right) = U^{A^{R_s}}_{x_{n,\tau}}(\gamma(\eta)) \qquad \eta \in K,$$

the limit theorem for the torus will then prove the limit theorem on R^d. □

6.2. Theorem. *Denote by M a compact manifold, $A_j(\ell+1)$ smooth vector fields, then $U_{x_{n,\cdot}}(\cdot)$ converges uniformly (as do its spatial derivatives). Furthermore $U_{x_{n,\tau}}(m_0)$ is the solution of the corresponding Stratonovich SDE.*

Proof. Following J.M. Bismut we will prove this theorem by an embedding.

6.3. Lemma. *Given a C^∞ compact manifold M, of dimension d there exists a C^∞ map of M into R^m $u = (u^1, \dots, u^m)$ such that its image $u(M)$ is a C^∞ submanifold of R^m and u is a diffeomorphism of M to $u(M)$.*

Proof. A chart ψ is a diffeomorphism of an open set of M to an open set of $\mathbb{R}^d$. By compactness we can find an atlas with a finite number $\psi_1 \dots \psi_q$ of charts.

We can construct a partition of unity subordinated to the open covering of this atlas. Then $p_i \psi_i = u_i$ are smooth C^∞ functions defined on M with values in $\mathbb{R}^d$. We denote $u = (u_1 \dots u_q)$.

Given $m_0 \in M$ there exists a neighbourhood of m_0 and d preferred coordinates $u^{j+1} \dots u^{j+d}$ such that all other coordinate functions can be written nearby m_0 as C^∞ functions of the preferred coordinate functions. □

6.4. *Proof of Theorem 6.2*

We identify M with $u(M)$. At each point $m_0 \in M$, the tangent plane $T_{m_0}(M)$ is a subspace of R^m. The orthogonal complement $T_{m_0}^{\perp}(M)$ is generated by the family of straight lines which are orthogonal to M. On a sufficiently small neighbourhood Ω of M there exists a unique normal to M passing through a given point ξ; the intersection $p(\xi)$ of the normal with M defines a map $p : \Omega \to M$. Then we define a vector $\tilde{A}_j$ on Ω by the condition

$$p'(\xi)\tilde{A}_j(\xi) = A_j(p(\xi))$$

$\tilde{A}_j(\xi)$ is orthogonal to $T_{p(\xi)}(M)$.

Using a partition of unity we extend $\tilde{A}_j$ to a C^∞ vector field $\hat{A}_j$ which is zero outside a compact set and such that

$$\hat{A}_j(\xi) = A_j(\xi) \quad \text{for} \quad \xi \in M.$$

By applying Theorem 6.1 to $\hat{A}_j$ we get Theorem 6.2. □

Remark. As the vector fields $\hat{A}_j$ have compact support we can also construct immediately a periodic C^∞ extension and apply directly Theorem 6.1.

7. Cauchy problem for SDE under Lipschitz hypothesis

The limit theorem leads immediately to an existence proof for the Cauchy problem for an SDE associated to C^∞ vector fields.

We want to establish this existence under weaker hypotheses: we shall assume that the vector fields A_j, defined on $\mathbb{R}^d$, satisfy the following global Lipschitz condition

$$\|A_j(\xi) - A_j(\xi')\| \leq c\|\xi - \xi'\|. \tag{7.1}$$

7.2. Theorem. *Under hypothesis (7.1), the Cauchy problem for the Itô SDE*

$$d\xi_x(t) = \sum_{j>0} A_j\,(\xi_x(t))\,dx^j(t) + A_0\,(\xi_x(t))\,dt, \qquad \xi_x(0) = a \tag{7.2.1}$$

has a unique solution.

Proof. Uniqueness was established in 1.5.

We transform (7.2.1) into the equivalent integral equation

$$\begin{aligned} \xi_x(t) &= a + \int_0^t \sum_{j\geq 1} A_j\,(\xi_x(\tau))\;dx^j(\tau) + \int_0^t A_0\,(\xi_x(\tau))\;d\tau, \\ &\quad \xi_x(t)\ \mathcal{N}^t - \text{measurable}, \end{aligned} \tag{7.2.2}$$

where the first integral is an Itô stochastic integral and where the second integral is a Lebesgue integral. Classically, the Picard iteration scheme is used to provide the existence of a solution to (7.2.2): this method works beautifully; for the sake of a unified approach we shall proceed from the C^∞ case.

We fix a C^∞ function v, defined on $\mathbb{R}^d$, with compact support, having integral 1. We choose a sequence $\varepsilon_s \to 0$ and we write $v_s(\xi) = \varepsilon_s^{-d} v(\varepsilon_s^{-1}\xi)$. Assume provisionally that

$$\sup_{\xi, j} \|A_j(\xi)\| < \infty. \tag{7.2.3}$$

Then

$$A_{j,s}(\xi) = (A_j * v)_s(\xi) = \int_{R^d} A_j(\xi - \lambda) v_s(\lambda)\;d\lambda$$

is a C^∞ vector field, with bounded derivatives; we write equation (7.2.1) for the $A_{j,s}$ and we transform this Itô SDE $(7.2.1)_s$ into a Stratonovich SDE; to this we can apply Theorem 6.1, which provides a solution $\xi_{x,s}(*)$ of $(7.2.1)_s$. Define

$$\delta_s(x, t) = \|\xi_{x,s}(t) - \xi_{x,s-1}(t)\|^2;$$

the energy identity applied to (7.2.2) gives

$$E\,(\delta_s(x, t)) \leq 2E\left[\int_0^t \sum_{j\geq 0} \|A_{j,s}\left(\xi_{x,s}(\tau)\right) - A_{j,s-1}\left(\xi_{x,s-1}(\tau)\right)\|^2\;d\tau\right].$$

Introduce $\gamma_s(t) = E\,(\delta_s(x, t))$ and use hypothesis (7.1) to see that there exists a constant c such that

$$\gamma_s(t) \leq c\epsilon_s^2 t + c\int_0^t \gamma_s(\tau)\;d\tau, \qquad \gamma_s(0) = 0,$$

which implies that

$$\sup_{0<t<1} \gamma_s(t) \leq (e^c - 1)\varepsilon_s^2;$$

we deduce that

$$\|\xi_{*,s}(t) - \xi_{*,s-1}(t)\|_{L^2(X)} \leq c_1 \varepsilon_s.$$

We choose $\{\varepsilon_s\}$ such that $\sum \varepsilon_s < \infty$, then $\lim \xi_{*,s}(t) = \xi_{*,\infty}(t)$ exists in $L^2(X)$; passing to the limit in (7.2.2) we find that $\xi_{x,\infty}$ satisfies (7.2.2). Finally we get rid of the extra hypothesis (7.2.3) by a truncation procedure as in in 6.1. □

7.3. Theorem. *The Itô map $x \to \xi_x$ takes Wiener measure $\mu \to \nu$; this measure ν is a solution of the martingale problem for the operator*

$$\Gamma = \frac{1}{2} \sum_{i,j} a^{i,j} \frac{\partial^2}{\partial \xi^i \partial^j} + \sum_i A_0^i \frac{\partial}{\partial \xi^i} \quad \text{where} \quad a^{i,j} = \sum_{1 \leq k \leq n} A_k^i A_k^j.$$

Proof. Given $f \in C^2(\mathbb{R}^d)$, apply Itô calculus to the semimartingale $t \to f(\xi_x(t))$. □

8. Critique of our approach

The lifting of an SDE on M to its diffeomorphism group has the following advantages:

i) elimination of the initial condition in the data of the Cauchy problem,
ii) an existence result at the level of the diffeomorphism group.

In Section 2 we have treated the driving vector fields A_j on M as elements of the Lie algebra of G the group of diffeomorphisms of M. In this framework we can write on G the following infinite-dimensional SDE

$$(8.1) \qquad dg_x(t) = \left(\sum_k A_k \circ dx^k(t) \right) g_x(t), \quad g_x(0) = \text{Identity}.$$

This SDE in infinite dimension has an appealing intuitive meaning; the limit of g_{x_n} constructed in 5.2 is solution of (8.1). The equation (8.1) will be considered in XI.1.1 for compact groups.

In this finite dimensional case the proof of the existence of a solution of (8.1) could be done along the lines of a Picard iteration solving the integral equation (7.2.2) *dependent upon the parameter a*. This approach would be easier than the proof of the limit Theorem 5.2.

The limit theorem induces an *embedding* of the theory of SDE into the theory of ODE: then constructions and theorems in the context of ODEs are automatically *transferred* by the limit theorem to SDEs. The limit theorem appears therefore as the corner stone of the theory of SDE: it gives much more information than a naked existence result for the Cauchy problem.

There exist many other approaches to the limit theorem (see the Notes at the end of this part). We have chosen a functional approach which emphasizes the underlying algebraic structure of the Lie algebra generated by the driving vector fields. The disadvantage of this approach is that it works at the C^∞ level; results in a C^k framework can be reached through some smoothing procedure as in Section 7: this indirect approach is laborious and will not lead to optimal differentiability results.

Chapter IX

Elliptic Estimates Through Stochastic Analysis

Contents: Variation of an ODE: linearized ODE determining the Jacobian matrix – The control map, its Jacobian – Jacobian of a stochastic flow – Higher derivatives of a stochastic flow – Itô functionals, their differentiability – Malliavin matrix of an Itô functional – Reduced variation of an Itô functional – Reduced variation and regularity in the backward variables – Bismut's identity – Hörmander's hypoellipticity theorem.

The first section will be devoted to the classical calculus variations for an ordinary differential equation and its associated Jacobian linear ODE.

Using infinite dimensional analysis on the diffeomorphism group we have estabished in Chapter VIII the existence of a smooth stochastic flow associated to an SDE with C^∞ coefficients. We want to revisit this question in a concrete way, making explicit computations in view of specific applications to parabolic regularity of PDE.

The transfer principle allows us to transport these results to smooth stochastic differential equations: the *Itô functional* defined by the solution of a smooth SDE is the limit of the *control map*: the transfer principle will be used in Section 2 where the differentiability of Itô's functional in the starting point will be computed explicitly.

In Section 3 we shall show that Itô functionals associated to a smooth SDE belong to D_∞.

Chapter III can be applied to study the regularity of the fundamental solution of the heat equation.

This will be done in the last section for the study of Hörmander hypoellipticity.

1. Variation of an ordinary differential equation

1.1. Notation

We shall work with non-autonomous vector fields $Z(\xi, \tau)$ defined on $\mathbb{R}^d$. We shall suppose that for fixed ξ, the partial function in τ is continuous on the closed intervals generated by a finite partition of [0, 1]. We shall suppose that Z and all its derivatives in ξ up to an appropriately high order are bounded uniformly $\xi \in \mathbb{R}^d$.

We will denote by

$$U_{\tau \leftarrow \tau_0}(\xi_0)$$

the solution of the following Cauchy problem:

(i)
$$\left[\begin{array}{lcl} \dfrac{d}{d\tau} U_{\tau \leftarrow \tau_0}(\xi_0) & = & Z\left(U^Z_{\tau \leftarrow \tau_0}(\xi_0), \tau\right) \\ U^Z_{\tau_0 \leftarrow \tau_0}(\xi_0) & = & \xi_0 \end{array}\right. .$$

We have then the classical *flow property*

(ii)
$$U^Z_{\tau_2 \leftarrow \tau_1} \circ U^Z_{\tau_1 \leftarrow \tau_0} = U^Z_{\tau_2 \leftarrow \tau_0}.$$

1.2. The equation for the Jacobian

We want to differentiate $U^Z_{\tau \leftarrow \tau_0}$ with respect to the *initial value*; more precisely, given ξ_0, $a \in \mathbb{R}^d$, we want to compute

$$D_a\left(U^Z_{\tau \leftarrow \tau_0}(\xi_0)\right) = \left\{\frac{d}{d\varepsilon} U_{\tau \leftarrow \tau_0}(\xi_0 + \varepsilon a)\right\}_{\varepsilon = 0}.$$

For fixed τ, ξ_0 the map

$$a \mapsto D_a\left(U^z_{\tau \leftarrow \tau_0}(\xi_0)\right)$$

is a linear map of $\mathbb{R}^d$ into $\mathbb{R}^d$. We denote by $J_{\tau \leftarrow 0}(\xi_0)$ the corresponding $d \times d$ matrix.

If we differentiate the differential system 1.1(i) in ε and if we interchange the order of differentiation in ε and τ, we get:

$$\frac{d}{d\tau} J^Z_{\tau \leftarrow \tau_0}(\xi_0) \cdot a = \left\langle (D_{.}Z)\left(U_{\tau \leftarrow \tau_0}(\xi_0), \tau\right),\ J_{\tau \leftarrow \tau_0}(\xi_0) \cdot a\right\rangle.$$

Denote by $M(\xi, \tau)$ the $d \times d$ matrix defined by

$$M^j_k(\xi, \tau) = \frac{\partial}{\partial \xi^k}\left(Z^i(\xi, \tau)\right),$$

then

$$\frac{d}{d\tau} J^z_{\tau \leftarrow \tau_0}(\xi_0) = M\left(U^Z_{\tau \leftarrow \tau_0}(\xi_0)\right) J^Z_{\tau \leftarrow \tau_0}(\xi_0).$$

Consider the matrix valued differential equation

(i)
$$\frac{dX}{d\tau} = M\left(U^Z_{\tau \leftarrow \tau_0}(\xi_0)\right) X,$$

then the Jacobian will be defined by the two properties

(ii)
$$\left\{\begin{array}{l} J^Z_{\tau \leftarrow \tau_0} \quad \text{satisfies (i)} \\ J^Z_{\tau_0 \leftarrow \tau_0}(\xi_0) = \text{Identity} \end{array}\right. .$$

By differentiating 1.1(ii) we get

(iii)
$$J^Z_{\tau_2 \leftarrow \tau_0} = J^Z_{\tau_2 \leftarrow \tau_1} \cdot J^Z_{\tau_1 \leftarrow \tau_0}.$$

1.3. Jacobian of the control map

We are given $(n+1)$-smooth vector fields $A_0, A_1, \ldots, A_n$ on $\mathbb{R}^d$. A *control* will be the data of an R^n-valued piecewise C^1 function f of τ. We associate to f the vector field

$$Z^f(\xi, \tau) = \sum_{q=1}^{n} \dot{f}^q(\tau) A_q(\xi) + A_0(\xi).$$

We shall denote by $U^f_{\tau \leftarrow \tau_0}$ the flow associated to Z_f. The map

$$\xi_0 \mapsto U^f_{\tau \leftarrow \tau_0}(\xi_0)$$

is the *control map*.

We shall apply to the control map the results of the preceding paragraph; we introduce the $(d \times d)$ matrix valued functions $M_0, M_1, \ldots, M_n$ defined by

$$\left(M_q\right)^j_k = \frac{\partial}{\partial \xi^k}\left(A^j_q\right).$$

We define then

$$M^f(\xi, \tau) = \sum_{q=1}^{n} \dot{f}^q(\xi) M_q(\xi) + M_0(\xi),$$

then the Jacobian $J^f_{\tau \leftarrow \tau_0}(\xi_0)$ of $U^f_{\tau \leftarrow \tau_0}(\xi_0)$ is given by solving the following matrix valued differential equation

(i) $$\frac{d}{d\tau}\left(J^f_{\tau \leftarrow \tau_0}(\xi_0)\right) = M^f\left(U^f_{\tau \leftarrow \tau_0}(\xi_0), \tau\right) J^f_{\tau \leftarrow \tau_0}(\xi_0).$$

We close by the following obvious observation:
(ii) *The Jacobian equation is obtained from the control equation by replacing the vector fields A_q by the matrices M_q.*

1.4. Differentiability of the control map

Given an $\mathbb{R}^d$-valued C^1 function h we define

$$D_h U^f_{\tau \leftarrow 0} = \lim \varepsilon^{-1}\left[U^{f+\varepsilon h}_{\tau \leftarrow 0} - U^f_{\tau \leftarrow 0}\right].$$

Proposition.

$$D_h U^f_{\tau \leftarrow 0}(\xi_0) = \int_0^\tau \sum_{q=1}^{n} J^f_{\tau \leftarrow \lambda}\left(A_q(U^f_{\lambda \leftarrow 0}(\xi_0)\right) \dot{h}^q(\lambda) d\lambda.$$

Proof. Denote

$$q(\tau) = D_h U^f_{\tau \leftarrow 0}(\xi_0).$$

We have

$$\frac{d}{d\tau} U^{f+\varepsilon h}_{\tau \leftarrow 0} = Z^{f+\varepsilon h}\left(U^{f+\varepsilon h}_{\tau \leftarrow 0}\right).$$

We differentiate in ε and interchange $\frac{d}{d\tau}$ with $\frac{d}{d\varepsilon}$ on the l.h.s.; we get

$$\begin{cases} \dfrac{d}{d\tau} q = Z^h(\xi(\tau)) + \partial Z^f \cdot q \\ q(0) = 0 \end{cases}.$$

As

$$Z^h(\xi(\tau)) = \sum_{q=1}^{n} A_q(\xi(\tau))\dot{h}^q(\tau)$$

we get the result. □

2. Differentiation of the stochastic flow

2.1. Notation

We consider $(n+1)$ vector fields $A_0, A_1, \ldots, A_n$ on $\mathbb{R}^d$ and the following Stratonovich SDE

(i)
$$d\xi_x = \sum_{q=1}^{n} A_q(\xi_x) \; o dx^q + A_0(\xi_x) d\tau$$

where $x \in X$, the probability space of the Brownian motion on $\mathbb{R}^n$.

We denote by

$$U^x_{\tau \leftarrow \tau_0}(\xi_0) = \xi_x(\tau) \qquad \tau > \tau_0$$

where ξ_x satisfies (i) and

(ii)
$$\xi_x(\tau_0) = \xi_0.$$

2.2. Theorem. *Assume that A_q are bounded (as are their first and second derivatives), then the map $U^x_{\tau \leftarrow \tau_0}$ is C^1 and its Jacobian $J^x_{\tau \leftarrow \tau_0}$ satisfies the following Stratonovich SDE*

(i)
$$d_\tau \left(J^x_{\tau \leftarrow \tau_0}(\xi_0)\right) = \left[\sum_{q=1}^{n} M_q\left(U^x_{\tau \leftarrow \tau_0}(\xi_0)\right) \; o dx^q + M_0\left(U^x_{\tau \leftarrow \tau_0}(\xi_0)\right) d\tau\right] J^x_{\tau \leftarrow \tau_0}(\xi_0).$$

Proof. We shall prove this theorem for C^∞-bounded vector fields: the general case will be reduced to this special case by a smoothing procedure as in VIII.7.

We denote by x_m the Stroock-Varadhan approximation of x. Then by the limit theorem

$$\lim_m U^{x_m}_{\tau\leftarrow\tau_0}(\xi_0) = U^x_{\tau\leftarrow\tau_0}(\xi_0), \quad \text{uniformly in} \quad \xi_0 \in K, K \quad \text{compact.}$$

We can differentiate the control map U^{x_m}; then 1.3(ii) shows that its Jacobian $J^{x_m}_{\tau\leftarrow\tau_0}$ satisfies the following ODE

$$\frac{d}{d\tau}\left(J^{x_m}_{\tau\leftarrow\tau_0}\right) = \left(M^{x_m} + M_0\right) J^{x_m}_{\tau\leftarrow\tau_0}.$$

2.3. Lemma.

$$\lim_m J^{x_m}_{\tau\leftarrow\tau_0}(\xi_0) = J^x_{\tau\leftarrow\tau_0}(\xi_0)$$

uniformly in ξ_0 belonging to a compact set.

Proof. The difficulty in applying the limit theorem is that the Lipschitz constant for a linear SDE is bounded by the sup of the solution; this bound is not an *a priori* bound. We have to proceed as in VIII.6.1.

3. Differentiability of Itô functionals

3.1. Itô functional

We shall call an Itô functional a functional of the form

(i) $$I(x) = U^x_{1\leftarrow 0}(\xi_0)$$

where U^x is the stochastic flow associated to the smooth SDE 2.1(i).

3.2. Partial derivatives

The space X is the probability space of the Brownian motion on $\mathbb{R}^n$; its Cameron-Martin space is

$$H^1\left([0,1];\mathbb{R}^n\right) = \bigoplus_{i=1}^n H^1([0,1],\mathbb{R}).$$

This direct sum decomposition induces a decomposition of D_τ into n components. More precisely, given an $\mathbb{R}^q$-valued smooth functional F on X, we will denote by $D_{\tau,q}F$ the $\mathbb{R}$-valued functional defined by the identity

$$D_h F = \sum_{q=1}^n \int_0^1 \left(D_{\tau,q}F\right) \dot{h}^q(\tau)d\tau.$$

3.3. Theorem. *The Itô functional 4.1(i) has derivative*

$$(D_\tau I)(x) = J^x_{1\leftarrow\tau}\left(A_q(\tau)\right).$$

Proof. We apply the transfer principle to 1.4.

3.4. Theorem. *Consider an SDE whose coefficients along with all their derivatives are bounded, then the associated Itô functional belongs to* D_∞.

Proof. We limit our proof to the case of the first derivative. Then using 3.3 we have to prove that

$$E\left[\int_0^1 E\left(\|J^x_{1\leftarrow\tau}A_q\|^2_{\mathbb{R}^n}\right)d\tau\right]^{p/2} < +\infty.$$

This inequality will result from the following lemma.

3.4.1. Lemma. *For all* $p < +\infty$, *we have*

$$E\left(\sup_\tau \|J^x_{1\leftarrow t}\|^p_{\mathrm{End}(\mathbb{R}^n)}\right) < +\infty.$$

Proof. We reverse the time defining $\tilde{x}(\tau) = x(1-\tau)$, then

$$J^x_{1\leftarrow\tau} = \exp\left(*\int_{[0,1-\tau]} - \left(M_k\, od\tilde{x}^k + M_0 d\tau\right)\right)$$

and the matrix exponential martingale of XI.1.2.2 gives the result. □

4. Malliavin covariance matrix of an $\mathbb{R}^d$-valued Itô functional

4.1. Notation

We shall consider a stochastic differential equation on $\mathbb{R}^d$ of the following form in the Stratonovich notation

(i) $$dv_x = \sum_{k=1}^n A_k\, odx^k + A_0 d\tau$$

where $A_0, \ldots, A_n$ are $(n+1)$ smooth vector fields on $\mathbb{R}^d$. Then $U^x_{\tau\leftarrow\tau_0}$ will denote the stochastic flow of diffeomorphisms of $\mathbb{R}^d$ associated to (i).

We denote by M_k the $(n+1)$ square $(d\times d)$-matrices defined by $(M_k)^B_\alpha = \partial_\alpha A^\beta_k$. Then the Jacobian J of U is defined by the following Stratonovich matrix valued SDE

(ii) $$\begin{cases} d_\tau \left(J^x_{\tau\leftarrow\tau_0}\right) = \left(\sum_{k=1}^{n} M_k \; odx^k + M_0 d\tau\right) J^x_{\tau\leftarrow\tau_0} \\ J_{\tau_0\leftarrow\tau_0} = \text{Identity} \qquad (0 \le \tau_0 \le \tau) \end{cases}.$$

We fix the starting point at v_0 and we consider the *Itô functional*

(iii) $$g_\tau(x) = U^x_{\tau\leftarrow 0}(v_0).$$

4.2. Computation of the Malliavin covariance matrix

The Itô functional g defined in 3.1 is $\mathbb{R}^d$-valued. The covariance matrix of g is the $(d \times d)$ square-valued matrix σ defined by

$$\sigma^{\alpha,\beta}_\tau = \left(Dg^\alpha_\tau \mid Dg^\beta_\tau\right)_{H^1([0,\tau];\mathbb{R}^n)} \qquad \alpha, \beta \in [1, d].$$

Using the notation of 3.2 we have

(i) $$\sigma^{\alpha,\beta}_\tau = \sum_{k=1}^{n} \int_0^\tau D_{\lambda,k}(g^\alpha) D_{\lambda,k}(g^\beta) d\lambda.$$

Furthermore by Section 3.3

$$\left(D_{\lambda,k} g\right)^\alpha = \left(J^x_{\tau\leftarrow\lambda}(A_k)\right)^\alpha.$$

Therefore the relation (i) can be written as

(ii) $$\sigma_\tau = \sum_{k=1}^{n} \int_0^\tau J^x_{\tau\leftarrow\lambda}(A_k) \otimes J^x_{\tau\leftarrow\lambda}(A_k) d\lambda$$

where we make the following convention. Given two vectors Z_1, Z_2 of $\mathbb{R}^d$ then: $Z_1 \otimes Z_2$ denotes the $(d + d)$-square matrix $Z^\alpha_1 \otimes Z^\beta_2$.

The quadratic form $q_\tau(z)$ associated to σ can be written

(iii) $$q_\tau(z) := \sigma^{\alpha,\beta}_\tau z_\alpha z_\beta = \sum_{k=1}^{n} \int_0^\tau \left[\langle J^x_{\tau\leftarrow\lambda}(A_k), z\rangle\right]^2 d\lambda$$

where $\langle Z, z\rangle := Z^\alpha z_\alpha$.

5. The reduced variation

5.1. Definition. Given the stochastic flow $U^x_{\tau\leftarrow 0}$ defined by 2.1.(i) and given an infinitesimal perturbation $\delta(\tau)$ of $U^x_{\tau\leftarrow 0}$ we write

(i) $$\tilde{R}^x_\tau(\delta) = J^x_{0\leftarrow\tau}\left(\int_0^\tau J^x_{\tau\leftarrow\lambda}(\delta(\lambda)) d\lambda\right).$$

We shall call R^x_τ the *reduced perturbation* of U^x. The reduced perturbation can also be described as in the following way:

The reduced perturbation is the perturbation of the initial condition v_0 which induces the given perturbation.

We call the *reduced variation* of the variation associated to h the reduced perturbation of the perturbation induced by h. We shall denote it by $R^x_\tau(h)$.

5.2. Theorem. *The reduced variation* satisfies *the following ordinary differential equation*

$$\frac{d}{d\tau} R^x_\tau(h) = \sum_{k=1}^{n} a_k(x,\tau)\dot{h}^k(\tau) + a_0(x,\tau)$$

where a_k are the R^d-valued adapted functionals defined by

(i) $$a_k(x,\tau) = J^x_{0\leftarrow\tau}\left(A_k\left(v_x(\tau)\right)\right).$$

The reduced variation determines the variation by the following formula

(ii) $$\frac{d}{d\varepsilon}\left\{U^{x+\varepsilon h}_{\tau\leftarrow 0}(v_0)\right\} = J^x_{\tau\leftarrow 0}\left(R^x_\tau(h)\right).$$

Proof. We have by the composition rule

$$J^x_{0\leftarrow\lambda} \circ J^x_{\tau\leftarrow\lambda} = J^x_{0\leftarrow\lambda}.$$

Therefore

$$R^x_\tau(h) = \int_0^\tau J^x_{0\leftarrow\lambda}\left(\dot{h}(\lambda)\right) d\lambda.$$

□

5.3. Remarks. (Advantages of the reduced variation)

(i) The reduced variation is built from the *adapted functionals a_k*.
(ii) In differential geometry all the reduced variations belong to the *same vector space*: the tangent space at the starting point.
(iii) The reduced variation of an SDE is given by an ODE.

5.4. The reduced covariance

5.4.1. Definition. *The reduced covariance matrix is:*

$$C^{i,j}(x) = \sum_{k=1}^{n} \int_0^1 a^i_k(x,\tau) a^j_k(x,\tau) d\tau.$$

The two covariance matrices are related by the identity

$$
\begin{aligned}
\sigma(x) &= \sum_{k=1}^{n} \int_0^1 J_{1\leftarrow\lambda}(A_k) \otimes J_{1\leftarrow\lambda}(A_k) d\lambda \\
&= \sum_{k=1}^{n} \int_0^1 J^x_{1\leftarrow 0}\,(a_k(x,\lambda)) \otimes J^x_{1\leftarrow 0}\,(a_k(x,\lambda))\, d\lambda \\
&= \left(J^x_{1\leftarrow 0} \otimes J^x_{1\leftarrow 0}\right) \mathcal{C}(x).
\end{aligned}
$$

5.4.2. Proposition. *The invertibility of σ is equivalent to the invertibility of $\mathcal{C}$. The two following statements are equivalent*

$$
(\det \sigma)^{-1} \in L^{\infty-0}, \qquad (\det \mathcal{C})^{-1} \in L^{\infty-0}.
$$

Proof. Using the same methodology as in 3.3.1 we see that for all p

$$
E\left(\| J^x_{1\leftarrow 0} \|^p\right) < +\infty \quad \text{and} \quad E\left(\| J^x_{0\leftarrow 1} \|^p\right) < +\infty.
$$

6. The reduced variation and regularity in the backward variables

We denote by $p_t(v_0, dv)$ the law of $v_x(\tau)$ where $v_x(\tau)$ is the diffusion associated to the SDE 2.1.(i). The *forward variable* is the variable dv. The main purpose of the stochastic calculus of variations is to produce estimates in the forward variable. Nevertheless estimates in the backward variable could be of interest.

6.1. Theorem. *Given a vector z at the point v_0, assume that for all τ we can find scalars $c_j(x,\tau)$ such that*

(i) $$
z = \sum_{i=1}^{n} c_j(x,\tau) a_j(x,\tau),
$$

where the vectors a_i have been constructed in Theorem 5.2, then

$$
\partial_z \left[\int_{\mathbb{R}^d} p_t(v_0, dv) f(v) \right] = \frac{1}{t}\, E\left(\dot{f}\,(v_x(t)) \int_0^t c(x,\tau) dx(\tau) \right).
$$

Proof. We have

$$
\begin{aligned}
\partial_z \left[\int_{\mathbb{R}^d} p_t\,(v_0, dv)\, f(v) \right] &= \frac{d}{d\varepsilon}\, \left[f\left(U^x_{t\leftarrow 0}\,(v_0 + \varepsilon z)\right)\right] \\
U^x_{t\leftarrow 0} &= U^x_{t\leftarrow \tau} \circ U^x_{\tau\leftarrow 0}.
\end{aligned}
$$

Define

$$
\zeta(\tau) = J^x_{\tau\leftarrow 0}(z).
$$

Then for every τ

$$\partial_z \int_{\mathbb{R}^d} p_t(v_0, dv)\, f(v) = E\left((df)(v_x(\tau)),\ J_{t\leftarrow\tau}(\zeta(t))\right).$$

The r.h.s. does not depend upon τ. We get therefore by integration in τ

$$= \frac{1}{t}\, E\left(\int_0^t \langle (df)(v_x(\tau)),\, J_{t\leftarrow\tau}(a_z(\tau))\rangle > d\tau\right).$$

Define an adapted vector field by the condition

(i) $$\dot{h}(x,\tau) = c_j(x,\tau)$$

$$\begin{aligned}
\partial_z\left[\int_{\mathbb{R}^d} p_t(v_0, dv)\, f(v)\right] &= \frac{1}{t} E\left(\left\langle\int_0^t J^x_{t\leftarrow\tau}\dot{h}\,d\tau,\ df\,(v_x(t))\right\rangle d\tau\right)\\
&= \frac{1}{t} E\left\{\frac{d}{d\varepsilon} f\left(U^{x+\varepsilon h}_{t\leftarrow 0}(v_0)\right)\right\}_{\varepsilon=0}\\
&= \frac{1}{t} E\left(f\,(v_x(t)\operatorname{div}(h))\right).
\end{aligned}$$

As h is adapted

(ii) $$\operatorname{div}(h) = \int_0^t \dot{h}(x,\tau)\,dx(\tau).$$

□

6.2. Remark. Condition (i) is reminiscent of an ellipticity condition. When Δ is elliptic, $p_t(v_0, dv) = p_t(v_0, v)\,dv$ where dv is the Lebesgue measure. We get then

$$\left(\int_{\mathbb{R}^d} \partial_z p_t(v_0, v)\, f(v)dv\right) = \int_{\mathbb{R}^d} f(v)q(v)p_t(v_0, v)\,dv$$

with

$$q(v) = E^{v_x(t)=v}\left(\int_0^t c(x,\tau)\,\frac{dx(\tau)}{t}\right)$$

which implies the Harnack type relation

$$\partial_z \log p_t(v_0, v) = q(v).$$

See XI.6.5 for a geometric version of this Harnack inequality.

6.3. Remark. The advantage of Theorem 6.1 is that it holds true under C^1-hypothesis of the coefficients of the SDE 2.1.(i), *written in Itô form*. The *cost is the ellipticity condition* 6.1.(i).

If we have C^2-hypotheses for the coefficients, then the vector field $J^x_{t\leftarrow 0}(z)$ will belong to $\mathbb{D}^p_1(X; H)$; the usual theory of divergence will be applicable to this vector field, under the *hypoellipticity* condition considered in Section 8.

7. Bismut identity

We shall take advantage of the *adaptedness* of the *reduced variation* by writing *divergences* as *Itô stochastic integrals*. The Bismut identity which follows provides therefore an approach to the stochastic calculus of variation *independent* of the theory developed in the first five chapters: in fact the relation between Itô stochastic integrals and divergences of adapted vector fields can be derived directly from the Girsanov formula. Although this approach is more straightforward it does not allow an essential modification of the hypotheses needed to apply the stochastic calculus of variations.

Bismut theorem. *Given a C^1 vector field Z on $\mathbb{R}^d$, bounded along with its derivative, we introduce*

$$\begin{aligned} z(x) &= J^x_{0\leftarrow 1}\left(Z_{v_x(1)}\right) \\ \zeta(x) &= C^{-1} z. \end{aligned}$$

Then for any smooth test function f we have

7.1.
$$E\left((\mathcal{L}_Z f)(v_x(1))\right) = E\left[f(v_x(1)) \left[\sum_{k=1}^{n} \left(\zeta(x) \Big| \int_0^1 a_k(x,\tau)dx^k(\tau) \right)_{\mathbb{R}^d} - \int_0^1 \left(D_{\tau,k}\zeta \mid a_k \right)_{\mathbb{R}^d} d\tau \right]\right].$$

Proof. If $g \in \mathbb{D}^p_1(X;\mathbb{R})$, we will have

7.2.
$$E\left(g(x) \int_0^1 a^\ell_k(x,\tau)dx^k(\tau) \right) = E\left(\int_0^1 \left(D_{\tau,k} a\right) a^\ell_k d\tau \right).$$

We will sharpen the difference between $k \in [0,n]$ and $\ell \in [1,d]$ by denoting k by α. With this notation the identity 7.2 will have for $g \in \mathbb{D}^p_1(X;\mathbb{R}^d)$ the following form

7.3.
$$E\left(\left(g \Big| \int_0^1 a_\alpha(x,\tau)dx^\alpha(\tau) \right)_{\mathbb{R}^d}\right) = E\left(\int_0^1 \left(D_{\tau,\alpha} g \mid a_\alpha\right)_{\mathbb{R}^d} d\tau \right)$$

If we take

$$g^\ell(x) = f(v_x(1))\, \zeta^\ell(x)$$

then after a summation on α the l.h.s. of 7.3 is equal to the first term of the r.h.s. of 7.1. Furthermore we have

$$D_{\tau,\alpha} g^\ell = \left[\left(D_{\tau,\alpha} f\right)(v_x(1)) \right] \zeta^\ell + f(v_x(1))\, D_{\tau,\alpha} \zeta^\ell(x).$$

The expectation of the second term of the r.h.s. after a summation on α can be written as

7.4.
$$E\left(f\left(v_x(1)\right)\sum_{\alpha,\ell}\int_0^1 D_{\tau,\alpha}\zeta^\ell a_\alpha^\ell d\tau\right) = E\left(f\left(v_x(1)\right)\int_0^1 \left(D_\tau\zeta|a\right)d\tau\right).$$

The r.h.s. of 7.4 appears in the r.h.s. of 7.1. Finally we have to prove

7.5.
$$E\left(\left(\mathcal{L}_Z f\right)\left(v_x(1)\right)\right) = \sum_{\ell,\alpha} E\left(\int_0^1 \left(D_{\tau,\alpha} f\left(v_x(1)\right)\right)\zeta^\ell a_\alpha^\ell d\tau\right).$$

We have by 3.3

$$\left(D_{\tau,\alpha} f\right)\left(v_x(1)\right) = \sum_k \frac{\partial f}{\partial v^k}\left(J_{1\leftarrow\tau}^x(A_\alpha)\right)^k$$

or

$$E\left(\left(\mathcal{L}_Z f\right)\left(v_x(1)\right)\right) = \sum_{k,\ell,\alpha} E\left(\frac{\partial f}{\partial v^k}\zeta^\ell\int_0^1 \left(J_{1\leftarrow\tau}(A_\alpha)\right)^k\left(J_{0\leftarrow\tau}(A_\alpha)\right)^\ell d\tau\right).$$

The identity $J_{1\leftarrow\tau} = J_{1\leftarrow 0}\circ J_{0\leftarrow\tau}$ implies that

$$\sum_\alpha\int_0^1 \left(J_{1\leftarrow\tau}(A_\alpha)\right)^k\left(J_{0\leftarrow\tau}(A_\alpha)\right)^\ell d\tau = J_{1\leftarrow 0}^{k,s}\sum_\alpha\int_0^1 \left(J_{0\leftarrow\tau}(A_\alpha)\right)J_{0\leftarrow\tau}^\ell(A_\alpha)d\tau = J_{1\leftarrow 0}^{k,s}C^{s,\ell}.$$

Finally we have to check that

$$Z^k = \sum_{s,\ell} J_{1\leftarrow 0}^{k,s}C^{s,\ell}\zeta^\ell.$$

This identity results from the definition of ζ and 7.5. is established. □

8. Hörmander hypoellipticity under degenerate hypotheses

We consider a differential operator Γ which can be written as a sum of squares of derivatives with respect to C^∞ vector fields:

$$\Gamma = \frac{1}{2}\sum_{k=1}^n \mathcal{L}_{A_k}^2.$$

The framework is a manifold M of dimension d where the limit theorem of Chapter VIII can be applied: no other hypotheses are needed if M is compact; the hypothesis that the A_k are C^∞-bounded is sufficient if $M = \mathbb{R}^d$.

Denote by $\wp$ the Lie algebra spanned by the A_k and their successive brackets. Given $m_1 \in M$ the evaluation $Z \to Z_{m_1}$ sends $\wp \to \wp_{m_1} \subset T_{m_1}(M)$, the tangent space to M at m_1. We write

$$\rho(m) = \text{dimension}(\wp_m), \qquad O = \{m \in M;\ \rho(m) = d\};$$

then O is an open subset of M.

Consider the Stratonovich stochastic differential equation

$$dm_x = \sum_k A_k(mx) \circ dx^k, \qquad m_x(0) = m_0 \tag{8.1}$$

Then the process (8.1) is the solution of the martingale problem associated to Γ; the law of $m_x(t)$ is the fundamental solution $p_t(m_0, dm)$ of the heat equation with initial source at m_0.

8.2. Theorem. *Define the stopping time $T(x) = \inf\{t;\ m_x(t) \in O\}$, assume*

$$T(x) = 0 \quad \textit{almost surely,}$$

then, $\forall t_0 > 0$, $p_{t_0}(m_0, *)$ is a measure which is absolutely continuous relatively to the Lebesgue measure of M.

Proof. We introduce the functional $f(x) = m_x(t_0)$; by 3.4 $f \in D_\infty(X)$. The absolute continuity of the law of f will result, according III.9 and 5.4, from the almost sure invertibility of the matrices

$$C^{i,j}(t_0, x) = \int_0^{t_0} \sum_{k=0}^{n} a_k^i a_k^j \, dt.$$

Denote by $\mathcal{K}_x(t_0)$ the kernel of the matrix $C^{i,j}$; for x fixed and for $t' < t$ we have $\mathcal{K}_x(t) \subset \mathcal{K}_x(t')$; consider the limit at 0 of this "decreasing function": $\mathcal{K}_x(0^+) = \lim_{t\to 0} \mathcal{K}_x(t)$.

The functional $x \to \mathcal{K}_x(0^+)$ is measurable with respect to the germ σ-field $\cap_{t>0}\mathcal{N}^t$; this germ σ-field is *trivial* according the Blumenthal zero-one law: this means that there exists a subspace $V \subset T_{m_0}(M)$, independent of x, such that $V = \mathcal{K}_x(0^+)$, almost surely.

We fix $z \in V$, and we consider the stopping time

$$\Theta(x) = \inf\{t;\ q_x(t) > 0\}, \quad \text{where} \quad q_x(t) = \sum_{i,j,k} a_k^i(x,t) a_k^j(x,t) z_i z_j;$$

then $\Theta(x) > 0$ almost surely and the semi-martingale $t \to q_*(t \wedge \Theta)$ is identically zero.

8.3. Lemma. *Consider a semi-martingale $s(t)$ defined on X:*

$$s(t) - s(o) = \int_0^t \alpha_k \, dx^k + \int_0^t \beta \, dt$$

assume that there exists a stopping time Θ such that $s(t \wedge \Theta) \equiv 0$, then

$$\alpha_k(t \wedge \Theta) \equiv 0.$$

Proof. Consider the $b(t) = \int_0^t \beta \, dt$, then b is Lipschitzian, the identity

$$b(t \wedge \Theta) = -\sum_k \int_0^{t\wedge\Theta} \alpha_k \, dx^k$$

will be an equality between a Lipschitzian function and a non-Lipschitzian function: such an equality is possible only if the l.h.s. and the r.h.s. are equal to zero. Using the energy identity for stochastic integrals, we obtain

$$0 = E\left(\int_0^{t\wedge\Theta} \sum_k \alpha_k \, dx^k\right)^2 = \sum_k \int_0^{t\wedge\Theta} E(\alpha_k^2) \, dt.$$

□

8.4. *Proof of 8.2.*

We consider the vector z_* as defining a linear form φ on $T_{m_0}(M) \approx \mathbb{R}^d$; denote by $g_x(t)$ the stochastic flow contructed in 2.1, then

$$q(x,t) = \sum_j \left\langle \varphi, \left(g_x^{-1}(t)\right)_* A_j \right\rangle^2 .$$

Therefore

$$s_x^j(t) = \left\langle \varphi, \left(g_x^{-1}(t)\right)_* A_j \right\rangle \tag{8.5}$$

is a semi-martingale satisfying the hypothesis of the Lemma 8.3. Consider the corresponding α_k^j computed by the Itô calculus; as we have on the group of diffeomorphisms the Stratonovich SDE

$$dg_x(t) = \left(\sum_k A_k \circ dx^k\right) g_x(t), \quad \text{then}$$

$$d\left(g_x^{-1}(t)\right) = -\left(g_x^{-1}(t)\right)\left(\sum_j A_j \circ dx^j\right). \tag{8.6}$$

The Itô calculus combined with VIII.2.1.4 gives

$$\alpha_k^j = \left\langle \varphi, \left(g_x^{-1}(t)\right)_* [A_k, A_j] \right\rangle .$$

Using Lemma 8.3 we get $\alpha_k^j(t \wedge \Theta) \equiv 0$.

We apply again Lemma 8.3 to the semi-martingale α_k^j; then the Itô calculus will prove

$$\alpha_{k,l}^j(t) = \left\langle \varphi, \left(g_x^{-1}(t)\right)_* [A_l, [A_k, A_j]] \right\rangle \quad \text{satisfies} \quad \alpha_{k,l}^j(t \wedge \Theta) \equiv 0.$$

We apply again Lemma 8.3 and Itô calculus and so on; finally we get

$$\text{(8.7)} \qquad \left\langle \varphi, \left(g_x^{-1}(t)\right)_* \wp \right\rangle = 0 \qquad \forall\, t \in [0, \Theta].$$

We take in (8.7) $t = \gamma(x) \leq \Theta(x)$ such that $m_x(\gamma(x)) \in O$ and we obtain $\varphi = 0$.

□

Notes and comments on Part IV

The theory of stochastic differential equations (SDE) was developed in the forties by Itô, modelled on the theory of ODE. This interrelation was emphasized in the sixties by Wong-Zakai [1], Stroock-Varadhan [1]. The limit theorem appeared in 1977 (Malliavin [7], [10]) and was used as a foundation of stochastic differential geometry and in particular for the stochastic flow of diffeomorphisms (Malliavin [9]). The proof used an infinite descent in the spirit of Fermat and also the machinery of extension of SDE to jet bundles. The limit theorem appeared then as a basic result with an unusual proof.

Many efforts were made to shape a proof of the limit theorem in a more familiar context. The mechanism of extension to jet bundles was replaced in Stroock [2], [3] by delicate bookkeeping.

Bismut [3] replaces the technique of descent by the *Kolmogorov criterion* for the existence of continuous version of a random field defined on $\mathbb{R}^d$; Bismut's method is to prove the continuity of a random field defined on a space including as one of its coordinates the smoothing parameter of the Brownian motion; a concise version of Bismut's method is presented in Kunita [2], book which is the basic reference for stochastic flows. Stroock observed that Kolmogorov's criterion can be interpreted as an embedding of a Sobolev space defined on $\mathbb{R}^d$ in a space of continuous functions. Ren [2], using a quasi-sure version of Kolmogorov's criterion, obtains a quasi-sure version of the limit theorem (see Malliavin-Nualart [2] and Denis [1]), and thus a quasi-sure theory of SDE, which is interesting for application to anticipative SDE (see Malliavin-Nualart [2]).

Caverhill-Elworthy [1] give a functional construction of stochastic flows by solving a SDE defined on the group of diffeomorphisms. The proof given in this book of the limit theorem is based on similar functional analytic considerations.

Stroock-Taniguchi [1] provide a proof based on the integration of ODE, driven by an *adapted control*; their proof shows perfectly the paramount role of the Lie brackets of the driving vector fields.

A forthcoming theory, presently developed by T. Lyons [3], [4], could provide a *universal limit theorem*; let us describe this theory for SDE driven by a two-dimensional Brownian motion; we associate to this Brownian motion the corresponding diffusion on the Heisenberg group (see Gaveau [1]); then the convergence of the approximating sequence for the Gaveau SDE will imply *pathwise* the convergence for *all SDE with smooth coefficients*. This universal property is formally analogous to the universal property of the Ben Arous [4] expansion of diffusions with real analytic coefficients.

A conceptual interest of the limit theorem is that it provides a *constructive approach* to SDE. An analogous constructive approach to processes associated to a Dirichlet form is made by Stroock-Zheng [1] where such processes are approximated by a sequence of Markov chains.

A fundamental fact, that stochastic flows dispel the difficulty coming from the non-adaptedness of the starting point, was undeclined in Bismut [1], Bismut-Michel [1], [2], Kunita [1], [2], Malliavin-Nualart [1]. The reversibility of the flow is discussed in Kunita [5], Föllmer [2], [4], Millet-Nualart-Sanz [1], [2].

Differentiation of solutions of an SDE with respect to the to the initial condition was performed in 1972 by Gikhman-Skorokhod [1]. The stochastic variation of an SDE was implemented in 1976 using the limit theorem (Malliavin [7], [10]).

A remarkable fact (Bouleau-Hirsch [5], [8], [7], Nualart-Vives [1]) is the existence of the Jacobian of an *elliptic* diffusion with Lipschitz coefficient.

The *reduced variation* was introduced in Malliavin [10]; Bismut [2] showed that the reduced variation makes possible to reduce the stochastic calculus of variations to the application of a Girsanov formula; reduced variation is used by Bismut [11] to obtain a Riemannian Harnack-type inequality, by Norris [4] and Elworthy-Li [1] to obtain heat kernel estimates.

The hypoellipticity with degeneracy appeared in Malliavin [6], [10]. The qualitative statement that the law is absolutely continuous with respect to Lebesgue measure can be replaced by a C^∞ quantitative statement (see Malliavin [8], Kusuoka-Stroock [1]) under appropriate quantitative hypotheses. This degenerate hypoellipticity was discussed from a PDE point of view in Morimoto [1], Hoshiro [1].

C^∞-results for degenerate Hörmander operators were obtained in Malliavin [8], Bismut [2], Kusuoka-Stroock [1], [3], [5], Bell-Mohammed [2].

Long-time decay estimates linked to spectral estimates were obtained by Malliavin [14], [19], [20], [21], [22], [23], Kusuoka-Stroock [6], Ikeda-Manabe [2], Ikeda-Shigekawa-Taniguchi [1], Prat [1], [2], Shigekawa [10], [11].

Estimates in infinite dimension are produced by projection estimates in finite dimension where the constants are dimension-free. This fact was used by Holley-Stroock [1] in order to get a Lyapunov function for the evolution of an Ising model with continuous spin.

Partial stochastic calculus of variations means to work on a *reducible* Gaussian space. This methodology leads to (i) estimates of harmonic measure of hypoelliptic operators (Ben Arous-Kusuoka-Stroock [1]); (ii) estimation of the conditional law in the filtering equation (Bismut-Michel [1], [2]).

Surprisingly, the same methodology can prove the impossibility of realizing a finite-dimensional filter (Ocone [2]). Approximate continuity of the filter is proved in Chaleyat-Maurel [1].

Large deviation problems are related to the differential structure of Wiener spaces: see Bismut [6], Ben Arous [1], [2], [3], Kusuoka-Stroock [7], [8].

The asymptotic of an oscillatory integral can be reached either by stationary phase estimates (see Ben Arous [1], Bogachev [1], Gaveau-Moulinier [1], [3], Ikeda-Manabe [3], Malliavin-Taniguchi [1]) or by smoothness property combined with a semigroup iteration (see Malliavin [19] to [23], Matsumoto-Ueki [1], Prat [2], [3], Shigekawa [10], Ueki [3]).

Application to non-parametric statistics in given by Nakahiro Yoshida in [1], [4].

Part V

Stochastic Analysis in Infinite Dimensions

Stochastic analysis plays an eminent role in infinite dimension; the main applications of the theory in finite dimension remain valid; furthermore, in infinite dimension, the methods of classical analysis (Fourier transform, pseudo-differential operators) collapse so that the theory of elliptic PDE can be approached essentially only through stochastic analysis. We cannot give here a global overview of such a rich field. We will restrict ourselves to some subjects which are directly linked to the previous chapters of this book.

The Ornstein-Uhlenbeck operator was a key ingredient in Chapter II; we will study its associated process from which we shall deduce another approach to the regularity of laws studied in Chapter III; this approach to the regularity of laws has greater potential than those considered in Chapter III and Chapter VIII and for this reason will be developed in an axiomatic framework.

The infinitesimal transformations leaving quasi-invariant the probability measure can be considered as being the "tangent vector fields to the probability space". In the case of a Gaussian probability space, those vector fields have been considered in part I: all smooth vector fields taking their values in the Cameron-Martin space are tangent vector fields. As in finite dimensions, the theory of stochastic differential equations is a nonlinear theory, infinite dimensional stochastic analysis must be nonlinear. The broad generality carried over by this nonlinearity requirement is a precious laboratory for the elaboration of the concept of "tangent space to the probability space". In the case of the path space over a Lie group, this elaboration is possible under minor adjustments to the Wiener space case. On the contrary, the case of the path space over a Riemannian manifold needs the new concept of *tangent processes*. The lifting to the frame bundle produces an intrinsic measure-preserving map from the probability space of Brownian motion on $\mathbb{R}^l$ to the probability space of Brownian motion on a Riemannian manifold: this is the *Itô map*. Then the Itô map behaves as a diffeomorphism: it induces an isomorphism at the level of tangent processes.

Chapter X

Stochastic Analysis on Wiener Spaces

Contents: The Ornstein-Uhlenbeck (OU) flow on a finite-dimensional Gaussian space – Lifting the OU flow to an abstract Wiener space, its axiomatic definition – Representation of the OU flow on the probability space of Brownian motion – Itô calculus of variations along the OU flow – Equilibrium processes and regularity of law.

1. The Ornstein-Uhlenbeck process

1.1. Elementary facts in the finite-dimensional case

We consider on $\mathbb{R}^n$ the normal law μ_n given by its density $(2\pi)^{-n/2}\exp\left(-\frac{1}{2}\|\xi\|^2\right)$. We denote by Δ the ordinary Laplacian

$$\Delta = \sum_{k=1}^{n} \partial_k^2 .$$

We introduce the Ornstein-Uhlenbeck generator

1.1.1. $$\mathcal{L} = \frac{1}{2}\,[\Delta - \xi \cdot \nabla];$$

then $\mathcal{L}$ is a *symmetric operator* in $L^2(\mu_n)$.

The associated symmetric bilinear form is:

1.1.2. $$-\,(\mathcal{L}u \mid v) = \int_{\mathbb{R}^n} \nabla u \cdot \nabla v \; d\mu_n .$$

The stochastic differential equation associated to $\mathcal{L}$ is

1.1.3. $$d\xi_x = dx - \frac{1}{2}\,\xi\, d\tau$$

where x is the n-dimensional Brownian motion.

1.1.4. Proposition. *The transition probability of the process 1.1.3 is given by the Mehler formula*

$$E_{\xi_0}\left(f\left(\xi_x(\tau)\right)\right) = \int_{\mathbb{R}^n} f\left(e^{-\tau}\xi_0 + (1-e^{-2\tau})^{1/2}\eta\right)\mu_n(d\eta).$$

1.1.5. Definition of the flow on R^+

We solve equation 1.1.3 with a random initial value given by μ_n.

Theorem. *The Ornstein-Uhlenbeck flow leaves the Gaussian measure invariant. It is furthermore reversible.*

Proof. The fact that the Gaussian measure is left invariant is equivalent to the identity

$$\int_{\mathbb{R}^n} \mu_n(d\xi)p_t(\xi, d\eta) = \mu_n(d\eta).$$

Resuming the computation of Chapter II we denote $e^{-t} = \cos\theta$, then we have to prove

$$\int_{\mathbb{R}^{2n}} \varphi(\xi\cos\theta + \eta\sin\theta)\mu(d\xi)\otimes\mu(d\eta) = \int_{\mathbb{R}^n}\varphi(z)\mu_n(d\zeta).$$

The rotation of $\mathbb{R}^{2n}$ is defined by the matrix

$$A = \begin{pmatrix} \cos\theta & \sin\theta \\ -\sin\theta & \cos\theta \end{pmatrix}.$$

It defines a linear endomorphism of $\mathbb{R}^{2n}$. Therefore the joint law ρ is *Gaussian*; it is fully determined by its covariance matrix. As $\xi_x(\tau)$ and $\xi_x(\tau+a)$ have the same law, it follows that ρ is invariant under interchange of the two $\mathbb{R}^n$ components. This proves reversibility. □

1.1.6. Proposition. *The Ornstein-Uhlenbeck semigroup is symmetric.*

Proof.

$$(P_\tau\varphi \mid \psi) = E\left(\varphi\left(\xi_x(\tau)\right)\psi\left(\xi_x(0)\right)\right).$$

We use now reversibility

$$= E\left(\varphi\left(\xi_{x'}(0)\right)\psi\left(\xi_{x'}(\tau)\right)\right) = (\varphi \mid P_\tau\psi).$$

□

1.1.7. Definition of the flow

We consider two independent copies of the forward flow $\xi_x^1(\tau)$, $\xi_x^2(\tau)$, starting from the same point ξ_0.

We define

$$\eta_y(\tau) = \xi_x^1(\tau) \qquad \text{if} \quad \tau > 0$$

$$\eta_y(\tau) = \xi_x^2(-\tau) \qquad \text{if} \quad \tau < 0$$

We get in this way a stationary stochastic process defined on $]-\infty, +\infty[$, leaving the measure μ_n invariant, and reversible. This will be the *Ornstein-Uhlenbeck flow*.

1.2. Lifting the OU flow to an abstract Wiener space

1.2.1. A setting of the problem

An abstract Wiener space X is given by the triplet

$$X^* \subset H \subset X$$

where H is the Cameron-Martin space. We pick an orthonormal basis $\{e_q\}$ of H such that $e_q \in X^*$. We define

$$\Phi(x) = \left\{\langle x, e_q\rangle\right\}.$$

Then Φ is an injective continuous map of X into $\mathbb{R}^{\mathbb{N}}$.

By the tightness of the capacities on X, we can find an increasing sequence of compact subsets $K_n \subset X$ such that

$$c_{2,1}\left(K_n^c\right) \to 0.$$

We write $\tilde{K}_n = \Phi(K_n)$. Then by the invariance of the capacity $c_{2,1}$ (cf. Chapter IV)

$$c_{2,1}\left(\tilde{K}_n^c\right) \to 0.$$

(i) We denote by $\Omega(\mathbb{R}^s)$ the probability space of the OU flow on R^s. We denote by $\Omega(\mathbb{R}^{\mathbb{N}})$ the probability space obtained as the product of an infinite number of copies of $\Omega(\mathbb{R})$. We write π_s for the projection $\pi_s : \mathbb{R}^{\mathbb{N}} \to \mathbb{R}^s$. Then it is possible using π_s to identify $\Omega(\mathbb{R}^s)$ to a probability subspace of $\Omega(\mathbb{R}^{\mathbb{N}})$; we identify $\pi_s^{-1}(\mathcal{B}_{\mathbb{R}^s})$ which is a sub σ-field of $\mathcal{B}_{\mathbb{R}^{\mathbb{N}}}$.

1.2.2. Main Lemma. *Almost surely in $\omega \in \Omega(\mathbb{R}^{\mathbb{N}})$ there exists $n(\omega)$ such that*

$$\pi_s\left(y_\omega(\tau)\right) \in \pi_s\left(\tilde{K}_{n(\omega)}\right) \quad \textit{for all} \quad \tau \in [0, 1].$$

We shall need the following classical result.

1.2.3. Lemma. *Consider the OU flow on $\mathbb{R}^n$. Let A be a Borel set in $\mathbb{R}^n$ and T_A the first entrance time of the process in A, $(T_A \geq 0)$, then*

$$E\left(e^{-T_A}\right) = c_{2,1}(A).$$

Proof. It is sufficient to prove the lemma when A is compact. Consider the function

$$u(\xi_0) = E_{\xi_0}\left(c^{-T_A}\right),$$

then

$$u(\xi_0) = 1 \qquad \text{for} \quad \xi_0 \in A.$$

Furthermore

$$\mathcal{L}u - u = 0 \quad \text{on} \quad A^c = 0.$$

Therefore u minimizes the following energy integral on O:

$$I(O) = \int_O \left(|\nabla v|^2 + v^2\right) d\mu_n$$

under the contraint $v \mid_{\partial O} = 1$; therefore u is the function which minimizes the energy integral $\int \left(\|\nabla w\|^2 + w^2\right) d\mu$ under the constraint $w = 1$ on A. Finally u is the equilibrium potential. Denote by γ the equilibrium measure of A. Then

$$\int d\gamma = \text{ charge } (A) = c_{2,1}(A) \quad \text{(see Chapter IV).}$$

Denote by U^ρ the potential of ρ, then

$$U^\rho = u, \qquad U^{\mu_n} = 1$$

$$\int U^\gamma d\mu_n = \int U^{\mu_n} d\gamma \quad \text{(symmetry of the Green kernel)}$$

$$E\left(e^{-T_A}\right) = \int d\gamma = c_{2,1}(A).$$

□

1.2.4. *Proof of Lemma 1.2.2*

(i) **Sublemma.** *Let K be a compact subset of X, then*

$$c_{2,1}\left(\pi_s(K)\right) = c_{2,1}\left(\pi_s^{-1}\left(\pi_s(K)\right)\right).$$

Remark. On the l.h.s. the capacity is computed on R^s, on the r.h.s. in $\mathbb{R}^N$.

Proof. Denote by ρ the equilibrium measure of $\pi_s(K)$. Then $\tilde{\rho} = \rho \otimes \mu_{\mathbb{R}^N - s}$ is carried by $\pi_s^{-1}\left(\pi_s(K)\right)$, we have $\Lambda^2_{\mathbb{R}^N}\tilde{\rho} = \left(\Lambda^2_{\mathbb{R}^s}\rho\right) \circ \pi_s$. Therefore the potential $\Lambda^2_{\mathbb{R}^N}\tilde{\rho}$ is equal to 1 on $\pi_s^{-1}\left(\pi_s(K)\right)$. □

(ii) On the other hand we have

$$\pi_s^{-1}\left(\pi_s(\tilde{K}_n)\right)^c \subset \tilde{K}_n^c,$$

then using (i)

$$c_{2,1}\left(\left(\pi_s(\tilde{K}_n)\right)^c\right) \le c_{2,1}(\tilde{K}_n^c).$$

Denote by $T_{s,n}$ the first entrance time of $\pi_s(y_\omega)$ in $\left(\pi_s(\tilde{K}_n)\right)^c$. Then $T_{s,n} \ge T_{s+1,n}$.

Set $T_{\infty,n} = \lim T_{s,n}$. Then by 1.2.3 and by (i)

$$E\left(e^{-T_{s,n}}\right) \le c_{2,1}(K_n^c).$$

We pass to the limit and obtain

$$E\left(e^{-T_{\infty,n}}\right) \le c_{2,1}(K_n^c).$$

Letting $n \to +\infty$, as $T_{\infty,n} \le T_{\infty,n+1}$, we have

$$\lim T_{\infty,n} = +\infty \qquad \text{a.s.}$$

which proves the main Lemma 1.2.2. □

1.2.5. Theorem. *Let K be a compact subset of X, denote by T the entrance time in K, then*

$$E(e^{-T}) = c_{2,1}(K).$$

Proof. We define

$$K_s = \pi_s(K)$$

$$T_s = \text{entrance time in} \quad K_s.$$

Then $T_s \le T_{s+1}$. As

$$K = \cap_s \pi_s^{-1}(\pi_s(K))$$

we get

$$\lim T_s = T$$

$$c_{2,1}(K) = \lim c_{2,1}(\pi_s(K)).$$

We then apply 1.2.3 to $\pi_s(K)$. □

1.2.6. Realization of the lifting

As $\pi_s(y_\omega(\tau))$ converges in the topology of $\mathbb{R}^{\mathbb{N}}$ to $y_\omega(\tau)$, we obtain that

$$y_\omega(\tau) \in \bigcap_s \pi_s^{-1}\left(\pi_s(\tilde{K}_n)\right) = \tilde{K}_n \quad \text{for} \quad n = n(\omega) \quad \text{and} \quad \tau \in [0, 1].$$

As Φ is a homeomorphism of K_n onto $\tilde{K}_n$, we get that

$$x_\omega(t) = \Phi^{-1}(y_\omega(t))$$

is a continuous path on X. □

1.2.7. Axiomatic definition of an OU process on X

Given an abstract Wiener space X, we denote by $\mathbb{P}(X)$ the vector space of *continuous* maps of $\mathbb{R}$ into X.

For any choice of $\ell_1 \ldots \ell_n \in X^*$, which are orthonormal in H, then the natural map $\psi : X \to \mathbb{R}^n$ induces a map

$$\tilde{\psi} : \mathbb{P}(X) \to \mathbb{P}(\mathbb{R}^n).$$

Definition. An OU process on X will be the data of a probability measure θ on $\mathbb{P}(X)$ such that $(\tilde{\psi})_* \theta$ is the measure defined by the OU process on $\mathbb{R}^n$ whatever the choice of the ℓ_i.

Theorem. *An abstract Wiener space has a unique OU flow.*

Proof. The existence results from 1.2.5. The uniqueness from the fact that polynomials in the ℓ, $\ell \in X^*$ are dense in $L^2(X)$. □

1.2.8. Proposition. *Consider the probability space of Brownian motion on $\mathbb{R}^k$. Given a smooth function $u(\xi_1, \ldots, \xi_k)$ defined on $\mathbb{R}^k$, and given $s_1 < s_2 < \ldots < s_k$, we define the following cylinder function:*

$$U(x) = u\left(\xi_x(s_1), \xi_x(s_2), \ldots, \xi_x(s_k)\right),$$

then

$$2\mathcal{L}U = \int_0^1 \left(D^{(2)}_{\tau,\tau} U\right) d\tau - \int_0^1 D_\tau U \, dx(\tau).$$

Remark. This formula has a *"formal meaning"* in all cases. We shall establish it in the cylindrical case.

Proof.

$$D^{(2)}_{\tau_1,\tau_2} u = \sum_{i,j} \frac{\partial^2 u}{\partial \xi_i \partial \xi_j} \mathbf{1}_{[0,s_1]}(\tau_1) \mathbf{1}_{[0,s_j]}(\tau_2)$$

$$\int_0^1 D^{(2)}_{\tau,\tau} u \, d\tau = \sum_{i,j} \frac{\partial^2 u}{\partial \xi_i \partial \xi_j} s_i \wedge s_j.$$

Consider the linear map $X \to \mathbb{R}^k$ defined by $x \to (x(s_1), \ldots, x(s_k))$. The image of the OU flow on X is the OU flow on $\mathbb{R}^k$ with the Euclidean metric defined by the matrix $(s_i \wedge s_j)$. □

1.3. Representation of the OU flow on the probability space of Brownian motion

In this paragraph, we want to give some constructive realizations of the OU flow.

1.3.1. Proposition. *Denote by $\Omega(\mathbb{R}^N)$ the probability space defined in 1.2.1; consider the series*

$$x_\omega(t,\tau) = \tau y_\omega^0(t) + \sqrt{2}\sum y_\omega^k(t)\,\frac{\sin k\pi\tau}{k\pi} \quad (\textit{where} \quad \omega \in \Omega(\mathbb{R}^{\mathrm{N}})).$$

This series converges uniformly in the variables (t,τ), $|t| < M$.
Furthermore, given $\ell_1, \ldots, \ell_q \in X^*$, $\ell_1, \ldots, \ell_q$ orthonormal in H, then

$$\{\ell_j\,(x_\omega(t,\cdot))\}$$

is isomorphic in law to the OU flow on $\mathbb{R}^q$.

Proof. Firstly we shall evaluate the maximum of the scalar OU flow; we have

$$\sup |y_\omega^1(\tau)| \le |y_\omega^1(0)| + \max_{0<\tau<1} |b_\omega(\tau)|.$$

Therefore

$$\text{(i)} \qquad P\left(\sup_{|\tau|<1} |y_\omega^1(\tau)| > a\right) = \mathcal{O}\left(\exp\left(-\frac{a^2}{4}\right)\right).$$

We define

$$P_r(\tau) = \sum_{k=2^r+1}^{2^{r+1}} \frac{\sin k\pi\tau}{k\tau}\, y_\omega^k(t).$$

By the theory of entire functions of exponential type we have (see Boas [1])

$$\text{(ii)} \qquad \|P_r'\|_{L^\infty} \le 2^{r+1}\|P_r\|_{L^\infty}.$$

We denote by Δ_r the set of equidistant points splitting the segment $[0, 1]$ in 2^{r+2} parts. Then (ii) implies

$$\text{(iii)} \qquad \|P_r\|_{L^\infty} < 2\sup_{\delta\in\Delta_r} |P_r(\delta)|.$$

We remark that $P_r(\delta)$ is, as a function of t, a *scalar* OU flow associated to the variance

$$\sum_{k=2^r+1}^{2^{r+1}} \frac{\sin^2 k\delta}{(k\pi)^2} < 2^{-r}.$$

We denote by $\|\ \|_{c(t)}$ the sup norm in t, $|c| < M$. Then by (i)

$$\begin{aligned}
\operatorname{Prob}\left(\|P_r(\delta)\|_{c(t)} > 2^{-r/2}r\right) &\le c\exp\left(-\frac{1}{20}\,r^2\right)\\
\operatorname{Prob}\left(\sup_{\delta\in\Delta_r} \|P_r(\delta)\|_{c(t)} > 2^{-r/2}r\right) &\le c2^{r+1}\exp\left(-\frac{1}{4}\,r^2\right).
\end{aligned}$$

This implies that by (iii)

$$\sum \operatorname{Prob}\left\{\|P_r\|_{L^\infty([-M,M]\times[0,1])} \geq 2^{-r/2}r\right\} < +\infty$$

which proves the uniform convergence $\tau \in [0,1]$, $|t| < M$.

Given $\ell_i \in X^*$, then $\frac{d^2}{d\tau^2}\ell_i$ is a measure and

$$\int_0^1 (\cos k\pi\tau)\dot{\ell}_i(\tau)d\tau = -\int_0^1 \frac{\sin k\pi\tau}{k\tau}\,\ddot{\ell}_i.$$

Therefore $\ell_1(x(t,\cdot))$ ($i \in [1,s]$) is an OU flow on $\mathbb{R}^s$ associated to the variance

$$\sum \frac{1}{(k\pi)^2}\left(\int_0^1 \sin k\pi\tau\,\frac{d^2}{d\tau^2}\,\ell_i\right)^2 = \sum_i\left(\int \cos(k\pi\tau)\dot{\ell}_1(\tau)dt\right)^2 = \|\ell_i\|^2_{H^1}.$$

1.3.2. Proposition. *Consider the white noise W in the two variables (t,τ), $t \in \mathbb{R}$, $\tau \in [0,1]$ associated to L^2 relatively to the Lebesgue measure. Define*

$$x_w(t,\tau) = \int_{]-\infty,t]\times[0,\tau]} \exp(-t+t')W(dt'\otimes d\tau'),$$

then $x_w(t,\tau)$ is a continuous function of (t,τ), the map

$$t \to x_w(t,\cdot)$$

defines the OU flow on the Wiener space of scalar Brownian motion.

Proof. See Ren [2], [3].

2. Itô calculus along the OU flow

2.1. Theorem. *Given $f \in D_2^p(X)$, then*

(i) $$f(x_\omega(t)) - \int_0^t (\mathcal{L}f)(x_\omega(t'))\,dt'$$

is an OU martingale M.

The increasing process $\langle M\rangle$ associated to M is given by

$$\langle M\rangle_t = \int_0^t \|Df(x_\omega(t'))\|_H^2 dt'.$$

Furthermore denoting by f^* the $(p,2)$ redefinition of f $(p>2)$;

(ii) the map $t \to f^*(x_\omega(t))$ is almost surely a continuous function of t, $t \in \mathbb{R}$.
(iii) for $f, g \in \mathbb{D}_2^p$ $(p>2)$ we have

$$(\mathcal{L}f \mid g) = (g \mid \mathcal{L}f).$$

Proof. We realize the space X as $\mathbb{R}^{\mathbb{N}}$. To the projection $\pi_s : \mathbb{R}^{\mathbb{N}} \to \mathbb{R}^s$ is associated a conditional expectation $E^{\pi_s} : \mathbb{D}_2^p(\mathbb{R}^{\mathbb{N}}) \to \mathbb{D}_2^p(\mathbb{R}^s)$.

Define $f_s = E^{\pi_s} f$. Then

$$\mathcal{L} f_s = E^{\pi_s}(\mathcal{L} f).$$

We can smooth f_s by a sequence of C^∞ convolution kernels φ_n with compact support such that

$$f_{s,m} := f_s * \varphi_m \to f_s \quad \text{a.e.}$$

We can apply to $f_{s,m}$ the Itô formula along the OU flow of $\mathbb{R}^s$.

We have

$$E\left(f_{s,m}\left(x_\omega(t)\right) g_{s,m}\left(x_\omega(0)\right)\right) = E\left(\int_0^t \left(\mathcal{L} f_{s,m}\right)\left(x_\omega(\xi) d\xi\right) g_{s,m}\left(x_\omega(0)\right)\right) + E(f_{s,m} g_{s,m})$$

$$E\left(f_{s,m}\left(x_\omega(0)\right) g_{s,m}\left(x_\omega(t)\right)\right) = E\left(f_{s,m}\left(x_\omega(0)\right) \int_0^t \left(\mathcal{L} g_{s,m}\right)\left(x_\omega(\xi) d\xi\right)\right) + E(f_{s,m} g_{s,m}).$$

By reversibility of the OU process, the two l.h.s. are equal. Therefore

$$E\left(\left(\int_0^1 \mathcal{L} f_{s,m}\left(x_\omega(ts)\right) ds\right) g_{s,m}\left(x_\omega(0)\right)\right) = E\left(f_{s,m}\left(x_\omega(0)\right) \int_0^1 \mathcal{L} g_{s,m}\left(x_\omega(ts)\right) ds\right).$$

The OU semigroup being continuous on L^p, we get when $t \to 0$

$$E\left[\left[\int_0^1 \mathcal{L} f_{s,m}\left(x_\omega(ts)\right) ds - \mathcal{L} f_{s,m}\left(x_\omega(0)\right)\right]^p\right] \to 0.$$

Using the Cameron-Martin formula on $\mathbb{R}^s$ we have for $p' < p$ that

$$\begin{aligned} \|f_{m.s} - f_s\|_{L^{p'}} &\to 0 \\ \|\mathcal{L} f_{s,m} - \mathcal{L} f_s\|_{L^{p'}} &\to 0, \end{aligned}$$

then

$$M_{s,m}(t) = f_{s,m}\left(x_\omega(t)\right) - \int_0^t \mathcal{L} f_{s,m}\left(x_\omega(t')\right) dt'$$

is an $L^{p'}$ martingale *for OU flow on X*

$$M(t) = \lim_s \lim_m M_{s,m}(t)$$

each of the two limits being taken in $L^{p'}$ $(p' < p)$ it follows that $M(t)$ is a martingale.

The expression given for the increasing process of $\langle M \rangle$ results from the following elementary identity:

$$\left\langle M_{s,m}\right\rangle_t = \int_0^t \|D f_{s,m}\|_{\mathbb{R}^s}^2 dt'.$$

Consider a redefinition net K_n of f, as $p \geq 2$, the $(p, 2)$ capacity is greater than the $(2, 1)$ capacity. Therefore

$$c_{2,1}(K_n^c) \to 0.$$

Using 1.2.3 we get that, denoting by T_n the entrance time in K_n^c. we have

$$E\left(e^{-T_n}\right) \to 0 \qquad \text{or}$$

$$T_n \mapsto +\infty \quad \text{a.s.} \qquad \text{when} \quad n \to +\infty.$$

Furthermore $f^*(x_\omega(t))$ is continuous $t \in [0, T_n]$. □

2.2. Theorem (of symbolic calculus). *Given $f_1 \dots f_s \in \mathbb{D}_2^p$ (with $p > 2$) and given a C^2 function u defined on $\mathbb{R}^s$, bounded along with its derivatives, we define*

$$g = u(f_1, \dots, f_s),$$

$$\textit{then} \quad g \in \mathbb{D}_2^p \quad \textit{and}$$

$$\mathcal{L}g = \sum_i \frac{\partial u}{\partial \xi_i} \mathcal{L}f_i + \sum_{i,k} \frac{\partial^2 u}{\partial \xi_1 \partial \xi_k} (Df_i \mid DF_k)$$

$$\|Dg\|^2 = \sum_{i,k} \frac{\partial u}{\partial \xi_i} \frac{\partial u}{\partial \xi_k} (Df_i \mid Df_k).$$

Proof. We apply Itô calculus to the semi-martingales $t \to f_j(x_\omega(t))$.

2.3. Lemma. *If $f_1, f_2 \in \mathbb{D}_2^p$ with $p > 2$, then*

$$(Df_1 \mid Df_2) = \mathcal{L}(f_1 f_2) - f_2\mathcal{L}(f_2) - f_2\mathcal{L}(f_1).$$

Proof. We apply 2.2 to the function

$$u_R(t_1, t_2) = t_1 t_2 \theta\left(\frac{t_1}{R}\right) \theta\left(\frac{t_2}{R}\right)$$

with $\theta(\xi) = 1$ if $|\xi| < 1$, $\theta(\xi) = 0$ if $\xi > 2$, θ smooth. Afterwards we let $R \to +\infty$.

3. Regularity of laws established via equilibrium processes

We shall proceed at an axiomatic level in order to emphasize the generality of the method proposed here; nevertheless our axiomatic presently has no other concrete applications than the OU processes on Wiener spaces.

The stochastic calculus of variations was built by Malliavin [10] around this approach; it was shaped by D. Stroock [1] into a calculus on the closed operator associated to the OU generator; Bouleau-Hirsch [7] worked along similar points of view developed in the context of Dirichlet forms. As we want to present a method, our exposition will not be fully formalized.

Chapter III was devoted to divergences. The elementary Laplacian on $\mathbb{R}^d$ can be defined by

$$\Delta f = \operatorname{div}(\operatorname{grad} f),$$

where the divergence is computed relative to Lebesgue measure. We can read this formula from right to left as saying that the divergences of vector fields which are equal to the gradient of a function are computable in terms of Laplacian. This point of view will be implemented in infinite dimensions in 3.2.

3.1. Equilibrium processes on a probability space

Definition. Given a probability space Ω, an *equilibrium process* on Ω is given by a probability measure Θ on the space $\mathcal{M}(R;\Omega)$ of the measurable maps from R to Ω satisfying the following properties:

(i) The measure Θ is invariant by the time reversal $\omega_t \to \omega_{-t}$.
(ii) $\exists$ an algebra $\mathcal{S}(\Omega)$ of random variables on Ω such that $\mathcal{S}(\Omega) \subset L^p(\Omega)$, $\forall p < \infty$, this inclusion having a dense image.
(iii) $E\left(\varphi(\omega_t)\right) = E(\varphi), \qquad \forall t \in R, \quad \forall \varphi \in \mathcal{S}$
(iv) $\forall \varphi \in \mathcal{S}$ the function $t \to \varphi(\omega_t)$ is a continuous semimartingale, homogeneous in law under the time shift.
(v) The infinitesimal generator $\mathcal{L}$ is a map $\mathcal{S}(X) \to \mathcal{S}(X)$ such that

$$M_\varphi(t) = \varphi(\omega_t) - \int_0^t \mathcal{L}\varphi(\omega_s)ds$$

is a martingale for $t > 0$.

3.2. Differential calculus on the probability space

This differential calculus will be built from the Itô calculus on the equilibrium process. In some sense we reverse the point of view of Chapter VII where from the differential calculus on the probability space of the Brownian motion, the Itô calculus was grasped through the Clark-Bismut-Ocone formula. Now the primitive notion is the Itô calculus on the equilibrium process which will induce a differential calculus on the probability space. We denote by $\mathcal{N}_t^+$ the forward σ-field generated by the ω_s; $s \leq t$; the backward σ-field $\mathcal{N}_t^-$ will be the forward σ-field of the time reversed process.

Proposition. *The operator $\mathcal{L}$ is formally self-adjoint.*

(3.2.1) $$E(\varphi\mathcal{L}\psi) = E(\psi\mathcal{L}\varphi).$$

Proof.

$$E(\varphi\mathcal{L}\psi) - E(\psi\mathcal{L}\varphi) = \lim_{\varepsilon\to 0} \varepsilon^{-1}\left[E\left(\varphi(\omega_0)\psi(\omega_\varepsilon)\right) - E\left(\psi(\omega_0)\varphi(\omega_\varepsilon)\right)\right];$$

by reversing the time $s \to \varepsilon - s$ we obtain that the two expectations appearing in the r.h.s. are equal. □

A *gradient vector field* will be associated to a function $\varphi \in \mathcal{S}(X)$ and denoted $\nabla\varphi$. It will correspond to the "infinitesimal martingale" along the equilibrium processes defined by

$$M_\varphi(\varepsilon) - M_\varphi(0), \quad \varepsilon > 0, \quad \varepsilon \to 0.$$

The notion of infinitesimal martingale was used in Debiard-Gaveau [1] to prove the differentiability of finely harmonic functions on $\mathbb{R}^d$.

The gradient is a linear operator. On the space of gradients a scalar product is defined by the formula

$$\nabla\varphi * \nabla\psi = \mathcal{L}(\varphi\psi) - \varphi\mathcal{L}\psi - \psi\mathcal{L}\varphi. \tag{3.2.2}$$

Then $\nabla\varphi * \nabla\varphi$ is a *non-negative* random variable, denoted by $\|\nabla\varphi\|^2$ associated to the non-decreasing process

$$\langle M_\varphi \rangle_t = \int_0^t \|\nabla\varphi\|^2 \, dt.$$

The linear combinations of gradient vector fields of the form

$$Z = \sum_{j=1}^{q} \psi_j \nabla\varphi_j, \qquad \psi_j, \varphi_j \in \Omega \tag{3.2.3}$$

constitutes the vector space of *tangent vectors.*

Given a tangent vector Z, we define a *derivation operator* $D_Z : \mathcal{S}(X) \to \mathcal{S}(X)$ by the formula

$$D_Z\varphi = \sum_{j=1}^{q} \psi_j \left(\nabla\varphi_j * \nabla\varphi\right). \tag{3.2.4}$$

Proposition.

$$D_Z(\theta_1\theta_2) = \theta_1 D_Z\theta_2 + \theta_2 D_Z\theta_1. \tag{3.2.5}$$

Proof. By Itô calculus on semimartingales we have

$$\nabla(\theta_1\theta_2) = \theta_1\nabla\theta_2 + \theta_2\nabla\theta_1.$$

□

Proposition. *The gradient vector field $Z = \nabla\varphi$ has divergence $\delta(Z) = -2\mathcal{L}\varphi$, that is to say the following identity holds true:*

$$E(D_Z\theta) = -2E(\theta\mathcal{L}\varphi), \qquad \forall\theta \in \mathcal{S}(X). \tag{3.2.6}$$

Proof. By the hypothesis 3.1.(iii) we have $E(\mathcal{L}\psi) = 0$, $\forall \psi \in \mathcal{S}(X)$. Applying this identity to (3.2.2), we get

$$E(D_Z\theta) = -E(\theta\mathcal{L}\varphi) - E(\varphi\mathcal{L}\theta).$$

We conclude by using (3.2.1). □

3.2.7. Theorem. *Every tangent vector field possesses a divergence.*

Proof. We use (3.2.6) together with the identity

$$\delta(\psi\nabla\varphi) = \delta(\nabla\varphi) - D_{\nabla\varphi}\psi.$$

3.3. Theorem. *Given $\varphi_k \in \mathcal{S}(X)$, $k = 1, \ldots, d$, assume the $\varphi_j \in L^\infty(\Omega)$ and that*

$$M_{i,j} = \nabla\varphi_i * \nabla\varphi_j \quad \textit{satisfy} \quad E\left(\det{}^{-p}(M_{*,*})\right) < \infty, \quad \forall p < \infty,$$

then the joint law of the φ_i has a density with respect to Lebesgue measure on $\mathbb{R}^d$ which is Hölder continuous for all exponents < 1.

Proof. The range in $\mathbb{R}^d$ of φ_* is a bounded set. By the Stone-Weierstrass theorem the polynomials P in the coordinates ξ_* are dense in the continuous functions. Assume that there exists a constant c such that for every polynomial P we have

$$\text{(i)} \qquad \left| E\left(\left[\frac{\partial}{\partial\xi_s}P\right](\varphi_1, \ldots, \varphi_d)\right)\right| \leq c\|\tilde{P}\|_{L^\infty(\Omega)} \text{ with } \tilde{P} = P(\varphi_1, \ldots, \varphi_d).$$

Then (i) will extend by density to all C^1-functions. Then by applying Chapter III we deduce that the law of φ_* has a density which is Hölder continuous.

In order to prove (i), we remark that $P(\varphi_1, \ldots, \varphi_d) \in \mathcal{S}(X)$; therefore as in III.3.1 we can reduce the desired estimate to computations of divergences done using the identities of paragraph 3.2, identities valid for $\tilde{P} \in \mathcal{S}(\Omega)$. □

Chapter XI

Path Spaces and Their Tangent Spaces

Contents: Stochastic calculus of variation on a Lie group: Reduced variation and adjoint representation – Path groups: left infinitesimal quasi-invariance of Wiener measure – Path group on a compact Lie group – Orthonormal frame bundles over a Riemannian manifold: Levi-Civita parallelism, structure equations – Lifting to the frame bundle of the Riemannian diffusion: the stochastic parallel displacement – The horizontal stochastic flow, determination of its Jacobian on a Riemannian diffusion – Invariance of Brownian motion by orthogonal transformations – Tangent processes – Stochastic analysis on the path space of a Riemannian manifold: Twisted differential, Clark-Bismut-Ocone representation – Harnack estimates via the reduced variation – Loop spaces.

The Cameron-Martin theorem can be considered as the seed containing the developments of the preceding chapters: in a differential geometry language, the Cameron-Martin space plays the role of the *tangent space* to Wiener space.

In a nonlinear situation we must answer the basic preliminary question: *What is the tangent space to the probability space*? In the case of path space on a Lie group we shall provide an answer quite close to the Wiener space case. For the path space on a Riemannian manifold, the *hypoellipticity in infinite dimensions* generates quite new phenomena, which makes it necessary to widen the concept of a tangent space.

It could be objected *a priori* to our project that the Itô map provides a canonical measure-preserving isomorphism between the Wiener space of a Brownian motion on $\mathbb{R}^\ell$ and the Wiener space of a manifold; in differential geometry language, the Itô map provides a *global chart*. We will make precise our point of view by going again to an elementary geometric example: the unit disk in the complex plane. Then the identity map φ of the unit disk into $\mathbb{C}$ provides a global chart. On the other hand, the unit disk carries the Poincaré metric $ds^2 = (1 - |z|^2)^{-2}|dz|^2$ invariant under this complex automorphism; the chart φ does not preserve the Poincaré metric. Coming back to our situation it can be said that the Itô map provides a global chart which does not preserve all the natural structures lying on path spaces.

In classical differential geometry, every explicit computation has to be made in a local chart using the facilities of differential calculus on $\mathbb{R}^d$. The Itô map will be for us the basic tool to lift stochastic calculus to path spaces.

Nevertheless, the ultimate objectives have to be thought of and formulated in terms of the intrinsic geometry of path spaces.

1. Tangent spaces to path groups

1.1. The Wiener measure

We denote by G a Lie group and by $\mathcal{G}$ its Lie algebra. To avoid technicalities of differential geometry we shall suppose that G is a group of matrices; this assumption is in fact unnecessary and all our results hold true for a general Lie group. Then the exponential map is the usual exponential of a matrix; the left invariant $\mathcal{G}$-valued differential form (i.e. the Maurer-Cartan differential form) has the following simple expression: $g^{-1}dg$.

Given $z \in \mathcal{G}$ we can associate to z the following two differential operators

$$\begin{aligned} \left(\partial_z^\ell f\right)(g) &= \left\{\frac{d}{d\varepsilon} f(\exp(\varepsilon z)g)\right\}_{\varepsilon=0} \\ \left(\partial_z^r f\right)(g) &= \left\{\frac{d}{d\varepsilon} f(g\exp(\varepsilon z))\right\}_{\varepsilon=0} . \end{aligned}$$

The operator ∂_z^r commutes with left translation and vice versa.

Choosing an Euclidean metric on $\mathcal{G}$, we define a left-invariant elliptic operator Δ on G by the following formula

$$\text{(1)} \qquad \Delta f = \frac{1}{2}\sum_{k=1}^{d} \left(\partial_{e_k}^r\right)^2 f$$

where $e_1, \ldots, e_d$ is an orthonormal basis of $\mathcal{G}$. To this operator the diffusion defined by the following Stratonovich SDE

$$\text{(ii)} \qquad df_x = g_x \circ \sum_{k=1}^{d} e_k dx^k .$$

is associated. In the Itô theory, this SDE takes the following equivalent form

$$dg_x = g_x \left[\sum_{k=1}^{d} e_k dx^k + \frac{1}{2}\left(\sum_{k=1}^{d} e_k^2\right) d\tau\right].$$

We shall denote by x the Brownian motion on the Euclidean space $\mathcal{G}$; then $x(\tau) = \sum e_k dx^k(\tau)$; in this way we will shorten (ii) into the following expression

(iii) $$dg_x = g_x \circ dx.$$

With this convention, the Cameron-Martin space will be

$$H = H^1([0,1]; \mathcal{G}).$$

Denote by $g_x(\tau)$ the solution of (ii) having for value at $\tau = 0$ the identity matrix e. Then granted the left invariance of Δ, the stochastic flow $U^x_{\tau\leftarrow 0}$ associated to (ii) has the following expression

$$U^x_{\tau\leftarrow 0}(\gamma) = \gamma g_x(\tau).$$

1.2. The adjoint representation

1.2.1. Definition. Given $g \in G$, we call $Ad(g)$ the linear endomorphism of $\mathcal{G}$ defined by

$$Ad(g) : z \to gzg^{-1}.$$

Then $Ad(\cdot)$ is a *representation* of G: we have that

$$Ad(g_1 g_2) = Ad(g_1)Ad(g_2) \qquad g_1, g_2 \in G.$$

1.2.2. Proposition. *For every* $p < +\infty$

$$E\left(\sup_{0<\tau<1} \|Ad\,(g_x(\tau))\|^p_{\mathrm{End}(\mathcal{G})}\right) < +\infty.$$

Proof. We consider the Hilbert-Schmidt norm of the matrix g

$$\|g\|_2^2 = \mathrm{trace}(g^* g).$$

Then there exists a constant c such that

$$\|Ad(g)\|_{\mathrm{End}(\mathcal{G})} \leq c_1 \|g\|_2.$$

We denote by $M(t) = \exp\left(\star \int_{[0,t]} A\right)$ the solution of the linear ODE $dM = MA\,dt$; by Itô's calculus, we have that

$$\left\| \exp\left(\star \int_{[0,t]} \sum_k e_k\, odx^k - \frac{1}{4}\left(e_k^2 + (e_k^*)^2 + 2e_k^* e_k\right) d\tau\right)\right\|_2^2$$

is a martingale. This implies that there exists a constant c_2 such that

$$E\left(\sup_{0<\tau<1} \|g_x(\tau)\|_2^2\right) < \exp\left(c_2 n^2\right).$$

□

1.3. The reduced variation

The *reduced variation* is the variation of the starting point which would produce at time τ the same variation as the variation induced by $h \in H = H^1(X;\mathcal{G})$. More precisely we define

$$dg_{x+\varepsilon h} = g_x \circ (dx + \varepsilon h d\tau).$$

Then the variation v_h is defined by the following equation

$$v_h(\tau) = \left\{\frac{\partial}{\partial\varepsilon}\, g_{x+\varepsilon h}(\tau)\right\}_{\tau=0}.$$

The reduced variation ρ_h is determined by the condition

$$\frac{d}{d\varepsilon}\, U^x_{\tau\leftarrow 0}\,(\exp(\varepsilon\rho_h)) = v_h(\tau)$$

$$\rho^h_x(\tau) g_x(\tau) = \left\{\frac{\partial}{\partial\varepsilon}\, g_{x+\varepsilon h}(\tau)\right\}_{\varepsilon=0}.$$

Theorem. *The reduced variation ρ_h is determined by the following ODE:*

$$\begin{aligned} \dot\rho^h_x(\tau) &= Ad\,(g_x(\tau))\,\dot h(\tau) \\ \rho^h_x(0) &= 0. \end{aligned}$$

Proof. We denote by x_n the Stroock-Varadhan sequence of piecewise linear approximations of x. We define

$$\frac{\partial}{\partial\tau}\, g_{x_n+\varepsilon h} = g_{x_n}\left(\dot x_n + \varepsilon\dot h\right).$$

We differentiate in ε and we denote by ρ^h_x the associated reduced variation

$$\text{(i)}\qquad \left\{\frac{\partial}{\partial\varepsilon}\, g_{x_n+\varepsilon h}(\tau)\right\}_{\varepsilon=0} = \rho^h_{x_n}(\tau) g_{x_n}(\tau).$$

We differentiate in τ and interchange $\frac{\partial}{\partial\varepsilon}$ and $\frac{\partial}{\partial\tau}$ to obtain

$$\begin{aligned} \left\{\frac{\partial}{\partial\varepsilon}\left[g_{x_n+\varepsilon h}\left(\dot x_n + \varepsilon\dot h\right)\right]\right\}_{\varepsilon=0} &= \dot\rho^h_{x_n} g_{x_n} + \rho^h_{x_n}\dot g_{x_n} \\ \text{(ii)}\qquad \rho^h_{x_n} g_{x_n}\dot x_n + g_{x_n}\dot h &= \dot\rho^h_{x_n} g_{x_n} + \rho^h_{x_n}\dot g_{x_n}. \end{aligned}$$

When $n \to \infty$, the term containing $\dot x_n$ would give rise to an SDE. We know by Chapter IX.5 that the reduced variation is governed by an ODE. Therefore the term containing $\dot x_n$ *must* disappear; in fact we have

$$\begin{aligned} \rho^h_{x_n} g_{x_n}\dot x_n &= \rho_{x_n}\dot g_{x_n} \qquad \text{or} \\ g_{x_n}\dot x_n &= \dot g_{x_n}. \end{aligned}$$

Therefore (ii) implies that

$$g_{x_n}\dot{h} = \dot{\rho}^h_{x_n} g \qquad \text{or}$$

(iii) $$\rho^h_{x_n}(\tau) = Ad\left(g_{x_n}(\tau)\right)\dot{h}.$$

By the limit Theorem VII.6 $\{g_{x_n}\}$ converges uniformly in τ to g_x. Passing to the limit in (iii) we prove the proposition. □

1.4. The path space

1.4.1. Definition of Wiener measure. We denote by $\mathbb{P}(G)$ the space of continuous map of [0, 1] into G, taking the value e at $\tau = 0$. Then the Itô map

$$\mathcal{J} : x \to g_x(\cdot)$$

sends the space X into $\mathbb{P}(G)$. The Wiener measure μ_x is carried by $\mathcal{J}$ into $\mu_{\mathbb{P}}$:

$$(\mathcal{J})_* \mu_x = \mu_{\mathbb{P}}.$$

1.4.2. Cylindrical functions

A function Φ on $\mathbb{P}(G)$ is a *cylindrical test function* if there exists $\tau_1, \ldots, \tau_q$ and a C^1*-function* φ defined on G^q, bounded along with its first left derivatives such that

$$\Phi(p) = \varphi\left(p(\tau_1), \ldots, p(\tau_q)\right) \quad \text{for } p \in \mathbb{P}(G).$$

Given $z \in H^1([0, 1]; \mathcal{G})$ we define

$$\partial^\ell_z \Phi = \partial^\ell_{z(\tau_1)}\varphi^{(1)} + \partial^\ell_{z(\tau_2)}\varphi^{(2)} + \ldots + \partial^\ell_{z(\tau_n)}\varphi^{(n)}$$

where we denote by $\varphi^{(1)}$ the function φ where all the variables are fixed except the first one and so on.

1.4.3. Theorem of left infinitesimal quasi-invariance. *Given* $z \in H^1([0, 1], \mathcal{G})$, *there exists a function* $k_z \in L^{\infty - 0}(\mathbb{P}(G); \mu_{\mathbb{P}})$ *such that for every cylindrical test-function*

$$E\left(\partial^\ell_z \Phi\right) - E\left(k_z \Phi\right)$$

where k_z *is defined by the following stochastic integral*

$$k_z(g_x) = \int_0^1 \left\langle \mathrm{Ad}\left(g_x^{-1}(\tau)\right)\dot{z}(\tau), \ \ dx(\tau)\right\rangle.$$

Proof. The theorem results from the computation of the reduced variation given in 1.3 combined with the Bismut identity IX.7.

1.5. Tangent space to the path group

The tangent space $T_p(P(G))$ to $P(G)$, *at the point* p, is defined as the space of infinitesimal deformations, $\varepsilon \to 0$, of the following form

$$\exp\big(\varepsilon z(\tau)\big)p(\tau) \quad \text{where} \quad z \in H^1([0,1],\mathcal{G}).$$

The tangent space $T_x(X)$ to the probability space of the Brownian motion X on $\mathcal{G}$, *at the point* $x \in X$, is defined as the space of infinitesimal deformations of the form

$$x(\tau) + \varepsilon h(\tau) \quad \text{where} \quad h \in H^1([0,1];\mathcal{G})$$

In the previous chapters we have omitted the point x in the definition of $T_x(X)$, taking advantage of its invariance under the translations; in the present nonlinear situation the previous abuse of notation could be misleading.

Main Theorem. *The Itô map realizes a diffeomorphim of X onto $P(G)$.*

More precisely, for almost all $x_0 \in X$ denote $p_0 = \mathcal{I}(x_0)$, then $\mathcal{I}$ is differentiable at x_0 and its differential

$$\mathcal{I}'(x_0) : T_{x_0}(X) \to T_{p_0}(P(G))$$

is an isomorphism which has the following expression:

$$\mathcal{I}'(x_0)(h) = k \quad \text{where} \quad \dot{k}(\tau) = \mathrm{Ad}\,(p_0(\tau))\,\dot{h}(\tau).$$

Furthermore, under the following hypothesis $\mathcal{H}$ the map $\mathcal{I}'(x_0)$ is unitary.

($\mathcal{H}$) For all $g \in G$, Ad(g) *is an orthogonal transformation.*

Proof. This theorem is a formalized statement of what was proved in paragraph 1.4.3. The hypothesis that $\mathrm{Ad}\,(p_0(*))$ is an orthogonal matrix implies that $\|\mathcal{I}'(x_0)(h)\|_{H^1} = \|h\|_{H^1}$.

Remark. The following facts are elementary results of the Lie group theory:

- If G is compact there exists always a Euclidean metric on $\mathcal{G}$ for which the hypothesis $\mathcal{H}$ is true.
- Conversely, if the hypothesis $\mathcal{H}$ is satisfied, then G is the direct product of a compact group with Euclidean space.

2. Transport of structures from the Wiener space to the path group

In all this paragraph we shall assume the hypothesis ($\mathcal{H}$); then $\mathrm{Ad}(g^{-1}) = {}^t\mathrm{Ad}(g)$.

Using quasi-invariance of $\mu_{\mathbb{P}}$ under the left action of $\mathbb{P}^1(\mathcal{G})$ we define, as in Chapter II, the space of smooth functions $\mathbb{D}_1^{\infty-0}(\mathbb{P}(G))$.

2.1. Definitions. We will define D_z^ℓ by the formula

$$\left(D_z^\ell f\right)(p) = \left[\frac{d}{d\varepsilon}\, f(\exp(\varepsilon z)p)\right]_{\varepsilon=0}$$

$$u \in \mathbb{D}_1^{\infty-0}(\mathbb{P}(G)) \quad \text{if and only if} \quad u \circ \mathcal{I} \in \mathbb{D}_1^{\infty-0}(X),$$

and will define D_τ^ℓ by the formula

$$\int_0^1 \left(D_\tau^\ell f\right) \dot{z}(\tau) d\tau = D_z^\ell f.$$

We denote by $\left(\delta p(\tau)\right) p^{-1}(\tau)$ the martingale part of the stochastic differential $[dp(\tau)]p^{-1}(\tau)$.

We shall call a *vector field* any map of $\mathbb{P}(G)$ into $\mathbb{P}^1(\mathcal{G})$. If the map is adapted then the stochastic integral will be defined by

$$\int_0^1 \left\langle \dot{Z}(p,\tau), (\delta p) p^{-1}(\tau)\right\rangle.$$

2.2. Theorem. *If $Z(p,\tau)$ is an adapted vector field on $\mathbb{P}(G)$, then*

$$E\left(\int_0^1 \left(D_\tau^\ell f\right) \dot{Z}(p,\tau) d\tau\right) = E\left(f(p)\int_0^1 \left\langle \dot{Z}(p,\tau), (\delta p)p^{-1}(\tau)\right\rangle\right).$$

Proof. We define a vector field $Z_1(x,\tau)$ on X by the formula

$$Z_1(x,\tau) = \int_0^\tau \operatorname{Ad}\left(g_x^{-1}(\lambda)\right) \dot{Z}\left(g_x(\lambda), \lambda\right) d\lambda.$$

Then we have

$$\int_0^1 \left\langle \dot{Z}_1(x,\tau), dx(\tau)\right\rangle = \int_0^1 \left\langle Z(p,\tau), (\delta p)p^{-1}(\tau)\right\rangle.$$

We are then reduced to 1.4.3. □

2.3. Theorem (Clark-Bismut-Ocone). *Given a function $f \in \mathbb{D}_1^2(\mathbb{P}(G))$, then*

$$f = E(f) + \int_0^1 \left\langle E^{\mathcal{N}_\tau}\left(D_\tau^\ell f\right),\ (\delta p)p^{-1}(\tau)\right\rangle.$$

Proof. We assume $E(f) = 0$. We define a martingale by

$$M(x,\tau) = E^{\mathcal{N}_\tau}\left(f(g_x)\right).$$

Then granted the Itô representation of a Brownian martingale, we have

$$f(g_x) = M(x,1) = \int_0^1 a(x,t) dx(\tau).$$

We replace dx by $\mathrm{Ad}\left(g_x^{-1}\right) dy$ to obtain

$$f(p) = \int_0^1 \left\langle b(p,\tau),\quad (\delta p) p^{-1}(\tau)\right\rangle .$$

For every adapted vector field $Z(p,\tau)$ we have

$$\begin{aligned} E\left(\int_0^1 Z(p,\tau)\left(D_\tau^\ell f\right)(p)(d\tau\right) &= E\left(f(p)\int_0^1 Z(p,\tau)(\delta p)p^{-1}(\tau)\right) \\ &= E\left(\int_0^1 \langle b(p,\tau),\quad Z(p,\tau)\rangle\, d\tau\right). \end{aligned}$$

Therefore

$$\int_0^1 \left\langle E^{\mathcal{N}_\tau} D_\tau^\ell f - b, Z\right\rangle d\tau = 0$$

for every adapted vector field; then by duality

$$E^{\mathcal{N}_\tau}\left(D_\tau^\ell f\right) = b.$$

□

Conclusion

Granted the Definitions 2.1, the theory on the path group looks the same as the theory on the classical Wiener space.

3. Some basic facts in Riemannian geometry

The purpose of this paragraph is to recall some classical geometric concepts; the proofs of the difficult results will be omitted. The reader can skip this paragraph in a first reading.

3.1. Laplacian on a Riemannian manifold

3.1.1. Riemannian manifold

We denote by M a C^3-Riemannian manifold: for each $m \in M$ an Euclidean metric is given on the tangent space $T_m(M)$. In a local chart m^i the Riemannian metric is given by a C^2-function γ with values in the symmetric positive matrices:

$$ds^2 = \sum \gamma_{i,j}(m) dm^i dm^j, \qquad i,j \in [1,\ell].$$

The gradient of a function f has for components

$$\sum_j \gamma_{1,j}\partial_j f, \ldots, \sum_j \gamma_{\ell,j}\partial_j f$$

where $\partial_j f = \dfrac{\partial f}{\partial m^j}$.

The volume element is defined by

$$dv = [\det(\gamma(m))]^{-1/2} dm^1 \ldots dm^2.$$

Given a vector field

$$Z = a^i(m)\frac{\partial}{\partial m^i},$$

its divergence is expressed by

$$\operatorname{div}_v(Z) = -\partial_i a - \frac{1}{2}\mathcal{L}_Z[\log(\det \gamma)],$$

where $(\mathcal{L}_Z\varphi)(m)$ is the derivative of φ along the vector $Z_{\dot{m}}$.

3.1.2. Riemannian diffusion

The Riemannian Laplacian is defined by

(i) $$\Delta_M f = \frac{1}{2}\operatorname{div}(\operatorname{grad} f).$$

Δ is a symmetric operator in $L^2(dv)$. The Dirichlet form associated to Δ has the following form

$$\mathcal{D}(f) = \int_M \|\nabla f\|^2 dv \qquad \text{where}$$

$$\|\nabla f\|^2 = \sum \gamma_{i,j}\partial_i f \partial_j f, \quad \operatorname{grad}(f) = \gamma^{\cdot\cdot j}\partial_j f.$$

A *Riemannian diffusion* will be the diffusion associated to the operator Δ defined in (i). In this chapter, we shall make the permanent assumption that the lifetime of all the considered diffusions is always infinite.

3.1.3. Geodesics

We denote by $\gamma^{i,j}$ the inverse matrix of $\gamma_{i,j}$. Given a C^1-curve $\psi(\tau)$ its length $L(\psi)$ is defined by

$$L^?(\psi) = \int_0^1 \gamma^{i,j}(\psi(\tau))\dot{\psi}_i(\tau)\dot{\psi}_i(\tau)d\tau.$$

A *geodesic* will be a C^2-curve which is an extremal of the length functional L. The Euler equation associated to L is a second order ordinary differential equation. This implies that given

$$m_0 \in M \quad \text{and} \quad z_0 \in T_{m_0}(M)$$

there exists a unique geodesic γ^0 satisfying the initial value problem

(i) $$\gamma(0) = m_0, \qquad \dot{\gamma}(0) = z_0.$$

The Riemannian distance on M will be defined by

$$d(m, m') = \inf L(\gamma)$$

where γ ranges over the C^1-curves satisfying $\gamma(0) = m$, $\gamma(1) = m'$.

We shall always assume that M is a *complete metric* space for this distance.

3.1.4. Normal chart

Given $m_0 \in M$ we define a map

$$\exp_{m_0} : T_{m_0}(M) \to M$$

in the following way: To every $z \in T_{m_0}(M)$ we associate the geodesic γ_z satisfying 3.1.3.(i). Then we define

$$\exp_{m_0}(z) = \gamma_z(1).$$

Theorem. *There exists a positive number r such that denoting*

$$\begin{aligned} B &= \left\{z \in T_{m_0}(M);\ \|z\| < r\right\} \\ \tilde{B} &= \{m \in M; d(m_0, m) < r\} \end{aligned}$$

then

(i) $$\exp_{m_0} \quad \text{is a } C^2\text{-diffeomorphism of } B \text{ to } \tilde{B}.$$

Therefore $\exp_{m_0}^{-1}$ *constitutes a local chart of* $\tilde{B}$. *The coefficients of the Riemannian metric expressed in this chart satisfy*

$$\gamma_{i,j}(0) = 1 \quad \text{if} \quad i = j, \quad \text{zero otherwise}$$

(ii) $$\left(\partial_s \gamma_{i,j}\right)(0) = 0 \quad \text{for all} \quad s, i, j.$$

(iii) Corollary. *The Riemannian Laplacian at the origin of the normal chart is equal to the Euclidean Laplacian on* $T_{m_0}(M)$.

Proof.

$$\begin{aligned} \partial_j[\log\det(g)](0) &= 0 \\ \left(\partial_j g_{i,k}\right)(0) &= 0. \end{aligned}$$

3.2. The bundle of orthonormal frames

3.2.1. Definitions. Given $m_0 \in M$, we call an *orthonormal frame* at m_0 a *Euclidean isomorphism* of $\mathbb{R}^\ell$ *onto* $T_{m_0}(M)$. We denote by $\mathcal{O}_{m_0}(M)$ the collection of all orthonormal frames of $T_{m_0}(M)$. We denote by $\mathcal{O}(M)$ the *disjoint union* of all the $\mathcal{O}_m(M)$

(i)
$$\mathcal{O}(M) = \bigcup_{m \in M} \mathcal{O}_m(M).$$

Given $r_0 \in \mathcal{O}(M)$ there exists a unique $m_0 \in M$ such that $r_0 \in \mathcal{O}_{m_0}(M)$. We denote by π the projection of $\mathcal{O}(M)$ onto M defined by

(ii)
$$\pi(r_0) = m_0.$$

Then we have $\pi^{-1}(m_0) = \mathcal{O}_{m_0}(M)$.

3.2.2. The $\mathcal{O}(\ell)$ action

We denote by $O(\ell)$ the orthogonal matrices of rank ℓ, by $SO(\ell)$ the matrices of $O(\ell)$ of determinant 1. Given $g_0 \in O(\ell)$, we define an action

$$m_g : \mathcal{O}(M) \to \mathcal{O}(M) \qquad \text{by}$$

$$m_g(r)(z) = r \circ g.$$

We have

$$m_{g_1} \circ m_{g_2} = m_{g_2 g_1}.$$

Furthermore the action of $O(\ell)$ on $\mathcal{O}_m(M)$ is simply transitive.

3.2.3. Sections of $\mathcal{O}(M)$

We shall call a *section* of $\mathcal{O}(M)$ any map

$$\sigma : M \to \mathcal{O}(M) \qquad \text{such that}$$

$$\pi \circ \sigma = \text{Identity on } M.$$

Proposition. *A local chart defines above its domain a natural section of $\mathcal{O}(M)$.*

Proof. Denote by φ this local chart and by g_{ij} the Riemannian metric in this chart. We denote by $\frac{\partial}{\partial \xi^1}, \dots, \frac{\partial}{\partial \xi^\ell}$ the canonical basis of $\mathbb{R}^\ell$; we apply to this basis the Schmidt orthogonalisation procedure. We get

$$\begin{aligned} e_1(\xi) &= \left[g^{1,1}(\xi)\right]^{-1/2} \frac{\partial}{\partial \xi^1} \\ e_k(\xi) &= \alpha_k(\xi) \left[\frac{\partial}{\partial \xi^k} - \sum_{j<k} \beta_{j,k} \frac{\partial}{\partial \xi^j} \right]. \end{aligned}$$

We define on the domain of φ a system of ℓ orthonormal vector fields E_j by the following formula

$$\varphi'(m) \cdot E_k(m) = e_k(\varphi(\xi)).$$

Given $m_0 \in M$, $\xi \in \mathbb{R}^\ell$ then the map

$$\xi \to \sum \xi^j E_j$$

defines a frame $\sigma(m_0)$. □

3.2.4. $\mathcal{O}(M)$ is a C^2-manifold

In fact, given a local chart ψ, denote by Ω its domain and by σ the associated section of $\mathcal{O}(\Omega)$. Then the map $O(\ell) \times \Omega \to \mathcal{O}(\Omega)$ defined by $(g, m) \to g \cdot \sigma_m$ gives the following identification

$$O(\ell) \times \Omega = \mathcal{O}(\Omega).$$

We take on the l.h.s. the product differentiable structure. This structure is compatible with the change of charts.

3.3. The canonical parallelism on $\mathcal{O}(M)$

We shall define parallel transport using the bundle of orthonormal frames.

3.3.1. The canonical horizontal vector fields

Given $r_0 \in \mathcal{O}(M)$ we denote $m_0 = \pi(r_0)$; then

$$r_0^{-1}\left(\exp_{m_0}^{-1}(m)\right) = \varphi_{r_0}(m)$$

defines a local chart around m_0. We denote by σ_{r_0} the section of $\mathcal{O}(M)$ associated in 3.2.3 to φ_{r_0}. We consider

$$\theta_{r_0} = \sigma_{r_0} \circ \varphi_{r_0}^{-1}.$$

Then θ_{r_0} is a C^2-map of a neighbourhood of zero in $\mathbb{R}^\ell$ into $\mathcal{O}(M)$. We define

$$A_j(r_0) = \theta'_{r_0}(0)\partial_j \qquad j \in [1, \ell].$$

In this way we associate to every point $r_0 \in \mathcal{O}(M)$ a tangent vector $A_j(r_0)$; the correspondence $r_0 \to A_j(r_0)$ defines a *vector field* A_j on $\mathcal{O}(M)$.

3.3.2. Horizontal space

We shall call the *horizontal space* H_{r_0} at r_0 the subspace of $T_{r_0}(\mathcal{O}(M))$ spanned by the $A_j(r_0)$.

3.3.3. Proposition. *We have*

(i) $$\pi'\left(A_j(r_0)\right) = r_0\left(\frac{\partial}{\partial \xi^j}\right)$$

(ii) $$(m_\gamma)_* A_j = \sum_i \gamma^{i,j} A_i$$

where m_γ are as defined in 3.2.2 and where $\gamma^{i,j}$ is the orthogonal matrix associated to γ.
(iii) The A_j are of class C^1.

3.3.4. Lemma. *Given $z_0 \in T_{m_0}(M)$ and given $r_0 \in \pi^{-1}(m_0)$ there exists a unique tangent vector such that*

$$\pi'(\tilde{z}) = z, \qquad \textit{and} \quad \tilde{z} \in H_{r_0}.$$

We shall call $\tilde{z}$ the horizontal lift of z at r_0.

Proof. We have

$$\pi'\left(A_j(r_0)\right) = r_0\left(\frac{\partial}{\partial \xi^j}\right),$$

therefore

$$\tilde{z} = \sum_j \left(r_0^{-1}(z)\right)^j A_j(r_0).$$

□

3.3.5. Proposition. *Given a C^1 curve $\gamma(\tau)$ on M and given $r_0 \in \pi^{-1}(m_0)$ there exists a unique curve $\tilde{\gamma}(\tau)$ on $\mathcal{O}(M)$ such that*

$$\tilde{\gamma}(0) = r_0,$$

$$\frac{d\tilde{\gamma}}{d\tau} \in H_{\tilde{\gamma}} \qquad \pi'\left(\frac{d}{d\tau}\tilde{\gamma}(\tau)\right) = \frac{d}{d\tau}\gamma(\tau).$$

We shall call $\tilde{\gamma}$ the lifted curve of γ starting from r_0.

To $r_1, r_2 \in \pi^{-1}(m_0)$ we associate two lifted curves $\tilde{\gamma}_1$, $\tilde{\gamma}_2$, then there exists $g \in O(\ell)$ such that

$$\tilde{\gamma}_2(\tau) = m_g \cdot \tilde{\gamma}_1(\tau) \quad \textit{for all} \quad \tau \in [0, 1].$$

Proof. Denote by φ a local chart of M around m_0. There exists τ_0 such that $\gamma(\tau)$ belongs to the domain Ω of φ for $\tau \in [0, \tau_0]$. Denote

$$\mathcal{O}(\gamma) = O(\ell) \times [0, \delta].$$

On $\mathcal{O}(\gamma)$ we define the vector field

$$Z(g, \tau) = (\tilde{z}(r, \tau), 1)$$

where $z(\tau) = \frac{d}{d\tau}\gamma$ and where $\tilde{z}(g, \tau)$ is the lift of z at the frame associated to r. Then the lifted curves are the integral curves of Z; given an initial point there exists a unique integral curve of Z starting from that point. □

We remark that by 3.3.3.(ii), m_g preserves the horizontal space. Therefore $m_g\left(\tilde{\gamma}(\tau)\right)$ is a lifted curve of γ. □

3.4. Theorem. *Given a C^1 curve γ on M, there exists an Euclidean isomorphism $t^{\gamma}_{\tau\leftarrow 0}$ of $T_{\gamma(0)}(M)$ with $T_{\gamma(\tau)}(M)$.*

We will call $t^{\gamma}_{\cdot\leftarrow 0}(z_0)$ the parallel transport of z_0 along γ.

Proof. We choose $r_0 \in \pi^{-1}(\gamma(0))$. Denote by $\tilde{\gamma}$ the lifted curve starting from r_0. We define

$$t^{\gamma}_{t\leftarrow 0} = [\tilde{\gamma}(\tau)] \circ [\tilde{\gamma}(0)]^{-1}.$$

This definition does not depend upon the choice of r_0; if we replace r_0 by $m_g r_0 = r_0 \circ g^{-1}$, the lifted curve will become $\tilde{\gamma}(\tau) \circ g_0^{-1}$ and

$$\tilde{\gamma}(\tau) \circ g_0^{-1} \circ g_0 \circ (\tilde{\gamma}(0))^{-1} = \tilde{\gamma}(\tau) \circ \tilde{\gamma}^{-1}(0).$$

□

3.5. Theorem. *A curve γ is a geodesic curve if and only if*

(i) $$t^{\gamma}_{\tau\leftarrow 0}(\dot{\gamma}(0)) = \dot{\gamma}(\tau).$$

3.5.1. Corollary. *Given a geodesic γ, choose $r_0 \in \pi^{-1}(\gamma(0))$ such that*

$$r_0(\dot{\gamma}(0)) = \frac{\partial}{\partial \xi^1},$$

then the lift $\tilde{\gamma}$ of γ starting at r_0 is the integral curve of the horizontal vector field A_1.

Proof. Consider an integral curve u of A_1, then $\gamma_1 = \pi \circ u$ has u for a lifted curve. Therefore γ satisfies the geodesic equation 3.5.(i).

As $\gamma(0) = \gamma_1(0)$, $\dot{\gamma}(0) = \dot{\gamma}_1(0)$, then given that the Euler equation is a second order differential equation we get $\gamma = \gamma_1$. □

3.6. The horizontal Laplacian

We define a semi-elliptic operator on $\mathcal{O}(M)$ by the following formula

$$\Delta_{\mathcal{O}(M)} = \frac{1}{2}\sum_{j=1}^{\ell} \mathcal{L}^2_{A_j}$$

where $A_1, \ldots, A_\ell$ are the canonical horizontal vector fields.

Theorem. *Given $f \in C^2(M)$, define $\tilde{f} = f \circ \pi$, then*

$$\Delta_{\mathcal{O}(M)} \tilde{f} = (\Delta_M f) \circ \pi$$

or symbolically

$$\Delta_{\mathcal{O}(M)} \pi^* = \pi^* \Delta_M.$$

Proof. We consider $r_0 \in \mathcal{O}(M)$, $m_0 = \pi(r_0)$. We define

$$u(\xi) = f\left(\exp_{m_0}(\xi)\right).$$

Then

$$(\Delta_M f)(m_0) = \left(\Delta_{\mathbb{R}^\ell} u\right)(0).$$

Consider the geodesic γ_j such that

$$\gamma_j(0) = m_0, \qquad r_0\left(\dot{\gamma}_j(0)\right) = \frac{\partial}{\partial \xi_1}.$$

Then

$$(\Delta_M f)(m_0) = \frac{1}{2} \sum_j \left\{ \frac{d^2}{d\tau^2}\, f\left(\gamma_j(\tau)\right)\right\}_{\tau=0}.$$

Denote by $\tilde{\gamma}_j$ the lift of γ_j starting from r_0. Then

$$\tilde{f}\left(\tilde{\gamma}_j(\tau)\right) = f\left(\gamma_j(\tau)\right).$$

We have

$$\Delta_M f(m_0) = \sum_j \left\{ \frac{d^2}{d\tau^2}\, \tilde{f}\left(\tilde{\gamma}_j(\tau)\right)\right\}_{\tau=0}.$$

By 3.5.1 we know that $\tilde{\gamma}_j(\tau)$ is the integral curve of the vector field A_j; therefore

$$\frac{d^2}{d\tau^2}\, \tilde{f}(\tilde{\gamma}_j) = \mathcal{L}^2_{A_j} \tilde{f}.$$

□

3.7. The canonical parallelism on $\mathcal{O}(M)$

We denote

$$D(\ell) = R^\ell \oplus so(\ell)$$

where $so(\ell)$ denotes the $\ell \times \ell$ antisymmetric matrices. We can identify $D(\ell)$ with the Lie algebra of the group of Euclidean motions of $\mathbb{R}^\ell$, the Lie algebra bracket being defined by

$$\left[(z, a), (z', a')\right] = (az' - a'z, aa' - a'a).$$

Theorem. *On $\mathcal{O}(M)$ there exists a canonical differential form $\theta = (\theta^1, \theta^2)$ taking its values in $D(\ell) = \mathbb{R}^\ell \times so(\ell)$ such that*

$$\theta_r \quad \text{is an isomorphism of } T_r(\mathcal{O}(M)) \text{ onto } D(\ell)$$

$$\langle \theta_r, A_j(r)\rangle = \left(\frac{\partial}{\partial \xi^j}, 0\right)$$

$$(m_g)^*\theta = (g \circ \theta_1, \mathrm{Ad}(g)\theta_2).$$

Proof. We split θ in its two components

$$\theta = (\theta_1, \theta_2)$$

where θ_1 is an $\mathbb{R}^\ell$-valued differential form, θ_2 is an $so(\ell)$-valued differential form.

Given $Z \in T_{r_0}(\mathcal{O}(M))$ we define

$$\langle \theta_1, Z\rangle = r_0^{-1}(\pi'(Z)).$$

Now we denote by Z_1 the horizontal vector such that $\pi' Z_1 = \pi' Z$. Define

$$Z_2 = Z - Z_1.$$

Then Z_2 is tangent to $\mathcal{O}_m(M)$. We define

$$\langle \theta_2, Z\rangle_{r_0} = \alpha$$

by the formula

$$\left\{\frac{d}{d\varepsilon} f\,(r_0 \circ \exp(-\varepsilon\alpha))\right\}_{\varepsilon=0} = \left(\mathcal{L}_{Z_2} f\right)(r_0)$$

which is to be satisfied for every test function f.

A vector field Z on $\mathcal{O}(M)$ is called *vertical* if we have $\pi'(Z) = 0$. Given $\alpha \in so(\ell)$ we define a canonical vector field Z^α on $\mathcal{O}(M)$ by

$$\theta(Z^\alpha) = (0, \alpha).$$

Then we have

$$(\mathcal{L}_{Z^\alpha} f)\,(r) = \left\{\frac{d}{d\varepsilon} f\big(r \circ \exp(-\varepsilon\alpha)\big)\right\}_{\varepsilon=0}.$$

We want to prove that

(i) $$(m_g)_* Z^\alpha = \mathrm{Ad}(g^{-1}) Z^\alpha.$$

In fact

$$\mathcal{L}_{(m_g)_* Z^\alpha} f(r) = (\mathcal{L}_{Z^\alpha} f)\left(m_g^{-1}(r)\right) = \left\{\frac{d}{d\varepsilon} f\big(r \circ \exp(-\varepsilon\alpha)\big) \circ g\right\}_{\varepsilon=0}.$$

We have

$$r \circ \exp(-\varepsilon\alpha) \circ g = r \circ g \circ \exp\left(-\varepsilon \mathrm{Ad}(g^{-1})\alpha\right).$$

This relation proves (i) and the desired action of $(m_g)^*$ on θ_2. The action on θ_1 results then from 3.3.3.(ii). □

Remark. The parallelism form θ on $\mathcal{O}(M)$ plays the same basic role in Riemannian geometry as the Maurer-Cartan differential form $g^{-1}dg$ does in Lie group theory.

3.8. The structure equation

Theorem. *There exists an antisymmetric map Ω of $\mathbb{R}^\ell$, taking its values in $so(\ell)$ such that*

$$\begin{aligned} \langle d\theta_1, Z \wedge Z' \rangle + \theta_2(Z)\theta_1(Z') - \theta_2(Z')\theta_1(Z) &= 0 \\ \langle d\theta_2, Z \wedge Z' \rangle + \theta_2(Z)\theta_2(Z') - \theta_2(Z')\theta_2(Z) &= \Omega(Z, Z'). \end{aligned}$$

We will denote

$$\left[\Omega\left(\frac{\partial}{\partial \xi^i}, \frac{\partial}{\partial \xi^j}\right)\right]_{\alpha,\beta} = R_{i,j,\alpha,\beta}.$$

We will call $\Omega_{i,j,\alpha,\beta}(r)$ the curvature tensor of M expressed in the frame r.

4. Calculus of variations on the horizontal flow

4.1. The horizontal flow

We consider the following stochastic differential equation

$$\begin{cases} dr_x = \sum_{j=1}^{\ell} A_j(r_x) \, odx^j \\ r_x(0) = r_0 \end{cases}.$$

We shall denote its solution by $U^x_{\tau\leftarrow 0}(r_0)$. Then $U^x_{\tau\leftarrow 0}$ defines a stochastic flow of diffeomorphisms which will be called the horizontal flow.

4.2. Action of $so(\ell)$ on the horizontal flow

Proposition. *Given $g \in so(\ell)$, we have*

$$m_g \, U^x_{\tau\leftarrow 0} \, m_g^{-1} = U^y_{\tau\leftarrow 0}$$

where

$$y(\tau) = g\big(x(\tau)\big).$$

Proof. 3.3.ii.

4.3. Corollary. *Consider an $so(\ell)$-valued adapted functional g, constant in the intervals $[\tau_k, \tau_{k+1}]$, $k \in [1, q]$ $0 < \tau_1 < \ldots < \tau_q < 1$.*

Define

$$y(\tau) = \int_0^\tau g(s)dx(s)$$

and define

$$\delta_s = g(\tau_k)g^{-1}(\tau_{k-1}).$$

Then

$$m_{g(\tau)}U^x_{\tau\leftarrow 0} = U^y_{\tau\leftarrow\tau_s} m(\delta_s) U^y_{\tau_s\leftarrow\tau_{s-1}} m(\delta_{s-1}) \ldots U^y_{\tau_1\leftarrow 0} m_{g(0)}.$$

Proof. We apply 4.2 in each of the intervals $[\tau_k, \tau_{k+1}]$. □

4.4. Theorem. *Fix $m_0 \in M$, $r_0 \in \pi^{-1}(m_0)$. Consider $\mathbb{P}_{m_0}$ with the probability measure defined by diffusion associated to Δ_M starting from m_0. Then the mapping*

$$\Phi : x \to \pi\left(U^x_{\cdot\leftarrow 0}(r_0)\right)$$

is a bijective probability preserving map.

Proof. The identity established in 3.6

$$\Delta_{\mathcal{O}(M)}\pi^* = \pi^*\Delta_M$$

shows that Φ is a probability preserving map. We only have to show the *injectivity* which has been proved in Malliavin [4].

4.5. The stochastic parallel transport (Itô [5])

Given a sample path $p(\tau)$ of the diffusion on M, we choose $r_0 \in \pi^{-1}(p(0))$. We take $r(\tau)$ to be the horizontal lift of $p(\tau)$ starting from r_0. Granted Theorem 4.4, this lift is uniquely defined. We define

$$t^{\gamma}_{\tau\leftarrow 0}(z) = r(\tau) \circ r^{-1}(0).$$

As in 3.4 the map t^{γ} is independent of the choice of r_0; *therefore there exists a canonical linear Euclidean isomorphism* $t^{\gamma}_{\tau\leftarrow 0}$ of $T_{p(0)}(M)$ *onto* $T_{p(\tau)}(M)$.

This isomorphism will be called the parallel displacement.

We denote by x_n the Stroock-Varadhan approximation of x associated to the partition $\{k2^{-n}\}$.

4.6. Proposition. *The differential equation of the horizontal flow $U^{x_n}_{\tau\leftarrow 0}$ can be expressed through the parallelism form θ as follows:*

$$\begin{aligned} \langle dr_{x_n}, \theta_1\rangle &= dx_n \\ \langle dr_{x_n}, \theta_2\rangle &= 0. \end{aligned}$$

Proof.

$$dr_{x_n} = A_j\left(r_{x_n}\right) o dx_n.$$

4.7. Jacobian matrices of the horizontal flow

Given a curve $\lambda \mapsto r_0(\lambda)$ on $\mathcal{O}(M)$, define

$$\Phi(\tau, \lambda) = U^{x_n}_{\tau\leftarrow 0}\left(r_0(\lambda)\right).$$

We get a map

$$\Phi : [0, 1]^2 \to \mathcal{O}(M).$$

We push back the parallelism form θ by Φ to obtain

$$\rho := \Phi^*(\theta) = a(\tau, \lambda)d\tau + b(\tau, \lambda)d\lambda.$$

We split a and b into their components:

$$\begin{aligned} a &= a_1 + a_2 \\ b &= b_1 + b_2 \end{aligned}$$

where $a_1, b_1 \in \mathbb{R}^\ell$ and $a_2, b_2 \in so(\ell)$.

Now we have by 4.6

4.7.1.
$$a_1(\tau, \lambda) = \dot{x}_n, \qquad a_2 = 0.$$

Furthermore

4.7.2.
$$b(\tau, 0) = \frac{\partial \Phi}{\partial \lambda}(\tau, 0)$$

is the desired variation.

We shall use the commutation of exterior differentiation with the inverse image:

$$\left\langle d\Phi^*(\theta), \frac{\partial}{\partial \tau} \wedge \frac{\partial}{\partial \lambda} \right\rangle = \left\langle \Phi^*(d\theta), \frac{\partial}{\partial \tau} \wedge \frac{\partial}{\partial \lambda} \right\rangle.$$

Then we express $d\theta$ by the structure equation 3.8. Taking into account the fact that $a_2 = 0$ (cf. 4.7.1) we get

$$\begin{aligned} \frac{\partial b_1}{\partial \tau} - \frac{\partial a_1}{\partial \lambda} &= b_2 a_1 \\ \frac{\partial b_2}{\partial \tau} - \frac{\partial a_2}{\partial \lambda} &= \Omega(a_1, b_1). \end{aligned}$$

Using 4.7.1 and 4.7.2 we have for $\lambda = 0$

$$\frac{\partial a_1}{\partial \lambda} = 0, \qquad \frac{\partial a_2}{\partial \lambda} = 0, \qquad a_1 = \dot{x}_n.$$

We get

$$\begin{aligned} \dot{b}_1 &= b_2 \dot{x}_n, \\ \dot{b}_2 &= \Omega\,(\dot{x}_n, b_1)\,. \end{aligned}$$

Denote by $\mathcal{M}_j$ the endomorphism of $D(\ell)$ defined by

$$\mathcal{M}_j(b_1, b_2) = \left([b_2]_j^*, \Omega(b_1, e_j)\right)$$

where $b_1 \in \mathbb{R}^\ell$, $b_2 \in so(\ell)$ and where e_* denotes the canonical basis of $\mathbb{R}^\ell$.

Theorem. *Consider the endomorphism $J^{x_n}_{\tau\leftarrow\sigma}$ of $D(\ell)$ constructed as the solution of the following linear differential equations, starting from the identity at $\tau = \sigma$:*

(4.7.3.)
$$\frac{d}{d\tau} J^{x_n}_{\tau\leftarrow\sigma} = \left[\sum_j \mathcal{M}_j \dot{x}^j_n\right] J^{x_n}_{\tau\leftarrow\sigma}.$$

Then the variation of the horizontal flow is given in the canonical parallelism of $\mathcal{O}(M)$ by

(4.7.4.)
$$\frac{d}{d\lambda} U^{x_n}_{\tau\leftarrow 0}(r_0(\lambda))\bigg|_{\lambda=0} = J^{x_n}_{\tau\leftarrow 0}\left(\frac{d}{d\lambda} r_0(\lambda)\right)\bigg|_{\lambda=0}.$$

Corollary. *Given a variation $x_n \to x_n + \varepsilon z$, where $z \in C^1([0,1];\mathbb{R}^\ell)$, then*

(4.7.5.)
$$\frac{d}{d\varepsilon} U^{x_n+\varepsilon z}_{\tau\leftarrow 0}\bigg|_{\varepsilon=0} = \int_0^\tau J^{x_n}_{\tau\leftarrow\sigma}\dot{z}(\sigma)\, d\sigma.$$

5. Tangent processes

5.1. Tangent processes on Wiener space and on path space and orthogonal invariance of Brownian motion

We denote by X the probability space of the $\mathbb{R}^d$-valued Brownian motion. To an adapted $so(\ell)$-valued process q we associate the $SO(d)$-valued process $\exp(\varepsilon q)$ and the process y defined by the following Itô stochastic integral

$$y_\varepsilon(x)(\tau) = \int_0^\tau \exp\left(\varepsilon q(x,s)\right) dx(s).$$

Then y is an $\mathbb{R}^\ell$-valued Brownian motion; therefore for every cylindrical function $\Phi(p) = \varphi(p(\tau_1), \ldots, p(\tau_n))$ we have

$$E(\Phi(y_\varepsilon(x)) = E(\Phi(x)).$$

Differentiating relatively to ε, we get

(5.1.1)
$$E\left[D_\zeta(\Phi)\right] = 0$$

where

$$D_\zeta\Phi = \sum_q \frac{\partial}{\partial\xi^\alpha}\varphi\left(p(\tau_1), \ldots, p(\tau_q), \ldots p(\tau_n)\right)\zeta^\alpha(\tau_q),$$

and where

$$\zeta^\alpha(\tau) = \int_0^\tau q^\alpha_\beta(\sigma)\, dx^\beta(\sigma).$$

The notion of *tangent processes on the Wiener space* will come from the merging of the previous variation with a variation of the Cameron-Martin type.

5.1.2. Definition. A *tangent process* is an $\mathbb{R}^\ell$-valued semi-martingale $\zeta^\alpha(\tau)$ such that its Itô differential satifies

$$d\zeta(\tau) = q^\alpha_\beta(\tau)\, dx^\beta + h^\alpha(\tau)\, d\tau \quad \text{where} \quad q^\alpha_\beta + q^\beta_\alpha = 0.$$

We denote by $T^0(X)$ the vector space of tangent processes; the subspace of tangent processes such that $q = 0$ will be denoted by $T^1(X)$: it will be called the space of Cameron-Martin-like tangent processes.

We denote by $\mathcal{C}^1$ the vector space of Itô semi-martingales, vanishing at time 0, which can be represented by an Itô integral formula and also by a Stratonovich integral. We shall assume in the next sections that

$$\zeta, q \in \mathcal{C}^1. \tag{5.1.3}$$

Theorem (Formula of integration by parts). *Given a tangent process ζ, then for every cylindrical function Φ*

$$E(D_\zeta \Phi) = E(\Phi k_\zeta), \quad \text{where} \quad k_\zeta(x) = \int_0^1 \sum_\alpha h^\alpha\, dx^\alpha. \tag{5.1.4}$$

Proof. We denote by ζ' the semi-martingale defined by $d\zeta' = d\zeta - h\, d\tau$; then $D_\zeta = D_{\zeta'} + D_h$. We apply to $D_{\zeta'}$ the identity (5.1.1); then we are reduced to a Cameron-Martin-like tangent process, which case was treated in Chapter VII.

Remark. Driver has proved conversely that a semi-martingale satisfying a formula of integration by part must be a tangent process.

5.1.5. Closability of the operator D_ζ

An unbounded densely defined linear operator A from the Banach space B_1 to the Banach space B_2 is said to be *closable* if the closure $\bar{\Gamma}$ of its graph $\Gamma = \{(b_1, b_2) \in B_1 \times B_2;\ \ b_2 = A(b_1)\}$ has the *graph property* which means that the restriction to $\bar{\Gamma}$ of the projection $\pi : B_1 \times B_2 \to B_1$ is injective: then $\bar{\Gamma}$ is the graph of an operator $\bar{A}$ called the *closure of the operator A*.

We take $A = D_\zeta$, its domain of definition being the cylindrical functions.

Assume that k_ζ defined in (5.1.4) satisfies $k_\zeta \in L^q$, then the operator D_ζ is closable in L^p.

The non-closability of D_ζ in L^p would imply the existence of $z \neq 0$, $z \in$ kernel $(\pi|_{\bar{\Gamma}})$: let us approximate z by a sequence $(\varphi_n, D_\zeta \varphi_n)$, where the φ_n are cylindrical functions, such that $\varphi_n \to 0$, $D_\zeta \varphi_n \to u \neq 0$, with convergence in L^p norm. We denote by v a generic cylindrical function, then

$$E\left(v D_\zeta \varphi_n\right) = -E\left(\varphi_n D_\zeta v\right) + E\left(\varphi_n v k_\zeta\right) \to 0;$$

therefore $E(uv) = 0$ for every cylindrical function; this implies $u = 0$. □

5.1.6. Definition. A *tangent process on the path space* is given by

$$Z_\tau(p) \in T_{p(\tau)}(M)$$

such that $z(\tau) = t^p_{0\leftarrow\tau} Z_\tau$ is a tangent process on X.

We denote by $T^0\left(P_{m_0}(M)\right)$ the vector space of tangent processes on $P_{m_0}(M)$.

The Cameron-Martin like tangent processes $T^1\left(P_{m_0}(M)\right)$ are the image under $t^p_{*\leftarrow 0}$ of $T^1(X)$.

5.2. Theorem (Cruzeiro-Malliavin isomorphism). *The Itô map $\mathcal{I}: X \to P_{m_0}(M)$, defined by $\mathcal{I}(x) = \pi\left(U^x_{\tau\leftarrow 0}(r_0)\right)$, induces by differentiation a map $\mathcal{I}'$ which is an isomorphism of the space of tangent processes on X satisfying (5.1.3) onto the space of tangent processes on $P_{m_0}(M)$, satisfying (5.1.3). More precisely, for every cylindrical function Φ on $P_{m_0}(M)$ we have the intertwining formula*

$$D_\zeta[\Phi \circ \mathcal{I}] = [D_{\zeta^*}\Phi] \circ \mathcal{I}$$

where

(5.2.1) $d\zeta^* = d\zeta + \rho \; odx, \quad d\rho = \Omega(\zeta^*, \; odx)$ *with* $\zeta(0) = 0, \; \rho(0) = 0.$

Proof. We apply to (4.7.5) the transfer principle and, denoting by Π the projection $\mathbb{R}^\ell \times so(\ell) \to \mathbb{R}^\ell$, we get by (4.8.4)

(i) $$\zeta^*(\tau) = \Pi \int_0^\tau J^x_{\tau\leftarrow\sigma} \; od\zeta(\sigma),$$

where

(ii) $$d_\tau \Pi J^x_{\tau\leftarrow 0} = \left[\sum_j \Pi \mathcal{M}_j \; odx^j\right] J^x_{\tau\leftarrow 0}.$$

We denote $\rho = (1-\Pi)J^x_{*\leftarrow 0}$. We compute the stochastic differential of the equation (i) and get

(iii) $$d\zeta^* = d\zeta + \rho \; odx.$$

We multiply the two members of 4.7.5 by $(1 - \Pi)$ and we differentiate:

(iv) $$d\rho = \Omega(\zeta^*, \; odx).$$

The two equations (iii) and (iv) prove (5.2.1); the initial conditions determine uniquely (ζ^*, ρ).

The correspondence $\zeta \to \zeta^*$ is injective; let ζ_0 an element of its kernel, then $d\rho = 0$, $\rho = 0$, $d\zeta = 0$, $\zeta = 0$.

Surjectivity results from the fact that, given ζ^* the matrix ρ is determined by the second equation (5.2.1) and then the first equation (5.2.1) determines ζ. □

Remark. If $\zeta \in T^1(X)$ and if the curvature $\Omega \neq 0$, then $\rho \neq 0$ and hence $\zeta^* \notin T^1\left(P_{m_0}(M)\right)$. The theory could not have been developed in the framework of Cameron-Martin like tangent spaces: the study of the path space of a Riemannian manifold obliges us to enlarge the notion of tangent space to the probability space.

6. Transport of structures to the path space of a Riemannian manifold

6.1. Theorem. *Every tangent process on $P_{m_0}(M)$ satisfying (5.1.3) satisfies an integration by parts formula.*

Proof. We transfer the problem by the Itô map from $P_{m_0}(M)$ to X; the given tangent process is transferred into a tangent process on X for which a formula of integration by parts is valid by (5.1.4). □

In paragraph 6.3. we shall give more details of this formula of integration by parts making explicit the transport of structures and the underlying stochastic contraction; in the next paragraph 6.2. we shall present some geometric objects which are needed.

6.2. Damped parallel transport

We denote by Ric the *Ricci tensor* of M defined by contraction of the curvature tensor: $\mathrm{Ric}(z) = \sum_j \Omega(e_j, z)e_j$, where e_* is the canonical basis of $\mathbb{R}^\ell$.

The covariant *derivative* is defined along a *smooth curve* γ by the formula

$$\nabla^\gamma_\tau(z) = \lim_{\varepsilon\to 0} \varepsilon^{-1}\left[t^\gamma_{\tau\leftarrow(\tau+\varepsilon)}\big(z(\tau+\varepsilon)\big) - z(\tau)\right].$$

Given $\zeta \in T_{p(\tau_0)}(M)$ we define

$$z_\tau \in T_{p(\tau)}(M) \qquad \tau < \tau_0$$

as the solution of

$$\begin{cases} \nabla^p z + \frac{1}{2}\mathrm{Ric}(z) = 0 \\ z(\tau_0) = \zeta \end{cases},$$

and we denote

$$\tilde{t}^p_{\tau\leftarrow\tau_0}(\zeta) = z(\tau) \qquad (\tau < \tau_0).$$

We have to associate to $p \in \mathbb{P}_{m_0}(M)$ the endomorphism R^p_τ of $T_{m_0}(M)$ defined by

$$R^p_\tau = t^p_{0\leftarrow\tau} \circ (\mathrm{Ric})_{p(\tau)} \circ t^p_{\tau\leftarrow 0}.$$

We consider the resolvent $Q^p_{\tau\leftarrow\tau_0}$ of the corresponding linear differential equation defined by

$$\begin{aligned} \frac{d}{d\tau}\, Q^p_{\tau\leftarrow\tau_0} &= -R^p_\tau Q^p_{\tau\leftarrow\tau_0} \\ \frac{d}{d\tau_0}\, Q^p_{\tau\leftarrow\tau_0} &= Q^p_{\tau\leftarrow\tau_0} R^p_{\tau_0}. \end{aligned}$$

Consider the adjoint $\left(Q^p_{\tau\leftarrow\tau_0}\right)^* = S^p_{\tau_0\leftarrow\tau}$ of the resolvent. Then

$$\frac{d}{d\tau}\left(Q^p_{\tau\leftarrow\tau_0}\right)^* = -\left(Q^p_{\tau\leftarrow\tau_0}\right)^* R^p_\tau \qquad \text{or}$$
$$\frac{d}{d\tau} S^p_{\tau_0\leftarrow\tau} = -S^p_{\tau_0\leftarrow\tau} R^p_\tau .$$

Then

$$\tilde{t}^p_{\tau\leftarrow\tau_0} = t^p_{\tau\leftarrow 0}\circ Q^p_{\tau\leftarrow\tau_0}\circ t^p_{0\leftarrow\tau_0}$$
$$\left(\hat{t}^p_{\tau\leftarrow\tau_0}\right)^* = t^p_{\tau_0\leftarrow 0}\circ S^p_{\tau_0\leftarrow\tau}\circ t^p_{0\leftarrow\tau}.$$

We denote

$$\left(\tilde{t}^p_{\tau\leftarrow\tau_0}\right)^* \text{ by } \hat{t}^p_{\tau_0\leftarrow\tau}.$$

Given a cylindrical function F we define its *twisted* gradient by

6.2.1.
$$\hat{D}_\tau F = \sum_{j=1}^k \mathbf{1}(\tau<\tau_j)\hat{t}^p_{\tau\leftarrow\tau_j}(\partial_j F) \quad \text{and its gradient by}$$
$$D_\tau F = \sum_{j=1}^k \mathbf{1}(\tau<\tau_j) t_{\tau\leftarrow\tau_j}(\partial_j F).$$

6.3. Theorem (Bismut, Driver). *For every $Z\in T^1\left(P_{m_0}(M)\right)$ and every cylindrical function F, we have the two following integration by parts formulae*

6.3.1.
$$E\left(\int_0^1 D_\tau F\cdot\nabla z_\tau d\tau\right) = E\left(F\int_0^1 \nabla_\tau Z\cdot dp(\tau)\right)$$

where

$$z(\tau) = \int_0^\tau \tilde{t}^p_{\tau\leftarrow s}(\nabla_s Z)ds.$$

6.3.2.
$$E\left(\int_0^1 \hat{D}_\tau F\cdot\nabla_\tau Z d\tau\right) = E\left(F\int_0^1 \nabla_\tau Z\cdot dp(\tau)\right)$$

where the stochastic integrals are Itô stochastic integrals transfered from X by

$$\int_0^1 u_\tau dp(\tau) = \int_0^1 t^p_{0\leftarrow\tau}(u_\tau)dx(\tau) \quad \text{where} \quad x=\mathcal{I}^{-1}(p).$$

Proof. We have

$$z(\tau) = \int_0^\tau \tilde{t}^p_{\tau\leftarrow s}(\nabla_s Z)ds.$$

Define $\zeta^*(\tau) = t^p_{0\leftarrow\tau}(z(\tau))$, then we want to compute the Itô stochastic differential of the associated tangent processes ζ given by (5.2.1). As we want to apply the formula (5.1.4) we can neglect the stochastic term in the Itô differential of ζ. We write $\cong$ for equality modulo such terms and we denote by $\bullet$ the stochastic contraction; then the first equation (5.2.1) appears in this formalism as

$$d\zeta \cong d\zeta^* - \frac{1}{2} d\rho \bullet dx,$$

the stochastic contraction is computed by the second equation (5.2.1):

$$d\rho \bullet dx = \sum_j \Omega(\zeta^*, e_j)e_j = -R^p(\zeta)\, d\tau.$$

We determine the congruence class of ζ^* by the formula

$$d\zeta(\tau) \cong d\zeta^* + \frac{1}{2} R^p_\tau(\zeta^*)\, d\tau.$$

Given a cylindrical function Φ on $P_{m_0}(M)$, define $f = \Phi \circ \mathcal{I}$; by applying Theorem 5.2 we get

$$\begin{aligned} E(D_{\zeta^*}\Phi) = E(D_\zeta f) &= E\left(f \int_0^1 \left(\dot{\zeta}^* + \frac{1}{2} R^p \zeta^*\right) dx\right) \\ &= E\left(\Phi \int_0^1 \nabla Z.dp(\tau)\right), \end{aligned}$$

which proves (6.3.1). We obtain (6.3.2) by taking the transpose of the damping term in the l.h.s. of (6.3.1). □

6.4. Theorem (Clark-Bismut-Ocone representation). *Given a smooth cylindrical function F we have*

$$F(p) = E(F) + \int_0^1 E^{\mathcal{N}_\tau}\left(\hat{D}_\tau F\right) \cdot dp(\tau).$$

Proof. Using the Itô map (which reduces everything to the classical Wiener space) we know that there exists an adapted functional $a(p,\tau)$ satisfying $a(p,\tau) \in T_{p(\tau)}(M)$, such that

$$F(p) - E(F) = \int_0^1 a(\tau) \cdot dp(\tau).$$

Then, using 6.3.1, we have for every adapted vector field Z

$$E\left(\int_0^1 D_\tau F \cdot \nabla^p_\tau z d\tau\right) = E\left(\int_0^1 a(\tau) \cdot \nabla^p_\tau Z d\tau\right)$$

which implies

$$E^{\mathcal{N}_\tau}\left(\hat{D}_\tau F\right) = a(\tau).$$

□

6.5. Harnack estimates via the reduced variation

Denote by $p_\tau(m_0, m)$ the heat kernel of M. We have the following estimate.

Theorem (Bismut). *Given $z \in T_{m_0}(M)$, we have for every bounded continuous function ψ*

$$\int_M \langle z, d_{m_0} \log p_\tau(m_0, m)\rangle \psi(m) dm = E\left(\psi(p(\tau)) \int_0^\tau a_z(s) dp(s)\right)$$

where

$$a_z(s) = \frac{1}{\tau}\, \tilde{t}^{p}_{s\leftarrow 0}(z).$$

Proof. We resume the computation of IX.6 replacing the usual reduced variation by a twisted variation. □

7. Loop spaces

The loops $L_{m_0}(M)$ are the paths $p \in P_{m_0}(M)$ such that, denoting $\Phi(p) = p(1)$, we have $\Phi(p) = m_0$. It can be proved easily that the map Φ is non-degenerate; therefore the loop space can be written as $[\Phi^*]^{-1}(m_0)$ where Φ^* denotes a re-definition in the sense of the quasi-sure analysis; the *restriction* of formulae of integration by parts from the path space to the loop space can be obtained through the quasi-sure analysis.

Also, genuinely new phenomena appear on loop spaces. We limit ourselves to the following Gross ergodicity theorem which has far reaching consequences:

Gross ergodicity theorem. *Assume that the group G is compact and simply connected. Consider a function φ defined on $L_e(G)$ belonging to a Sobolev space and such that its derivative in the direction of all Cameron-Martin like tangent vectors vanishes, then φ is constant.*

Under the same hypothesis on $P_e(G)$, we could use the representation formula 2.3 and would obtain $\varphi = E(\varphi)$. On loop space the formula 2.3 is no longer valid and new ideas are needed in order to prove the Gross ergodicity theorem.

Notes and comments on Part V

The theory of analytic functions of one complex variable led Emile Borel in 1915 to a class of "holomorphic" functions defined on a subset $O \subset C$, without any interior points for the usual topology but open for the fine topology; the corresponding differential calculus on O was built in 1975 by Debiard-Gaveau [1] using the Itô calculus along the Brownian motion localized on O.

The original development in 1976 of the stochastic calculus of variations (Malliavin [10]) was inspired by Debiard-Gaveau [1] and was carried out along the following two principles: α) Gaussian measures have to be characterized as the invariant measure of some Ornstein-Uhlenbeck process; β) Itô calculus along this OU process provides the differential calculus needed on the probability space.

Itô calculus for processes with an infinite dimensional state space was not at this time a standard object; a specific theory was built in Stroock [2]. In an opposite direction Stroock [1] substitutes for the Itô differential calculus along the OU process a differential calculus on Sobolev space: a basic fact is the existence of a symbolic calculus on this Sobolev space. This approach was amplified in Kusuoka-Stroock [1], incorporating the basic ideas of Shigekawa [1], [2].

In the approach of Bouleau [1] and Bouleau-Hirsch [7], Itô calculus associated to Dirichlet forms also plays a basic role; it must be emphasized that Bouleau-Hirsch obtain absolute continuity of law under hypotheses involving only the first derivative.

For future investigations, the approach to the regularity of laws through some kind of Itô calculus could have great potential for being free from any filtration on the probability space or from any *a priori* inequalities.

The introduction of the *Itô map* via the frame bundle machinery was done in 1974 (Malliavin [2]); the stochastic calculus of variations of the Itô map was done in 1977 (Malliavin [11]); it was then observed that the natural Cameron-Martin spaces are not preserved through the Itô map; this annoying fact left the subject untouched for seven years until Bismut [11] succeeded in obtaining for a Riemannian path space his *formula of integration by parts* (see Dohrn-Guerra [1] for a physicist's approach); this formula, which is fully proved in Bismut's book, nevertheless stood as a mystery for eight years until Driver [1] gave the integrated version of the Bismut integration by parts formula; an easier approach to Driver's result was provided by Enchev-Stroock [1]. The introduction of *tangent processes* in Cruzeiro-Malliavin [1] restores to the Itô map its computational value:

it becomes possible again to compute derivatives on the path space from derivatives on the classical Wiener space. More generally, a basic problem for non-linear stochastic analysis in infinite dimension is the following: "What could be the tangent space to the probability space?": see Albeverio-Kondratiev-Röckner [1], [2], Malliavin [25], Airault-Malliavin [6].

Differential geometry on Riemannian path space is developed in Cruzeiro-Malliavin [1] along the lines of a *geometric limit theorem* which describes the moving frame on the path space as the limit of its cylindrical approximations; all the summations appearing in this infinite-dimensional differential geometry can be then renormalized by stochastic integrals. In this geometric framework Shigekawa-Weitzenböck-type identities can be proved and give L^2 *a priori* inequalities for anticipative stochastic integrals (Cruzeiro-Fang [1]).

The Ornstein-Uhlenbeck processes on Riemannian path space have several approaches: Dirichlet forms (Driver-Röckner [1]), two-parameter processes (Norris [5]), martingale problem (Kazumi [2]).

The theory of abstract Wiener-Riemann manifolds, modelled on abstract Wiener spaces (see Kusuoka [5], [6], [7], [8], [9]; Aida [3]), is quite different from path space theory where the filtration plays a paramount role.

The *integration by parts formula for loop groups* was obtained in 1990 through quasi-sure analysis from the corresponding formula for path groups (Malliavin-Malliavin [5]); an explicit form of the modulus of integrated quasi-invariance can be found in Gross [3], Airault-Malliavin [2]. Geometry on loop groups is developed in Shigekawa-Taniguchi [1].

A chaos expansion on loop groups was obtained in Gross [3] (see also Pontier-Üstünel [1]); this chaos expansion becomes an unexpected source of results for finite-dimensional compact groups: for instance, a *coherent states transform* (Hall [1]) gives a unitary isomorphism between $L^2(G; p_t)$ and $L^2\left(G^C; p^C_{t/2}\right)$, where G is a simply connected compact Lie group, G^C its complexification, and where p_t, p^C_t denote the correponding heat kernels (see Gross-Malliavin [1] for an infinite dimensional proof).

The logarithmic Sobolev inequality on loop groups with an extra potential term is obtained in Gross [4]; corresponding results for Riemannian path spaces are in Hsu [3], Aida-Elworthy [1]. The spectral gap on Riemannian path space is proved in Fang [10].

The interplay between integration on loop groups and theory of representations of affine Lie algebras is still in its infancy (see Malliavin-Malliavin [6], and De Concini-Malliavin-Malliavin [1]).

The heat operator, which can be intrinsically defined, provides a Brownian motion on loop groups; it becomes possible to replace, on loop groups, the Wiener measure by the heat measure, that is the law of this Brownian motion at a fixed time: Driver-Lohrenz [1]; some interrelation between heat measure and Wiener measure is obtained in Airault-Malliavin [2]. Driver very recently solved the long-standing problem (see Malliavin [25]) of proving an integration by parts formula on the loop group with its heat measure.

Bibliography

The bibliography below contains considerably more papers than those that are cited in the text. They are included in order to provide the reader with a means of orientation in the fields related to the material of this book.

Adams, R.A.

[1] Sobolev spaces. Academic Press (1975).

Aida, S.

[1] D^∞-cohomology group and D^∞-maps on submanifolds of the Wiener space. J. Funct. Anal. **107** (1992) 289–301.

[2] Certain gradient flows and submanifolds in Wiener spaces. J. Funct. Anal. **112** (1993) 346–372.

[3] On the Ornstein-Uhlenbeck operators on Wiener-Riemannian manifolds. J. Funct. Anal. **116** (1993) 83–110.

[4] Sobolev spaces over loop groups. J. Funct. Anal. **127** (1995) 155–172.

[5] Support theorem for diffusion processes on Hilbert spaces. Pub. Res. Inst. Math. Sci. Kyoto **26** (1990) 947–965.

Aida, S., and Driver, B.

[1] On the uniqueness of ground states on loop spaces over a homogeneous space. Preprint

Aida, S., and Elworthy, D.

[1] Differential calculus on path spaces and loop spaces. I. Logarithmic Sobolev inequalities on path spaces. C.R. Acad. Sci., Paris, Sér. I **391** (1995) 97–102.

Aida, S., Kusuoka, S., and Stroock, D.

[1] On the support of Wiener functionals. Asymptomatic problems in probability theory. Pitman Res. Notes Math. Ser. **294** (1993) 3–34.

Aida, S., Masuda, T., and Shigekawa, I.

[1] Logarithmic Sobolev inequalities and exponential integrability. J. Funct. Anal. **196** (1994) 83–101.

Aida, S., and Shigekawa, I.

[1] Logarithmic Sobolev inequalities and spectral gaps: perturbation theory. J. Funct. Anal. **126** (1994) 448–473.

Aida, S., and Stroock, D.
[1] Moment estimates derived from Poincaré and logarithmic Sobolev inequalities. Math. Res. Letters **1** (1994) 75–86.

Airault, H.
[1] Perturbation singulière et solution stochastique de problèmes de D. Neumann-Spencer. J. Math. Pures Appl. **55** (1976) 233–268.
[2] Projection of the infinitesimal generator of a diffusion. J. Funct. Anal. **85** (1989) 353–391.
[3] Differential calculus on finite-codimensional submanifolds of the Wiener space. J. Funct. Anal. **100** (1991) 291–316.
[4] Formules de Cameron-Martin – Girsanov – Kusuoka – Buckdahn, Application à l'étude de rétractes. C.R. Acad. Sci., Paris, Sér. I **316** (1993) 727-732.

Airault, H., and Malliavin, P.
[1] Intégration géométrique sur l'espace de Wiener. Sémin. Ec. Norm. Supér. 1985, and Bull. Sci. Math. **112** (1988) 3–52.
[2] Integration on loop groups, II, Heat equation for the Wiener measure. J. Funct. Anal. **104** (1992) 71–109.
[3] Integration and probability. Grad. Texts in Math. 157, Springer (1995) 322 pp.
[4] Semi-martingales with values in a euclidean vector bundle and Ocone's formula on a Riemannian manifold. In: Proc. Symp. Pure and Appl. Math. **57** (M. Cranston and M. Pinsky ed.) Cornell 1993, Amer. Math. Soc. (1995) 175–193.
[5] Some heat operators on $\mathbb{P}(R^n)$. Ann. Math. Blaise Pascal **3** (1996) 1–11.
[6] Integration by parts formula and dilatation vector fields on elliptic probability spaces. Probab. Theory Relat. Fields **106** (1996) 447–494.

Airault, H., and Van Biesen, J.
[1] Processus d'Ornstein-Uhlenbeck sur une sous-variétè de codimension finie de l'espace de Wiener. Bull. Sci. Math. **115** (1991) 195–210.

Akiyama, H.
[1] Geometric aspects of Malliavin calculus on vector bundles. J. Math. Kyoto Univ. **26** (1986) 673–696.

Albeverio, S.
[1] Mathematical physics and stochastic analysis, Saint-Chéron Round Tables 1992 (M. Yor ed.). Bull. Sci. Math. **117** (1993) 125–152.

Albeverio, S., Daletskii, Y., Kondratiev, Y., and Streit, L.
[1] Non-Gaussian infinite-dimensional analysis. J. Funct. Anal. **138** (1996) 311–350.

Albeverio, S., Fukushima, M., Hansen, W., Ma, Z., and Röckner, M.
[1] An invariance result for capacities on the Wiener space. J. Funct. Anal. **106** (1992) 35–49.

Albeverio, S., Høegh-Krohn, R., and Testard, D.
[1] Factoriality of representation of the group of paths. J. Funct. Anal. **57** (1984) 49–55.

Albeverio, S., Kondratiev, Y., and Röckner, M.
[1] Differential geometry of Poisson spaces. C.R. Acad. Sci., Paris, Sér. I **323** (1996).
[2] Canonical Dirichlet operator and distorted Brownian motion on Poisson spaces. C.R. Acad. Sci., Paris, Sér. I **323** (1996) 1179–1184.

Albeverio, S., and Kusuoka, S.
[1] A basic estimate for the two-dimensional stochastic holonomy along Brownian bridges. J. Funct. Anal. **127** (1994) 132–155.

Albeverio, S., Léandre, R., and Röckner, M.
[1] Construction of a rotational invariant diffusion on free loops. C.R. Acad. Sci., Paris, Sér. I **316** (1993) 287–292.

Albeverio, S., Ma, Z.M., and Röckner, M.
[1] Partitions of unity in Sobolev spaces over infinite dimensional state space. J. Funct. Anal. **143** (1997) 247–267.

Albeverio, S., and Röckner, M.
[1] Classical Dirichlet forms on topological vector spaces, closability and a Cameron-Martin formula. J. Funct. Anal. **88** (1988) 395–436.

Albeverio, S., Röckner, M., and Zhang, T.S.
[1] Girsanov transform for symmetric diffusions with infinite dimensional state space. Ann. Prob. **21** (1993) 961–978.

Al-Hussaini, A.N.
[1] A simplified proof of the representation of functionals of diffusion. Appl. Math. Optimization **20** (1989) 63–70.

Amine, S.
[1] Law of large numbers and the central limit theorem for distributions on the Wiener space. Stoch. Anal. and Rel. Topics, (A.S.Üstünel ed.) Birkhäuser. Prog. Probab. 31 (1992) 187–196, and
Loi des grands nombres et théorème de -la limite centrale pour les distributions sur l'espace de Wiener. C.R. Acad. Sci., Paris, Sér. I **316** (1993) 381–385.

Arai, A.
[1] Comparison and nuclearity of spaces of differential forms on topological vector spaces. J. Funct. Anal. **111** (1993) 278–294.

Arai A., and Mitoma, I.
[1] De Rham-Hodge-Kodaira decomposition in ∞ dimension. Math. Ann. **291** (1991) 51–73.

Arnold, L. and Imkeller, P.
[1] Lyapunov exponent of stochastic flows and Malliavin Calculus. Preprint.

Asch, J. and Potthof, J.
[1] Itô's lemma without non-anticipatory conditions. Probab. Theory Relat. Fields **88** (1991) 17–46.

Bakry, D.
[1] Remarques sur le processus d'Ornstein-Uhlenbeck en dimension infinie, Sém. Probab. XVI. Springer, Lecture Notes in Math. 920 (1982) 134–137.
[2] Transformation de Riesz pour les semi-groupes positifs, Sém. Probab. XIX. Springer, Lecture Notes in Math. 1123 (1985) 179–206.
[3] Etude des transformations de Riesz dans les variétés à courbure de Ricci minorée, Sém. Probab. XXI. Springer, Lecture Notes in Math. 1247 (1987) 137–172.
[4] L'hypercontractivité et son utilisation en théorie des semi-groupes. In: Ecole d'été de Saint-Flour. Springer, Lecture Notes in Math. 1581 (1994) 1–114.

Baldi, P. and Sanz, M.
[1] Modulus of continuity for stochastic flows, Barcelona seminar (D. Nualart and M. Sanz ed.). Birkhäuser, Prog. Probab. 32 (1993) 1–20.

Bally, V.
[1] On the connection between the Malliavin covariance matrix and Hörmander condition. J. Funct. Anal. **96** (1991) 219–255.
[2] The Euler scheme for SDE. Math. Comput. Simul. **38** (1995) 35–41.

Bally, V., Gyöngy, I., and Pardoux, E.
[1] White noise driven parabolic PDE with measurable drift. J. Funct. Anal. **120** (1994) 484–510.

Bally, V., and Talay, D.
[1] The law of the Euler scheme for stochastic differential equations, I. Convergence rate of the distribution function. Probab. Theory Relat. Fields **104** (1996) 43–60.

Bass, R.F., and Cranston, M.
[1] The Malliavin calculus for pure jump processes and applications to local time. Ann. Probab. **14** (1986) 490–532.

Baxendale, P., and Stroock, D.W.
[1] Large deviation for stochastic flows of diffeomorphisms. Probab. Theory Relat. Fields **80** (1988) 169–215.

Bell, D.
[1] A quasi-invariance theorem for measures on Banach spaces. Trans. Am. Math. Soc. **290** (1985) 851–855.
[2] On the relationship between differentiability and absolute continuity of measures in $\mathbb{R}^n$. Probab. Theory Relat. Fields **72** (1986) 417–424.
[3] The Malliavin calculus. Pitman Monogr. Surv. Pure Appl. Math. 34 (1987) 105 pp.

Bell, D., and Mohammed, S.E.
[1] The Malliavin calculus and stochastic delay equation. J. Funct. Anal. **99** (1991) 75–99.
[2] Hypoelliptic parabolic operators with exponential degeneracies. C.R. Acad. Sci., Paris, Sér. I **317** (1993) 1059–1064.
[3] An extension of the Hörmander theorem for infinitely degenerate parabolic operators. In: Proc. Symp. Pure and Appl. Math. **57** (M. Cranston and M. Pinsky ed.) Cornell 1993, Amer. Math. Soc. (1995) 553–562, and Duke Math. J. **78** (1995) 453–475.

Belopolskaya, Ya.I., and Dalecky, Yu.L.
[1] Stochastic equations and differential geometry. Kluwer (1993).

Ben Arous, G.
[1] Méthode de Laplace et de la phase stationnaire sur l'espace de Wiener. Stoch. **25** (1988) 125–153.
[2] Développement asymptotique du noyau de la chaleur hypoelliptique hors du cut-locus. Ann. Scient. de l'Ecole Normale Supérieure, 4ème série **21** (1988) 307–331.
[3] Développement asymptotique du noyau de la chaleur hypoelliptique sur la diagonale. Ann. Inst. Fourier **39** (1989) 73–99.
[4] Flots et séries de Taylor stochastiques. Probab. Theory Relat. Fields **81** (1989) 29–77.
[5] Schilder's large deviation principle without topology. Pitman Res. Notes Math. Ser. 284 (1993) 107–121.

Ben Arous, G., and Castell, F.
[1] A probabilistic approach to semi-classical expansion. J. Funct. Anal. **137** (1996) 243–280.

Ben Arous, G., and Gradinaru, M.
[1] Singularity of hypoelliptic Green functions. Ann. Math. Blaise Pascal **3** (1996) 23–32.

Ben Arous, G., Gradinaru, M., and Ledoux, M.
[1] Normes hölderiennes et support des diffusions. Ann. Inst. Henri Poincaré, Probab. Stat. **30** (1994) 415–436.

Ben Arous, G., Kusuoka, S., and Stroock, D.
[1] The Poisson kernel of certain degenerate elliptic operators. J. Funct. Anal. **56** (1984) 171–208.

Ben Arous, G., and Léandre R.
[1] Décroissance exponentielle du noyau de la chaleur sur la diagonale (parts I and II). Probab. Theory Relat. Fields **90** (1991) 175–202 and 377–402.
[2] Annulation plate du noyau de la chaleur hypoelliptique. C.R. Acad. Sci., Paris, Sér. I **312** (1991) 463–464.

Berger, M.
[1] A Malliavin-type anticipative stochastic calculus. Ann. Probab. **16** (1988) 231–245.

Berger, M., and Mizel, V.
[1] An extension of the stochastic integral. Ann. Probab. **10** (1982) 435–450.

Biane, P.
[1] Segal-Bargman transform, functional calculus on matrix spaces and the theory of semi-circular systems. J. Funct. Anal. **144** (1997)

Bichteler, K., and Fonken, D.
[1] A simple version of the Malliavin calculus in dimension one. Springer, Lecture Notes in Math. 939 (1982) 6–12.
[2] A simple version of the Malliavin calculus in dimension N. Birkhäuser, Prog. Probab. 5 (1983) 97–110.

Bichteler, K., Gravereaux, J.B., and Jacod, J.
[1] Malliavin calculus for processes with jumps. Gordon & Breach, Stoch. monogr. 2 (1987) 157 pp.

Bichteler, K., and Jacod, J.
[1] Calcul de Malliavin pour les diffusions avec sauts: existence d'une densité dans le cas unidimensionnel, Sém. Probab. XVII. Springer, Lecture Notes in Math. 986 (1983) 132–157.

Bismut, J.M.
[1] A generalized Itô formula and some other properties of stochastic flow. Z. Wahrsch. verw. Gebiete **55** (1981) 85–109.
[2] Martingales, the Malliavin calculus and hypoellipticity under general Hörmander's conditions. Z. Wahrsch. verw. Gebiete, **56** (1981) 469–505.
[3] Mécanique aléatoire. Springer, Lecture Notes in Math. 866 (1981) 563 pp.
[4] Mécanique aléatoire. Springer, Lecture Notes in Math. 929 (1982) 1–100.
[5] An introduction to the stochastic calculus of variations. In: Stoch. Diff. Systems (M. Kohlmann and N. Christopeit ed.). Springer, Lecture Notes Control Inf. Sci. 43 (1982) 33–72.
[6] Calcul des variations stochastiques et grandes déviations. C.R. Acad. Sci., Paris, Sér. I **296** (1983) 1009–1012.
[7] Calcul des variations stochastiques et processus de sauts. Z. Wahrsch. verw. Gebiete **63** (1983) 147–235.
[8] The Atiyah-Singer theorems for classical elliptic operators: a probabilistic approach. J. Funct. Anal. **57** (1984) 56–99.
[9] The Atiyah-Singer theorems: a probabilistic approach, II, The Lefschetz fixed point formulas. J. Funct. Anal. **57** (1984) 329–348.

[10] The calculus of boundary processes. Ann. Sci. Ec. Norm. Supér. **17** (1984) 507–622.
[11] Large deviations and the Malliavin calculus. Birkhäuser, Prog. Math. **45** (1984) 216 pp.
[12] Jump processes and boundary processes. Proc. Symp. on Stoch. Anal. (K. Itô ed.) Katata/Kyoto 1982, North-Holland. Math. Library **32** (1984).
[13] Index theorems and equivariant cohomology on the loop space. Commun. Math. Phys. **98** (1985) 207–240.
[14] Last exit decompositions and regularity at the boundary of transition probabilities. Z. Wahrsch. verw. Gebiete **69** (1985) 65–98.
[15] Probability and geometry. In: CIME lectures. Springer, Lecture Notes in Math. 1206 (1985) 1–60.
[16] An infinitesimal Lefschetz formula: A heat equation proof. J. Funct. Anal. **62** (1985) 435–457.
[17] The Witten complex and the degenerate Morse inequalities. J. Differ. Geom. **23** (1986) 207–240.
[18] Index theorem equivariant cohomology and the filtering equation. Proc. of the VIIIth International Congres in Math. Phys. (M. Mebkhout, R. Sénéor ed.) World Scientific (1987) 17–56.
[19] Demailly asymptotic Morse inequalities: a heat equation proof. J. Funct. Anal. **72** (1987) 263–278.

Bismut, J.M., and Michel, D.
[1] Diffusions conditionnelles I. Hypoellipticité partielle. J. Funct. Anal. **44** (1981) 174–211.
[2] Diffusions conditionnelles II. Générateur conditionnel. Application au filtrage. J. Funct. Anal. **45** (1982) 274–292.

Boas, R.P.
[1] Entire functions. Academic Press 1954.

Bogachev, V.I.
[1] Polynômes et intégrales oscillantes sur les espaces de dimension infinie. C.R. Acad. Sci., Paris, Sér. I **311** (1990) 807–812.
[2] Smooth measures, the Malliavin calculus and approximation in infinite dimensional spaces. Acta Univ. Carol., Math. Phys. **31** (1990) 9–23.
[3] Functionals of random processes and infinite dimensional oscillatory integrals. Izv. Ross. Akad. Nauk, Ser. Mat. **40** (1993) 235–266.
[4] Deterministic and stochastic differential equations in infinite dimension. Acta Appl. Math. **40** (1995) 25–93.

Bogachev, V., Krylov, N., and Röckner, M.
[1] Regularity of invariant measure: The case of non-constant diffusion part. J. Funct. Anal. **138** (1996) 223–242.

Bogachev, V.I., and Mayer-Wolf, E.
[1] Some remarks on Rademacher's theorem in infinite dimension. Preprint

Bogachev, V.I., and Röckner, M.
[1] Regularity of invariant measure on finite- and infinite-dimensional spaces and applications. J. Funct. Anal. **133** (1995) 168–223.
[2] Mehler formula and capacities for Ornstein-Uhlenbeck process with general drift. Osaka Math. J. **32** (1995) 237–274.
[3] Hypoellipticity and invariant measure of infinite-dimensional diffusions. C.R. Acad. Sci., Paris, Sér. I **318** (1994) 553–558.

Bogachev, V.I., and Smolyanov, O.G.
[1] Fonctionnelles analytiques des processus. Usp. Mat. Nauk **45** (1990) 3–83.

Bouleau, N.
[1] Décomposition de l'énergie par niveau de potentiel, Colloque J. Deny. In: Springer, Lecture Notes in Math. 1096 (1984) 149–172.
[2] Suite de points d'un espace de Hilbert distribuée selon la promesure gaussienne canonique. C.R. Acad. Sci., Paris, Sér. I **315** (1992) 1297–1301.

Bouleau, N., and Hirsch, F.
[1] Formes de Dirichlet générales et densité des variables aléatoires réelles sur l'espace de Wiener. J. Funct. Anal. **69** (1986) 229–259.
[2] Propriétés d'absolue continuité dans les espaces de Dirichlet et application aux équations différentielles stochastiques, Sém. Probab. XX. Springer, Lecture Notes in Math. 1204 (1986) 131–161.
[3] Calculs fonctionnels dans les espaces de Dirichlet et application aux équations différentielles lipschitziennes. C.R. Acad. Sci., Paris, Sér. I **302** (1986) 263–266.
[4] Sur une extension du théorème de Girsanov. C.R. Acad. Sci., Paris, Sér. I **309** (1989) 955–958.
[5] Quasi-everywhere regularity of solutions of Brownian stochastic differential equations and their flows. J. Math. Pures Appl. **69** (1990) 95–128.
[6] Algèbre des structures de Dirichlet. C.R. Acad. Sci., Paris, Sér. I **310** (1990) 15–18.
[7] Dirichlet forms and analysis on the Wiener space. de Gruyter 1991, 320 pp.
[8] On the derivability, with respect to the initial data, of the solution of a stochastic differential equation with Lipschitz coefficient. Springer, Lecture Notes in Math. 1393 (1989) 39–57.

Buckdahn, R.
[1] Anticipative Girsanov transformation. Probab. Theory Relat. Fields **89** (1991) 211–238.
[2] Linear Skorokhod stochastic differential equations of diffusion type. Probab. Theory Relat. Fields **90** (1991) 223–240.
[3] Skorokhod stochastic differential equations of diffusion type. Probab. Theory Relat. Fields **93** (1992) 297–323.
[4] Quasi-linear partial differential equations without non-anticipation requirement. Preprint

Buckdahn, R., and Föllmer, H.
[1] A conditional approach to the anticipating Girsanov transformation. Probab. Theory Relat. Fields **95** (1993) 311–330.

Buckdahn, R., Malliavin, P., and Nualart, D.
[1] Multidimensional linear stochastic differential equations in the Skorokhod sense. Preprint

Buckdahn, R., and Nualart, D.
[1] Linear stochastic differential equations and Wick products. Probab. Theory Relat. Fields **99** (1994) 501–526.

Caballero, M.E., and Fernández, B.
[1] Critère de Hörmander: un cas non adapté. C.R. Acad. Sci., Paris, Sér. I **317** (1993) 303–306.

Caballero, M.E., Fernández, B., and Nualart, D.
[1] Smoothness of distributions for solutions of anticipating stochastic differential equations. Stoch. Stoch. Rep. **52** (1995) 303–322.

Calderon, A., and Zygmund, A.
[1] On singular integrals. Am. J. Math. **78** (1956) 289–303, and Acta Math. **88** (1952) 85–139.

Cameron, R.H., and Martin, T.
[1] A transformation of Wiener integrals. Ann. Math. **45** (1944) 380–286, Trans. Am. Math. Soc. **58** (1945) 184–219, and **66** (1949) 253–283.

Cameron, R.H., and Storvick, D.A.
[1] Feynman integral of variations of a functional, Third Nagoya Lévy Sem. 1990, (K. Itô, T. Hida ed.). World Scientific (1991) 144–157.

Carlen, E., and Krée, P.
[1] Sur la méthode des inégalités H^p pour les intégrales stochastiques itérées. C.R. Acad. Sci., Paris, Sér. I **314** (1992) 393–399.

Carlen, E., and Pardoux, E.
[1] Differential calculus and integration by parts on Poisson spaces. In: Stoch., Algebra and Anal. in classical and quantum dynamics, (S. Albeverio et al. ed.) (1990) 63–73.

Carlen, E., and Stroock, D.
[1] An application of the Bakry-Emery criterion to infinite-dimensional diffusions, Sém. Probab. XX. Springer, Lecture Notes in Math. 1204 (1985) 341–348.

Carmona, R., and Nualart, D.
[1] Random non-linear wave equation: smoothness of the solution. Probab. Theory Relat. Fields **79** (1988) 469–508.
[2] Random non-linear wave equation: propagation of singularities. Ann. Probab. **16** (1988) 730–751.
[3] Traces of random variables on Wiener spaces and the Onsager-Machlup functional. J. Funct. Anal. **107** (1992) 402–438.

Cattiaux, P.
[1] Régularité au bord pour les densités et les densités conditionnelles d'une diffusion réfléchie hypoelliptique. Stoch. **20** (1987) 309–340.
[2] Hypoellipticité et hypoellipticité partielle pour les diffusions avec une condition frontière. Ann. Inst. Henri Poincaré, Probab. Stat. **22** (1986) 67–112.
[3] Stochastic calculus and degenerated boundary value problem. Ann. Inst. Fourier **42** (1992) 541–624.
[4] Calcul stochastique et opérateur dégénéré du second ordre I. Bull. Sci. Math. **114** (1990) 421–462,
[5] and II, ibidem **115** (1991) 81–122.

Cattiaux, P., Roëlly, S., and Zessin, H.
[1] Une approche Gibbsienne des diffusions Browniennes infini-dimensionnelles. Probab. Theory Relat. Fields **104** (1996) 147–180.

Caverhill, A.P., and Elworthy, K.D.
[1] Flows of stochastic dynamical systems: the functional-analytic approach. Z. Wahrsch. verw. Gebiete **65** (1983) 245–267.

Chaleyat-Maurel, M.
[1] Robustesse du filtre et calcul des variations stochastiques. J. Funct. Anal. **68** (1986) 55–71.

Chaleyat-Maurel, M., and Michel, D.
[1] Hypoellipticity theorems and conditional laws. Z. Wahrsch. verw. Gebiete **65** (1984) 573–597.

[2] Des résultats de non-existence de filtres en dimension finie. Stoch. **13** (1984) 83–102.
[3] Une propriété de continuité en filtrage non-linéaire. Stoch. **19** (1986) 11–41.
[4] A Stroock-Varadhan support theorem in non-linear filtering. Probab. Theory Relat. Fields **84** (1990) 119–134.

Chaleyat-Maurel, M., Michel, D., and Pardoux, E.
[1] Un théorème d'unicité pour l'équation de Zakai. Stoch. Stoch. Rep. **29** (1990) 1–12.

Chaleyat-Maurel, M., and Nualart, D.
[1] The Onsager-Machlup functional for a class of anticipating processes. Probab. Theory Relat. Fields **94** (1992) 247–270.

Chen, Mu Fa, and Zhou, X.Y.
[1] Applications of Malliavin calculus to stochastic differential equations with time-dependent coefficients. Acta Math. Appl. Sin. **7** n°3 (1991) 193–216.

Chung, D.M.
[1] Conditional Feynmann integral for the Fresnel class on functions on an abstract Wiener space, Third Nagoya Lévy Sem. 1990, (K. Itô and T. Hida ed.). World Scientific (1991) 172–186.

Clark, J.M.C.
[1] The representation of functionals of the Brownian motion by stochastic integrals. Ann. Math. Stat. **41** (1970) 1282–1295.

Coquio, A.
[1] Forme de Dirichlet sur l'espace canonique de Poisson et applications aux équations différentielles stochastiques. Ann. Inst. Henri Poincaré, Probab. Stat. **29** (1993) 1–36.
[2] Calcul de Malliavin, existence et régularité de la densité d'une probabilité invariante d'une diffusion sur une variété Riemannienne. Grenoble (1994). Preprint

Coquio, A., and Gravereaux, J.B.
[1] Calcul de Malliavin et régularité de la densité d'une probabilité invariante d'une chaîne de Markov. Ann. Inst. Henri Poincaré, Probab. Stat. **28** (1992) 431–478.

Cranston, M.
[1] Gradient estimates on manifolds using compling. J. Funct. Anal. **99** (1991) 110–124.

Cranston, M., Kendall, W., and March, P.
[1] The radial part of the Brownian motion II: its life on the cut-locus. Probab. Theory and Relat. Fields **96** (1993) 353–368.

Cruzeiro, A.B.
[1] Equations différentielles sur l'espace de Wiener et formules de Cameron-Martin non-linéaires. J. Funct. Anal. **54** (1983) 206–227.
[2] Equations différentielles ordinaires, non-explosion, et mesures quasi-invariantes. J. Funct. Anal. **54** (1983) 193–205.
[3] Unicité de solutions d'équations différentielles sur l'espace de Wiener. J. Funct. Anal. **58** (1984) 335–347.
[4] Estimation capacitaire sur l'espace de Wiener. Bull. Sci. Math. **110** (1986) 139–147.
[5] Processus sur l'espace de Wiener associés à des opérateurs elliptiques à coefficients dans certains espaces de Sobolev. J. Funct. Anal. **72** (1987) 346–367.

Cruzeiro, A.B., and Fang, S.
[1] Une inégalité L^2 pour les intégrales stochastiques anticipatives sur une variété Riemannienne. C.R. Acad. Sci., Paris, Sér. I **321** (1995) 1245–1250, and J. Funct. Anal. **143** (1997) 400–414

Cruzeiro, A.B., and Malliavin, P.
[1] Renormalized differential geometry on path spaces: Structural equation, curvature. J. Funct. Anal. **139** (1996) 119–181, and C.R. Acad. Sci., Paris, Sér. I, **320** (1995) 603–606, and **319** (1994) 859–864.

Cruzeiro, A.B., and Zambrini, J.C.
[1] Malliavin calculus and Euclidean quantum mechanics I. The symbolic calculus. J. Funct. Anal. **96** (1991) 62–95.
[2] Feynman functional calculus and stochastic calculus of variation. In: Proc. Lisbonne conf. (A.B. Cruzeiro ed.). Birkhäuser, Prog. Probab. 26 (1991) 83–95.
[3] Ornstein-Uhlenbeck processes as Bernstein processes, Barcelona seminar (D. Nualart and M. Sanz ed.). Birkhäuser, Prog. Probab. 32 (1992) 40–62.
[4] Malliavin calculus and Euclidean quantum mechanics II: Variational principles for infinite-dimensional processes. J. Funct. Anal. **130** (1995) 450–476.

Daleckii, J.I., and Paramanova, S.N.
[1] Integration by parts with respect to measures in function spaces. Theory Probab. Math. Stat. **18** (1979) 34–49.

Da Prato, G., Fuhrman, M., and Malliavin, P.
[1] Asymptotic ergodicity for the Zakai filtering equation. C.R. Acad. Sci., Paris, Sér. I **321** (1995) 613–615.

Da Prato, G., Malliavin, P., and Nualart, D.
[1] Compact families of Wiener functionals. C.R. Acad. Sci., Paris, Sér. I **315** (1992) 1287–1291.

Da Prato, G., and Zabczyk, J.
[1] Stochastic equations in infinite dimension. Cambridge Univ. Press (1992) 452 pp.
[2] Regular densities of invariant measure in Hilbert spaces. J. Funct. Anal. **130** (1995) 427–449.
[3] Ergodicity for infinite-dimensional systems, London Math. Soc. Lecture Notes, Cambridge Univ. Press, 1996.

Davis, M.H.A.
[1] The Wiener space derivative for functionals of diffusions on manifolds. Nonlinearity **1** (1988) 241–251.

Debiard, A., and Gaveau, B.
[1] Différentiabilité des fonctions finement harmoniques. Invent. Math. **29** (1975) 111–123.

De Concini, C., Malliavin, M.P., and Malliavin, P.
[1] Intégration invariante associée à un anneau #fermé de représentations. C.R. Acad. Sci., Paris, Sér. I **323** (1996) 1265–1270.

Decreusefond, L., Hu, Y., and Üstünel, A.S.
[1] Une inégalité d'interpolation sur l'espace de Wiener. C.R. Acad. Sci., Paris, Sér. I **317** (1993) 1065–1067.

Decreusefond, L., and Üstünel, A.
[1] Application du calcul des variations stochastiques au mouvement brownien fractionnaire. C.R. Acad. Sci., Paris, Sér. I **321** (1995) 1605–1608.

Denis, L.
[1] Analyse quasi-sure de l'approximation d'Euler et du flot d'une équation differentielle stochastique. C.R. Acad. Sci., Paris, Sér. I (1992).

[2] Convergence quasi partout pour les capacités définies par un semi groupe Markovien. C.R. Acad. Sci., Paris, Sér. I **315** (1992) 1033–1036.
[3] Quasi-sure analysis for the Euler approximation and the flow relation to an SDE in Dirichlet forms. In: Dirichlet forms, Beijing 1993 (Ma Zhi Ming, M. Röckner, and J.A. Yan eds.). de Gruyter (1995) 103–113.

Deuschel, J.D., and Stroock, D.
[1] Hypercontractivity and spectral gaps of symmetric diffusion with applications to the stochastic Ising model. J. Funct. Anal. **92** (1990) 30–48.

Diebolt, J.
[1] Régularité de la loi de la solution d'une équation différentielle excitée par un processus gaussien physiquement réalisable. C.R. Acad. Sci., Paris, Sér. I **305** (1987) 143–146.

Dohrn, D., and Guerra, F.
[1] Stochastic behavior in classical and quantum Hamiltonian systems. Springer, Lecture Notes in Phys. 93 (1979) 241–249.

Donati-Martin, C.
[1] Equations différentielles stochastiques dans $\mathbb{R}$ avec conditions au bord. Stoch. Stoch. Rep. **35** (1991) 143–173.

Doob, J.L.
[1] Classical potential theory and its probabilistic counterparts. Springer 1984, 846 pp.

Doss, H.
[1] Liens entre équations différentielles stochastiques et ordinaires. Ann. Inst. Henri Poincaré, Probab. Stat. **13** (1977) 99–125.
[2] Démonstration probabiliste de certains développements asymptotiques quasi-classiques. Bull. Sci. Math. **109** (1985) 179–208.

Doss-Bachelet, C.
[1] Diffusion de fonctions convexes sur une variété Riemannienne, contrôle sur le fibré des repères. Bull. Sci. Math. **116** (1992) 427–452.

Doss, H., and Stroock, D.
[1] Nouveaux résultats concernant les petites perturbations des systèmes dynamiques. J. Funct. Anal. **101** (1991) 370–391.

Driver, B.
[1] A Cameron-Martin type quasi-invariance theorem for Brownian motion on a compact manifold. J. Funct. Anal. **109** (1992) 272–376.
[2] Towards calculus and geometry on path spaces. In: Proc. Symp. Pure and Appl. Math. **57** (M. Cranston and M. Pinsky ed.) Cornell 1993, Amer. Math. Soc. (1995).
[3] On the Kakutani-Itô-Segal-Gross and Segal-Bargmann-Hall isomorphism. J. Funct. Anal. **133** (1995) 69–128.
[4] The non-equivalence of Dirichlet forms on path spaces. Pitman Res. Notes Math. Ser. 310 (1994) 75–88.
[5] Lie brackets of adapted vector fields on Wiener spaces. Preprint 1995.

Driver, B., and Gross, L.
[1] Hilbert spaces of holomorphic functions on a complex Lie group. Preprint.

Driver, B., and Lohrenz, T.
[1] Logarithmic Sobolev inequalities for pinned loop groups. J. Funct. Anal. **140** (1996) 381–448.

Driver, B., and Röckner, M.
[1] Construction of diffusions on path and loop spaces of compact Riemannian manifolds. C.R. Acad. Sci., Paris, Sér. I **315** (1992) 603–608.

Duc, N., Nualart, D., and Sanz, M.
[1] Application of Malliavin calculus to a class of stochastic differential equations. Probab. Theory Relat. Fields **84** (1990) 549–571.

Eberle, A.
[1] Weak Sobolev spaces and Markov uniqueness of operators. C.R. Acad. Sci., Paris, Sér. I **320** (1995) 1249–1254.

El Kharroubi, A.
[1] Lois de probabilité marginales de la solution de certaines équations différentielles stochastiques, C.R. Acad. Sci., Paris, Sér. I **296** (1983) 1013–1016.

Elliott, R.J., and Kohlmann, H.
[1] Martingale representation and the Malliavin calculus, Appl. Math. Optimization **20** (1989) 105–112.
[2] Integration by parts and densities for jump processes, Stoch. **27** (1989) 83–98.
[3] Integration by parts, homogeneous chaos expansions and smooth densities, Ann. Probab. **17** (1989) 194–207.
[4] The existence of smooth densities for the prediction filtering and smoothing problems, Acta Appl. Math. **14** (1989) 269–286.

Elworthy, K.D.
[1] Stochastic methods and differential geometry, Sém. Bourbaki 1980–1981. Springer, Lecture Notes in Math. 901 (1981) 95–110.
[2] Stochastic differential equations on manifolds, Lond. Math. Soc. Lect. Note Ser. 70. Cambridge Univ. Press (1982) 326 pp.
[3] Geometric aspects of diffusion on manifold. In: Ecole d'été de Saint-Flour. Springer, Lecture Notes in Math. 1362 (1987) 276–425.
[4] Manifolds and graphs with mostly positive curvature, In: Proc. Lisbon conference (A.B. Cruzeiro ed.) Birkhäuser. Prog. Probab. 26 (1991) 96–110.
[5] Stochastic differential geometry, Saint-Chéron Round Tables 1992, (M. Yor ed.). Bull. Sci. Math. **117** (1993) 7–22.
[6] Stochastic flow on Riemannian manifolds in diffusion problems and related problems in analysis, vol. III (A. Pinsky and V. Vishutz, eds.) Birkhäuser 1992, 37–72.

Elworthy, D., Le Jan, Y., and Li, X.M.
[1] Integration by parts formulae for degenerate diffusion measures on path space and diffeomorphism groups. Comptes Rendus Acad. Sci., Paris, Sér. I **323** (1996) 921–926.

Elworthy, D., and Li, X.M.
[1] Formulae for the derivatives of heat semi-group. J. Funct. Anal. **125** (1994) 252–287.
[2] Derivative flows of SDE: Moment exponent and geometric properties, In: Proc. Symp. Pure and Appl. Math. **57** (M. Cranston and M. Pinsky ed.) Cornell 1993, Amer. Math. Soc. (1995), 565–573.
[3] A class of integration by parts in stochastic analysis I. In: Itô's Stochastic Calculus and Probability Theory (N. Ikeda, S. Watanabe, M. Fukushima, and H. Kunita eds.). Springer, Tokyo 1996, 15–37.

Elworthy, K.D., and Rosenberg, S.
[1] The Witten Laplacian on negatively curved simply connected manifold. Tokyo J. Math. **16** (1993) 513–524.

Elworthy, D., and Yor, M.
[1] Conditional expectation for derivatives of certain stochastic flows, Sém. Probab. XXVII. Springer, Lecture Notes in Math. 1557 (1993) 159–172.

Enchev, O.
[1] Non-linear transformations on the Wiener space. Ann. Probab. **21** (1993) 2169–2188.

Enchev, O., and Stroock, D.W.
[1] Anticipative diffusion and related change of measure. J. Funct. Anal. **116** (1993) 449–473.
[2] Rademacher's theorem for Wiener functionals. Ann. Probab. **23** (1993) 25–33.
[3] Towards a Riemannian geometry on the path space over a Riemannian manifold. J. Funct. Anal. **134** (1996) 392–416.

Epperson, J., and Lohrenz, T.
[1] Diffusions on finite energy loop spaces. C.R. Acad. Sci., Paris, Sér. I **317** (1993) 609–614.
[2] Brownian motion and heat semi-groups on path spaces. Pac. J. Math. **161** (1993) 233–253.

Erdös, L.
[1] Estimates on stochastic oscillatory integrals and Schrödinger operators with martingale fields. Duke Math J. **76** (1994) 541–565.

Estrade, A., and Pontier, M.
[1] Relèvement horizontal d'une martingale cadlag, Sém. Probab. XXVI. Springer, Lecture Notes in Math. 1526 (1992) 127–145.

Fabes, E.
[1] Gaussian upper bounds on fundamental solutions of parabolic equations: the method of Nash. In: CIME lectures on Dirichlet forms 1992. Springer, Lecture Notes in Math. 1563 (1993) 245 pp.

Fang, S.
[1] Une inégalité isopérimétrique sur l'espace de Wiener. Bull. Sci. Math. **112** (1988) 345–355.
[2] Pseudo-théorème de Sard et connexité sur l'espace de Wiener. Bull. Sci. Math. **113** (1989) 483–490.
[3] Sur la continuité approximative forte des fonctionnelles de Itô. Stoch. Stoch. Rep. **26** (1991) 193–204.
[4] Non dégénérescence des pseudo-normes de Sobolev sur l'espace de Wiener. Bull. Sci. Math. **115** (1991) 223–234.
[5] Sur une méthode de Lyons-Zheng appliquée à l'ergodicité de l'action de H dans l'espace de Wiener. C.R. Acad. Sci., Paris, Sér. I **313** (1991) 615–619.
[6] Sur un résultat d'entropie de l'espace de Wiener. Potential Anal. **1** (1992) 167–176.
[7] Le calcul différentiel quasi-sûr et son application à l'estimation du noyau de la chaleur. Trans. Am. Math. Soc. **339** (1993) 221–241.
[8] Théorème limite pour une équation différentielle stochastique anticipante. Stoch. Stoch. Rep. **39** (1992) 95–106.
[9] Inégalité du type Poincaré sur l'espace des chemins Riemanniens. C.R. Acad. Sci., Paris, Sér. I **318** (1994) 257–260.
[10] Equation $\bar{\partial} u = f$ sur un espace de Wiener complexe. In: Stoch. Anal. and Rel. Problems (B. Øksendal and A. Üstünel eds.), Gordon & Breach. Stochastic monogr. 8 (1993) 101–118.
[11] Stochastic anticipative integrals on a Riemannian manifold. J. Funct. Anal. **131** (1995) 228–252.

Fang, S., and Franchi, S.
[1] Platitude de la structure riemannienne sur le groupe des chemins et identité d'énergie pour les intégrales stochastiques. C.R. Acad. Sci., Paris, Sér. I **321** (1995) 1371–1376.

Fang, S., and Malliavin, P.
[1] Stochastic analysis on the path space of a Riemannian manifold, I. Markovian stochastic calculus. J. Funct. Anal. **118** (1993) 249–274.

Fang, S., and Ren, J.
[1] Quelques propriétés des fonctions holomorphes sur un espace de Wiener. C.R. Acad. Sci., Paris, Sér. I **315** (1992) 447–450.
[2] Sur le squelette et les dérivées de Malliavin des fonctions holomorphes sur un espace de Wiener complexe. J. Math. Kyoto Univ. **33** (1993) 749–764, and **34** (1994) 637–640.

Federer, H.
[1] Geometric measure theory. Grundlehren der math. Wiss., Springer, 1969, 676 pp.

Florchinger, P.
[1] Malliavin calculus with time dependent coefficients and application to non-linear filtering. Probab. Theory Relat. Fields **86** (1990) 203–223.
[2] Existence of a smooth density for the filter in non-linear filtering with infinite-dimensional noise. Syst. Control Lett. **16** (1991) 131–137.

Florchinger, P., and Léandre, R.
[1] Estimation de la densité d'une diffusion très dégénérée, étude d'un exemple. J. Math. Kyoto Univ. **33** (1983) 115–142.
[2] Décroissance non-exponentielle du noyau de la chaleur. Probab. Theory Relat. Fields **95** (1993) 237–262.

Florit, C., and Nualart, D.
[1] A local criterion for smoothness of density and application to the supremum of the Brownian sheet. Stat. Probab. Lett. **22** (1995) 25–21.

Föllmer, H.
[1] Calcul de Itô sans probabilités, Sém. Probab. XX. Springer, Lecture Notes in Math. 850 (1981) 143–150.
[2] Time reversal on Wiener spaces. In: Proc. BiBoS Symp. on Stoch. Proc., Springer, Lecture Notes in Math. 1158 (1986) 119–129.
[3] Random fields and diffusion processes, Ecole d'Eté de Saint- Flour, 1985. Springer, Lecture Notes in Math. 1326 (1988) 125–158.
[4] Martin boundaries on Wiener spaces. In: Diff. Proc. and Rel. Probl., in Anal., vol.I, (M. Pinsky ed.) Birkhäuser. Prog. Probab. 22 (1989) 3–16.

Föllmer, H., and Imkeller, P.
[1] A paradox arising in transformation on Wiener spaces. Ann. Inst. Henri Poincaré, Probab. Stat. **29** (1993) 569–586.

Föllmer, H., and Wakolbinger, A.
[1] Time reversal of infinite-dimensional diffusion. Stoch. Proc. Appl. **22** (1986) 59–77.

Frangos, N., and Imkeller, P.
[1] On the existence of occupation densities for stochastic integral processes in the second Wiener chaos. In: Stoch. Anal., Liber Amicorum for Moshe Zakai. Academic Press (1991) 189–212.

Frenkel, L.A.
[1] Orbital theory for affine Lie algebras. Invent. Math. **77** (1984) 301–352.

Fukushima, M.
[1] Capacitory maximal inequalities and an ergodic theorem. In: Proc. 4th Japan/USSR Symp. on Probab. Th., Springer, Lecture Notes in Math. 1021 (1983) 130–136.
[2] A note on capacities in infinite dimension. In: Proc. 5th Japan/USSR Symp. on Probab. Th. Kyoto (S. Watanabe and Prokorov eds.). Springer, Lecture Notes in Math. 1299 (1988) 80–85.
[3] Basic properties of Brownian motion and a capacity on the Wiener space. J. Math. Soc. Japan **36** (1984) 161–176.
[4] A Dirichlet form on the Wiener space and properties on Brownian motion. Springer, Lecture Notes in Math. 1096 (1984) 290–300.
[5] $(r-p)$-capacities and Hunt processes in infinite dimension. In: Proc. of the 6th URSS-Japan Symp. on Probab., Kiev 1991, (A.N. Shiryaev ed.). World Scientific (1992) 96–103.
[6] Two topics related to Dirichlet forms: quasi-everywhere convergence and additive functionals. In: CIME Lectures on Dirichlet forms. Springer, Lecture Notes in Math. 1563 (1994) 21–53.

Fukushima, M., and Okada, M.
[1] On conformal martingales and pluripolar sets. J. Funct. Anal. **55** (1984) 377–388.
[2] On Dirichlet forms for plurisubharmonic functions. Acta Math. **159** (1987) 171–218.

Fukushima, M., Oshima, Y., and Takeda, M.
[1] Dirichlet forms and symmetric Markov processes. de Gruyter 1994, 369 pp.

Gao, F., and Ren, J.
[1] Large deviations for stochastic flows and their applications. Preprint

Gaveau, B.
[1] Principe de moindre action, propagation de la chaleur et estimées sous-elliptiques sur certains groupes nilpotents. Acta Math. **139** (1977) 95–153.
[2] Systèmes dynamiques associés à certains opérateurs hypoelliptiques. Bull. Sci. Math. **102** (1978) 203–229.
[3] Noyaux des probabilités de transition de certains opérateurs d'Ornstein-Uhlenbeck. C.R. Acad. Sci., Paris, Sér. I **293** (1981) 460–472 et Proc. Japan Acad. Sér. A **59** (1983) 8–10.
[4] Analyse sur certaines sous-variétés de l'espace de Wiener. C.R. Acad. Sci., Paris, Sér. I bf 301 (1985) 881–884.

Gaveau, B., and Mazet, E.
[1] Opérateurs elliptiques et mesures sur l'espace des lacets invariantes par le groupe des difféomorphismes du cercle. Ann. Inst. Henri Poincaré, Probab. Stat. **35** (1981) 105–111.

Gaveau, B., and Moulinier, J.M.
[1] Intégrales oscillantes stochastiques: estimation asymptotique de fonctionnelles caractéristiques. J. Funct. Anal. **54** (1983) 161–176.
[2] Régularité de mesures dans l'espace de Hilbert et perturbation du processus d'Ornstein-Uhlenbeck par un champ de vecteurs. C.R. Acad. Sci., Paris, Sér. I **299** (1984) 559–562.
[3] Intégrales oscillantes stochastiques et front d'onde des solutions d'équations semi-elliptiques. C.R. Acad. Sci., Paris, Sér. I **301** (1985) 825–828.
[4] Régularité des mesures et perturbations stochastiques de champ de vecteurs sur des espaces de dimension infinie. RIMS Kyoto **21** (1985) 593–616.

Gaveau, B., and Trauber, P.H.
[1] Constructions de diffusions et de mesures invariantes par le groupe de jauge sur l'espace des connexions. J. Funct. Anal. **38** (1980) 324–341.
[2] L'intégrale stochastique comme opérateur de divergence dans l'espace fonctionnel. J. Funct. Anal. **46** (1982) 230–238, and C.R. Acad. Sci., Paris, Sér. I **29** (1980) 283–286.
[3] Une construction de la quantification euclidienne du champ de Yang-Mills. J. Funct. Anal. **42** (1981) 356–367.

Gaveau, B., and Vauthier, J.
[1] Intégrales stochastiques oscillantes: l'équation de Pauli. J. Funct. Anal. **44** (1981) 388–400.

Getzler, E.
[1] Degree theory for Wiener maps. J. Funct. Anal. **68** (1986) 213–233, 388–403.
[2] Dirichlet forms on loop spaces. Bull. Sci. Math. **113** (1989) 151–174.
[3] An extension of Gross's log Sobolev inequality for the loop space of a compact Lie group. In: Probab. model in Math. Phys., (G. Morrow ed.). World Scientific (1991) 73–97.
[4] The Thom class of Mathai and Quillen and probability theory. In: Proc. Lisbon conf. (A.B. Cruzeiro ed.), Birkhäuser. Prog. Probab. 26 (1991) 111–123.
[5] The degree of the Nicolai map. J. Funct. Anal. **74** (1987) 121–138.
[6] Cyclic homology and the path-integral of the Dirac operator. Preprint

Gikhman, J.I., and Skorohod, A.V.
[1] Stochastic differential equations. Grundlehren der math. Wiss. 218, Springer (1972), 574 pp.

Girsanov, I.V.
[1] On transforming a certain class of stochastic processes by absolutely continuous substitution of measures. Theory Probab. Appl. **5** (1960) 285–301.

Gong, F.Z., and Ma, Z.M.
[1] Quasi-sure analysis on Itô's Wiener space. C.R. Acad. Sci., Paris, Sér. I **321** (1995) 1609–1614, and J. Funct. Anal. (1996).

Goodman, V.
[1] A divergence theorem for a Hilbert space. Trans. Am. Math. Soc. **164** (1972) 411–426.

Graveraux, J.B.
[1] Calcul de Malliavin et probabilité invariante d'une chaîne de Markov. Ann. Inst. Henri Poincaré, Probab. Stat. **24** (1988) 159–166.

Graveraux, J.B., and Jacob, J.
[1] Opérateur de Malliavin sur l'espace de Wiener-Poisson. C.R. Acad. Sci., Paris, Sér. I **300** (1985) 81–84.

Grorud, A.
[1] Un crochet non-symétrique en calcul stochastique anticipatif. In: Proc. Stoch. Anal. and Rel. Topics, Silivri 1988 (A.S. Üstünel ed.). Springer, Lecture Notes in Math. 1444 (1990) 183–192.

Grorud, A., Nualart, D., and Sanz, M.
[1] Hilbert-valued anticipating S.D.E.. Ann. Inst. Henri Poincaré Probab. Stat. **30** (1994) 133–161.

Gross, L.
[1] Integration and non-linear transformation in Hilbert space. Trans. Am. Math. Soc. **94** (1960) 404–440.
[2] Potential theory on the Wiener space. J. Funct. Anal. **1** (1967) 123–181.
[3] Logarithmic Sobolev inequalities on loop groups. J. Funct. Anal. **102** (1991) 268–313.
[4] Uniqueness of ground states for Schrödinger operators over the loop space. J. Funct. Anal. **112** (1993) 373–441.

Gross, L., and Malliavin, P.
[1] Hall's transform and Segal-Bargmann maps. In: Itô's Stochastic Calculus and Probability Theory (N. Ikeda, S. Watanabe, M. Fukushima, and H. Kunita ed.). Springer, Tokyo 1996, 73–117.

Gundy, R.
[1] Sur les transformations de Riesz pour le semi-groupe de Ornstein-Uhlenbeck. C.R. Acad. Sci., Paris, Sér. I **303** (1986) 967–970.

Gutierrez, C.
[1] On Riesz transforms for gaussian measures. J. Funct. Anal. **120** (1994) 107–134.

Hargé, G.
[1] Régularité de certaines fonctionnelles sur l'espace de Wiener. C.R. Acad. Sci., Paris, Sér. I **316** (1993) 593–596.
[2] Continuité approximative de certaines fonctionelles sur l'espace de Wiener. J. Math. Pures Appl. **74** (1995) 53–93.

Haussmann, U.
[1] On the integral representation of functionals of Itô processes. Stoch. **3** (1979) 17–28.
[2] A stochastic maximum principle for optimal control of a diffusion. Pitman Res. Notes Math. Ser. 151 (1986).

Haussmann, U., and Pardoux, E.
[1] Time reversal of diffusions, Ann. Probab. **14** (1986) 1188–1205.

Havin, V.P., and Mazya, V.P.
[1] Non-linear potential theory. Russ. Math. Surv. **27** (1972) 71–148.

Hirsch, F.
[1] Propriété d'absolue continuité pour les équations différentielles stochastiques dépendant du passé. J. Funct. Anal. **76** (1988) 193–215.

Hirsch, F., and Song, S.
[1] Propriétés de Markov des processus à plusieurs paramètres et capacités. C.R. Acad. Sci., Paris, Sér. I **319** (1984) 483–488.
[2] Une inégalité maximale pour certains processus de Markov à plusieurs paramètres. C.R. Acad. Sci., Paris, Sér. I **320** (1995) 719–724, and 867–870.

Holley, R., and Stroock, D.
[1] Diffusions on an infinite-dimensional torus. J. Funct. Anal. **42** (1981) 29–63.
[2] Generalized Ornstein-Uhlenbeck process as limit of interacting systems. In: Proc. Stoch. integrals (D. Williams ed.) Durham 1980. Springer, Lecture Notes in Math. 851 (1981) 152–168.
[3] Logarithmic Sobolev inequalities and stochastic Ising model. J. Stat. Phys. **46** (1987) 1159–1196.

Hoshiro, T.
[1] Hypoellipticity for infinite degenerate elliptic and parabolic operators. J. Math. Kyoto Univ. **28** (1988) 615–632.

Hsu, E.P.
[1] Quasi-invariance of the Wiener measure on the path space over a compact Riemann manifold. J. Funct. Anal. **134** (1995) 417–450.
[2] Flows and quasi-invariance of the Wiener measure on path spaces. In: Proc. Symp. Pure and Appl. Math. **57** (M. Cranston and M. Pinsky ed.) Cornell 1993, Amer. Math. Soc. (1995).
[3] Logarithmic Sobolev inequalities on path spaces over Riemannian manifolds. C.R. Acad. Sci., Paris, Sér. I **320** (1995) 1009–1014.

Hu, Y.Z., and Meyer, P.A.
[1] Sur les intégrales multiples de Stratonovitch, Sém. Probab. XXII. Springer, Lecture Notes in Math. 1321 (1988) 72–81.

Huang, Z.
[1] Stochastic calculus of variations on gaussian spaces and white noise analysis. In: Third Nagoya Lévy Sem. 1990 (K. Itô and T. Hida eds.). World Scientific (1991) 227–242.

Huang, Z., and Ren, J.
[1] Quasi-sure stochastic flow. Stoch. **33** (1990) 149–158.

Hu-Ying
[1] A generalized Haussman formula. Stoch. Anal. Appl. **11** (1993) 49–60.

Ikeda, N.
[1] On the asymptotic behavior of the fundamental solution of the heat equation on certain manifolds. In: Proc. Symp. on Stoch. Anal. (K. Itô ed.) Katata/Kyoto 1982. North-Holland, Math. Library 32 (1984) 169–195.
[2] Limit theorem for a class of random current. In: Proc. Taniguchi Symp. on Probab. method in Math. Phys. (N. Ikeda and K. Itô eds.). Kinokuniya, Tokyo (1987) 181–193.
[3] Probabilistic methods in the study of asymptotics. In: Ecole d'été de Saint-Flour XVIII. Springer, Lecture Notes in Math. 1427 (1990) 195–325.

Ikeda, N., and Kusuoka. S.
[1] Short time asymptotics for fundamental solutions of diffusion equations, (Japanese/French Sem., Paris, 1987). Springer, Lecture Notes in Math. 1322 (1988) 37–49.

Ikeda, N., Kusuoka, S., and Manabe, S.
[1] Lévy's stochastic area and related problems. In: Proc. Symp. Pure and Appl. Math. **57** (M. Cranston and M. Pinsky ed.) Cornell 1993, Amer. Math. Soc. (1995) 284–305.
[2] Lévy's stochastic area for Gaussian processes. Comm. Pure Appl. Math. **47** (1994) 329–360.

Ikeda, N., and Manabe, S.
[1] Integral of differential forms along the path of diffusion processes. Pub. RIMS Kyoto Univ. **15** (1979) 827–852.
[2] Asymptotic formulae for stochastic oscillatory integrals, asymptotic problems in probability theory: Wiener functionals and asymptotics. Pitman Res. Notes Math. Ser. **284** (1993) 136–155.
[3] Van Vleck-Pauli for Wiener integrals and Jacobi fields: Itô's Stochastic Calculus and Probability Theory (N. Ikeda, S. Watanabe, M. Fukushima, H. Kunita eds.). Springer Tokyo 1996, 141–158.

Ikeda, N., and Matsumoto, H.
[1] Short time asymptotics for the trace of Schrödinger operators with magnetic fields. Bull. Sci. Math. **116** (1992) 53–66.

Ikeda, N., and Ochi, Y.
[1] Central limit theorem and random currents. Springer, Lecture Notes Control Inf. Sci. **78** (1986) 195–205.

Ikeda, N., Shigekawa, I., and Taniguchi, S.
[1] The Malliavin calculus and long time asymptotics of certain Wiener integrals. Proc. Cent. Math. Anal. Aust. Natl. Univ. **9** (1985) 146–213.

Ikeda, N., and Watanabe, S.
[1] Heat equation and diffusion on Riemannian manifolds with boundary, Proc. Intern. Symp. SDE Kyoto 1976. John Wiley (1978) 75–94.
[2] Stochastic flows of diffeomorphisms, Stoch. Anal. Appl. (M. Pinsky ed.) Dekker. Adv. Probab. Relat. Top. 7 (1984) 179–198.
[3] An introduction to Malliavin's calculus. In: Proc. Symp. on Stoch. Anal. (K. Itô ed.) Katata/Kyoto 1982. North-Holland, Math. Library 32 (1984) 1–52.
[4] Malliavin calculus of Wiener functionals and its applications, (K.D. Elworthy ed.). Pitman Res. Notes Math. Ser. 150 (1986) 132–178.
[5] Stochastic differential equations and diffusion processes. North-Holland, Math. Library 24 (1981) 464 pp., 2nd ed. (1989) 554 pp.

Imkeller, P.
[1] Existence and continuity of occupation densities of stochastic integral processes. Ann. Probab. **22** (1993) 469–473.
[2] Occupation densities of Stratonovich stochastic differential equations with boundary condition, Stoch. Stoch. Rep. **45** (1993) 249–264.
[3] Enlergement of the Wiener filtration by an absolutely continuous random variable via Malliavin's calculus. Probab. Theory Relat. Fields **106** (1996) 103–115.

Imkeller, P., and Nualart, D.
[1] Integration by parts on the Wiener space and the existence of occupation densities. Ann. Probab. **22** (1994) 469–493.

Imkeller, P., and Yan, J.A.
[1] Multiple intersection local time of planar Brownian motion as a particular Hida distribution. J. Funct. Anal. **140** (1996) 256–273.

Itô, K.
[1] Stochastic Integrals. Proc. Jap. Acad., Ser. A **20** (1944) 519–526.
[2] Stochastic differential equations on a manifold. Nagoya Math. J. **1** (1950) 35–47.
[3] Multiple Wiener integrals. J. Math. Soc. Japan **3** (1951) 157–169.
[4] On a formula concerning stochastic differentials. Nagoya Math. J. **3** (1951) 55–65.
[5] The Brownian motion and tensor fields on Riemannian manifolds. Proc. Int. Cong. Math., Stockholm, (1962) 536–539.
[6] Complex multiple Wiener integral. Jap. J. Math. **22** (1965) 63–86.
[7] Extension of stochastic integrals. In: Proc. Int. Symp. on SDE, Kyoto 1976. John Wiley (1978) 95–109.
[8] Infinite-dimensional Ornstein-Uhlenbeck processes. In: Proc. Symp. on Stoch. Anal. (K. Itô ed.) Katata/Kyoto 1982. North-Holland, Math. Library 32 (1984) 197–224.
[9] Foundations of stochastic differential equations in infinite-dimensional spaces, CBMS-NDF Regional Conferences in Appl. Math., SIAM (1984).
[10] Malliavin calculus on a Segal space. In: Proc. Stoch. Anal. (Métivier and Watanabe ed.) Paris 1987. Springer, Lecture Notes in Math. 1322 (1988) 50–72.
[11] Positive generalized functionals on $\mathbb{R}^\infty$. In: Proc. Colloq. White Noise Analysis, Bielefeld 1989. World Scientific (1990) 166–179.
[12] On Malliavin calculus. In: Proc. Singapore Probab. (Louis H.Y. Chen ed.). de Gruyter (1992) 47–72.

[13] An elementary approach to Malliavin fields. In: Asymptotics Problems in Probab. Theory, vol. 2 (D. Elworthy and N. Ikeda eds.). Pitman Res. Notes Math. Ser. 284 (1993) 35–89.
[14] On Malliavin tensor fields. Commun. Pure Appl. Math. **47** (1994) 377–403.
[15] A measure-theoretic approach to Malliavin calculus. In: New Trends in Stochastic Analysis, Proc. Taniguchi Symp. on Stoch. Anal., Charington 1994 (D. Elworthy, S. Kusuoka and I. Shigekawa eds.). World Scientific 1996.

Itô, K., and McKean Jr., H.P.
[1] Diffusion processes and their sample paths. Springer, Grundlehren der math. Wiss. 125 (1965) 321 pp.

Itô, K., and Watanabe, S.
[1] Introduction to stochastic differential equations. In: Proc. Int. Symp. on SDE, Kyoto 1976. John Wiley (1978) 1–32.

Jacod, J.
[1] Malliavin calculus for discontinuous processes. In: Probab. Th. and Math. Stat., vol.II (Vilnius, 1985). VNU Sci. Press (1987) 17–25.

Jacquot, S.
[1] Comportement asymptotique de l'intervalle spectral d'opérateur de Kolmogoroff sur l'espace de Wiener. Stoch. **37** (1991) 213–224.

Janson, S.
[1] Gaussian Hilbert spaces. Preprint (1996) 300 pp.

Jolis, M., and Sanz-Sole, M.
[1] On generalized multiple stochastic integrals and multiparameter anticipative calculus. In: Stoch. Anal. and Rel. Topics II (A.S. Üstünel ed.). Springer, Lecture Notes in Math. 1444 (1990) 141–182.
[2] Doob-Meyer decomposition and integrator properties of the Wong-Zakai anticipative integrals. In: Stoch. Anal. and Rel. Topics (B. Øksendal and A. Üstünel, eds.). Gordon & Breach, Stochastic Monogr. (1993) 177–202.

Julia, O., and Nualart, D.
[1] The distribution of a double stochastic integral with respect to two independent Brownian sheets. Stoch. **25** (1988) 171–182.

Kallianpur, G., and Perez-Abreu, V.
[1] The Skorokhod integrals and the derivative operator for cylindrical Brownian motion. Appl. Math. Optim. **25** (1992) 11–29.

Kaneko, H.
[1] On (r, p)-capacities for Markov processes. Osaka J. Math. **23** (1984) 325–336.
[2] A stochastic resolution of the Monge-Ampère equation. Osaka J. Math. **24** (1987) 307–319.

Karatzas, I., Li, J., and Ocone, D.
[1] An extension of the Clark formula. Stoch. **37** (1991) 127–131.

Katznelson, Y., and Malliavin, P.
[1] Image des points critiques d'une application régulière. In: Proc. Stoch. Anal., Paris 1987, (M. Mélivier and S. Watanabe ed.). Springer, Lecture Notes in Math. 1322 (1988) 85–92.
[2] Un contre exemple au théorème de Sard en dimension infinie. C.R. Acad. Sci., Paris, Sér. I **306** (1988) 37–41.

Kazumi, T.
[1] Refinements in terms of capacities of certain limit theorems on an abstract Wiener space. J. Math. Kyoto Univ. **32** (1992) 1–33.
[2] Le processus d'Ornstein-Uhlenbeck sur l'espace des chemins et le problème des martingales. J. Funct. Anal. (1977).

Kazumi, T., and Shigekawa, I.
[1] Measures of finite (r, p) energy and potential on a separable metric space, Sém. Probab. XXVI. Springer, Lecture Notes in Math. 1526 (1992) 415–444.
[2] Differential calculus on a submanifold of an abstract Wiener space, I. Covariant derivative. In: Stochastic analysis in infinite-dimensional spaces (H. Kunita and H.H. Kuo, eds.). Longman, 1994, 117–140.
[3] Differential calculus on a submanifold of an abstract Wiener space, II. Weitzenböck formula. In: Dirichlet forms, Beijing 1993 (Ma Zhi Ming, M. Röckner, and J.A. Yan eds.). de Gruyter (1995) 235–253.

Kendall, W.S.
[1] Stochastic differential geometry: An introduction. Acta Appl. Math. **9** (1987) 29–60.

Kleptsina, M., and Veretennikov, A.Yu
[1] On filtering and properties of conditional laws of Itô-Volterra processes. Stat. and Control of Stoch. Proc., (Moscow, 1984). Optim. Software, New York (1985) 179–196.

Komatsu, T., and Takashima, K.
[1] The Hausdorff dimension of quasi-all Brownian paths. Osaka J. Math. **21** (1984) 613–619.
[2] On the existence of intersectional local time except sets of zero capacity. Osaka J. Math. **21** (1984) 913–929.

Kôno, N.
[1] Propriétés quasi-partout de fonctions aléatoires gaussiennes. In: Séminaire d'Analyse des Fonctions Aléatoires, Univ. de Strasbourg (1983).
[2] 4-dimensional Brownian motion in recurrent with positive capacity. Proc. Japan Acad. **60** (1984) 57–59.

Koski, T.
[1] A note on a probabilistic decomposition of linear delta modulators of a Wiener process. Ann. Acad. Sci. Fenn., Ser.A I Math. **13** (1988) 209–218.

Kossyak, A.V.
[1] The irreducible regular Gaussian representation of the groups of the interval and of the circle diffeomorphism. J. Funct. Anal. **125** (1994) 493–547.

Krée, M.
[1] Propriété de trace en dimension infinie d'espace du type de Sobolev. Bull. Soc. Math. Fr. **105** (1977) 141–163.

Krée, M., and Krée, P.
[1] Continuité de la divergence dans les espaces de Sobolev relatifs à l'espace de Wiener. C.R. Acad. Sci., Paris, Sér. I **296** (1983) 833–836.

Krée, P.
[1] Calcul d'intégrales et de dérivées en dimension infinie. J. Funct. Anal. **31** (1979) 150–186.

Krylov, N.V.
[1] On the first quasi-derivative of solutions of Itô stochastic equations. Russian Acad. Sci. Izvestiya Math. **40** (1993) 377–403.

Kunita, H.
[1] On the decomposition of solutions of stochastic differential equations. In: Proc. Symp. Stoch. integrals, Univ. Durham, 1980. Springer, Lecture Notes in Math. 851 (1981) 213–255.
[2] Stochastic flows and stochastic differential equations. Cambridge Univ. Press (1988).
[3] Convergence of stochastic flows connected with stochastic ordinary differential equations. Stoch. **17** (1986) 215–251.
[4] Generalized solution of SPDE. J. Theor. Probab. **7** (1994) 279–308.
[5] On backward stochastic differential equations, Stoch. **6** (1982) 293–313.

Kunita, H., and Watanabe, S.
[1] On square integrable martingale. Nagoya J. Math. **30** (1967) 209–245.

Kuo, H.H.
[1] Integration by parts for abstract Wiener measure. Duke Math. J. **41** (1974) 373–379.
[2] Potential theory associated with the Ornstein-Uhlenbeck process. J. Funct. Anal. **21** (1976) 63–75.
[3] Brownian motion, diffusion and infinite-dimensional calculus. In: Proc. Stoch. Anal. and Rel. Topics, Silivri 1986. Springer, Lecture Notes in Math. 1316 (1988) 130–169.
[4] Fourier-Mehler transform in white noise analysis. In: Proc. Third Nagoya Lévy Sem. 1990 (K. Itô and T. Hida, eds.). World Scientific (1991) 257–272.

Kusuoka, S.
[1] Analytic functionals of Wiener processes and absolute continuity. In: Proc. Funct. Anal., Markov Proc., Katata/Kyoto 1981 (M. Fukushima ed.). Springer, Lecture Notes in Math. 923 (1982) 1–47.
[2] The non-linear transformation of Gaussian measure on a Banach space and its absolute continuity. J. Fac. Sci. Univ. Tokyo Sect. IA **29** (1982) 567–590.
[3] On the absolute continuity of the law of a system of multiple Wiener integrals. J. Fac. Sci. Univ. Tokyo Sect. IA **30** (1983) 191–197.
[4] The Malliavin calculus and its applications. Sugaku **36** (1984) 97–109.
[5] Degree theorem in certain Wiener Riemannian manifolds. In: Proc. Stoch. Anal., Japanese/French Sem., Paris 1987, (M. Métivier and S. Watanabe ed.). Springer, Lecture Notes in Math. 1322 (1988) 93–108.
[6] On the foundations of Wiener-Riemann manifolds. Stoch. Anal., Pitman Res. Notes Math. 200 (1989) 130–164.
[7] Analysis on Wiener spaces, I. Nonlinear maps. J. Funct. Anal. **98** (1991) 122–168.
[8] Analysis on Wiener spaces, II. Differential forms. J. Funct. Anal. **103** (1992) 229–274.
[9] De Rham cohomology on Wiener-Riemann manifolds. In: Proc. ICM Kyoto 1990, Springer, 1075–1082.
[10] Stochastic analysis as infinite dimensional analysis. Sugaku (1993) 289–298.
[11] Some remarks on Getzler's degree. Springer, Lecture Notes in Math. 1299 (1988) 239–249.

Kusuoka, S., and Stroock, D.
[1] Applications of the Malliavin calculus, I. In: Proc. Symp. on Stoch. Anal. (K. Itô ed.) Katata/Kyoto 1982. North-Holland, Math. Library 32 (1984) 271–306.
[2] The partial Malliavin calculus and its application to nonlinear filtering. Stoch. **12** (1984) 83–142.
[3] Applications of the Malliavin calculus, II. J. Fac. Sci. Univ. Tokyo Sect. IA **32** (1985) 1–76.

[4] Some boundedness properties of certain stationary diffusion semigroups. J. Funct. Anal. **60** (1985) 243–264.
[5] Applications of the Malliavin calculus, III. J. Fac. Sci. Univ. Tokyo Sect. IA **34** (1987) 391–442.
[6] Longtime estimates for the heat kernel associated with a uniformly subelliptic operator. Ann. Math. **127** (1988) 165–189.
[7] Precise asymptotics of certain Wiener functionals. J. Funct. Anal. **99** (1991) 1–74.
[8] Asymptotics of certain Wiener functionals with degenerate extrema. Commun. Pure Appl. Math. **47** (1994) 477–501.

Kusuoka, S., and Taniguchi, S.
[1] Pseudoconvex domains in almost complex Wiener spaces. J. Funct. Anal. **117** (1993) 62–117.

Lascar, B.
[1] Propriétés locales d'espaces du type de Sobolev de dimension infinie. C.R. Acad. Sci., Paris, Sér. I **280** (1975) 1587–1590, and Commun. Partial Differ. Equations **1** (1976) 561–584.

Léandre, R.
[1] Majoration en temps petit de la densité d'une diffusion dégénérée. Probab. Theory Relat. Fields **74** (1987) 289–294.
[2] Applications quantitatives et géométriques du calcul de Malliavin. In: Proc. Stoch. Anal., Japanese/French Sem., Paris 1987, (M. Métivier and S. Watanabe ed.). Springer, Lecture Notes in Math. 1322 (1988) 109–133.
[3] Calcul des variations sur un brownien subordonné, Sém. Probab., XXII. Springer, Lecture Notes in Math. 1321 (1988) 414–433.
[4] Intégration dans la fibre associée à une diffusion dégénérée. Probab. Theory Relat. Fields **76** (1987) 341–356.
[5] Integration by parts formulas and rotationally invariant Sobolev calculus on free loop spaces. J. Geom. Phys. **11** (1993) 517–528.

Léandre, R., and Russo, F.
[1] Estimation de Varadhan pour des diffusions à deux paramètres. Probab. Theory Relat. Fields **84** (1990) 429–451.
[2] Estimation de densité pour une équation des ondes faiblement perturbée par un bruit blanc. C.R. Acad. Sci., Paris, Sér. I **311** (1990) 737–740.
[3] Small stochastic perturbation of a one-dimensional wave equation. C.R. Acad. Sci., Paris, Sér. I **311** (1990) 737–739, and Stoch. Anal. and Rel. Fields (A.S. Üstünel, ed.), Birkhäuser, Prog. Probab. 31 (1992) 285–333.

Ledoux, M.
[1] L'algèbre de Lie des gradients itérés d'un générateur Markovien. C.R. Acad. Sci., Paris, Sér. I **317** (1993) 1049–1052.

Lee, Y.
[1] Integral transform of analytic functions on abstract Wiener space. J. Funct. Anal. **47** (1982) 153–193.

Le Jan, Y.
[1] Flots de diffusion sur $\mathbb{R}^d$. C.R. Acad. Sci., Paris, Sér. I **294** (1982) 697–699.
[2] On isotropic brownian motion. Z. Wahrsch. verw. Gebiete **70** (1985) 609–620.
[3] Equilibre statistique pour les produits aléatoires de difféomorphismes indépendants. Ann. Inst. Henri Poincaré, Probab. Stat. **23** (1987) 111–120.
[4] On the Fock space representation of functionals of the occupation time and their renormalization. J. Funct. Anal. **80** (1988) 88–108.

Le Jan, Y., and Watanabe, S.
[1] Stochastic flow of diffeomorphisms. In: Proc. Symp. on Stoch. Anal. (K. Itô ed.) Katata/Kyoto 1982. North-Holland, Math. Library 32 (1984) 307–332.

Lepingle, D., Nualart, D., and Sanz, M.
[1] Dérivation stochastique de diffusion réfléchie. Ann. Inst. Henri Poincaré, Probab. Stat. **25** (1989) 283–305.

Lescot, P.
[1] Un critère de régularité des lois pour certaines fonctionnelles de Wiener. C.R. Acad. Sci., Paris, Sér. I **316** (1993) 1313–1318.
[2] Un théorème de désintégration en analyse quasi-sure, Sém. Probab. XXVII. Springer, Lecture Notes in Math. 1557 (1993) 256–275.
[3] Sard theorem for hyper-Gevrey functionals on the Wiener space. J. Funct. Anal. **122** (1995) 191–221.

Liao, M.
[1] Stochastic flow on the boundary of Lie group. Stoch. **39** (1992) 213–238.
[2] Stochastic flows on the boundaries of $SL(n;\mathbb{R})$. Probab. Theory Relat. Fields **96** (1993) 261–281.
[3] Brownian motion on a symmetric space. Trans. Am. Math. Soc. **341** (1996) 253–274.

Lyons, T.J.
[1] The critical dimension at which quasi-every Brownian path is self-avoiding. Adv. Appl. Probab., Suppl. (1986) 87–99.
[2] The interpretation and solution of ordinary differential equations driven by rough signals. In: Proc. Symp. Pure and Appl. Math. **57** (M. Cranston and M. Pinsky, eds.) Cornell 1993, Am. Math. Soc. (1995) 115–128.
[3] Differential equations driven by rough signals. Math. Res. Letters **1** (1994) 451–464.

Lyons, T.J., and Quian, Z.M.
[1] A class of vector fields on path spaces. J. Funct. Anal. (1997).

Lyons, T.J., and Zheng, W.A.
[1] A crossing estimate for the canonical process on a Dirichlet space and a tightness result. Astérisque **157** (1988) 249–271.
[2] On conditional diffusion processes. In: Proc. Royal Soc. Edinburgh **115**A (1990) 243–255.

Malliavin, M.P., and Malliavin, P.
[1] Asymptotique de la diffusion horizontale au-dessus d'un espace Riemannien symétrique. Springer, Lecture Notes in Math. 404 (1974) 155–210.
[2] Holonomie stochastique au-dessus d'un espace Riemannien symétrique. C.R. Acad. Sci., Paris, Sér. I **279** (1974) 185–188.
[3] Ibidem, (1975) 793–795.
[4] An infinitesimally quasi-invariant measure on the group of diffeomorphisms of the circle. In: Special Functions, ICM-90 Satellite Conference Proc. (M. Kashiwara and T. Miwa eds.). Springer Tokyo (1991) 234–244.
[5] Integration on loop groups, I. Quasi invariant measures. J. Funct. Anal. **93** (1990) 207–237.
[6] Integration on loop groups, III. Asymptotic Peter-Weyl orthogonality. Ibidem **108** (1992) 13–46.

Malliavin, P.
[1] Asymptotic of the Green's function of a Riemannian manifold and Itô's stochastic integral. Proc. Nat. Acad. Sci. **71** (1974) 381–383.

[2] Formule de la moyenne, calcul des perturbations et théorie d'annulation pour les formes harmoniques. J. Funct. Anal. **17** (1974) 274–291.
[3] Diffusions et géométrie différentielle globale. Centro Matem. Estivo Varenna (1975) 208–279.
[4] Parametrix trajectorielle. C.R. Acad. Sci., Paris, Sér. I **281** (1975) 241–245.
[5] Champs de Jacobi stochastiques. C.R. Acad. Sci., Paris, Sér. I **285** (1977) 789–792.
[6] Hypoellipticité avec degénérescence et analyse stochastique en dimension infinie. C.R. Acad. Sci., Paris, Sér. I **284** (1977).
[7] Un principe de transfert et son application au calcul stochastique. C.R. Acad. Sci., Paris, Sér. I **284** (1977) 187–189.
[8] C^k-hypoellipticity with degeneracy. In: Stoch. Anal. (A. Friedman and M. Pinsky eds.). Academic Press (1978) 199–214 and 327–340.
[9] Géométrie différentielle stochastique. Sém. Math. 64, Les Presses de l'Univ. de Montréal (1978) 132 pp.
[10] Stochastic calculus of variation and hypoelliptic operators. In: Proc. Int. Symp. on SDE, Kyoto 1976. John Wiley (1978) 195–263. Summary in: C.R. Acad. Sci., Paris, Sér. I **283** (1976) 515–518, and **284** (1977) 1455–1456.
[11] Stochastic jacobian fields. In: Proc. Conf. partial differential equations and geometry (Christofer, Byrhes, eds.). M. Dekker (1979) 203–235, and C.R. Acad. Sci., Paris, Sér. I **285** (1977) 789–792.
[12] Diffusion on the loops. In: "Conference in the honour of A. Zygmund", Chicago (Beckner, ed.). Wadsworth Math. Ser. vol. II (1981) 764–782.
[13] Calcul des variations stochastiques subordonnées à l'équation de la chaleur. C.R. Acad. Sci., Paris, Sér. I **295** (1982) 167–172.
[14 Intégrales oscillantes, C.R. Acad. Sci., Paris, Sér. I **295** (1982) 295–300.
[15] Analyse différentielle sur l'espace de Wiener. In: Proc. Intern. Congress, Math., Warsaw 1983, (P. Malliavin ed.). PWN, Warszawa (1984) 1089–1096.
[16] Implicit functions in finite corank on the Wiener space. In: Proc. of the Tanniguchi Symp. Stoch. Anal., Katata/Kyoto 1982 (K. Itô, ed.). North-Holland (1984) 369–386.
[17] Calcul des variations intégrales stochastiques et complexes de de Rham sur l'espace de Wiener. C.R. Acad. Sci., Paris, Sér. I **299** (1984) 347–350.
[18] Régularité des lois conditionelles et calcul des variations stochastiques. C.R. Acad. Sci., Paris, Sér. I **289** (1979) 357–360.
[19] Intégrales oscillantes. C.R. Acad. Sci., Paris, Sér. I **300** (1985) 141–143.
[20] Intégrales oscillantes, analyticité transverse d'opérateurs hypoelliptiques C^3 sur des fibrés principaux: spectre equivariant et courbure. C.R. Acad. Sci., Paris, Sér. I **301** (1985) 767–770.
[21] Intégrales oscillantes, analyticité réelle des lois conditionnelle de fonctionnelles additives. C.R. Acad. Sci., Paris, Sér. I **302** (1986) 73–78.
[22] Intégrales oscillantes. C.R. Acad. Sci., Paris, Sér. I **302** (1986) 359–362.
[23] Intégrales oscillantes. C.R. Acad. Sci., Paris, Sér. I **302** (1986) 481–486.
[24] Functional integration and partial differential equations. Adv. Math. Beijing **15** (1986) 131–174, (in chinese).
[25] Hypoellipticity in infinite dimension. In: Diff. Proc. and Rel. Probl. in Anal. (M. Pinsky ed.). Birkhäuser, Prog. Probab. 22 (1989) 17–33.
[26] Smooth σ-fields. In: Stoch. Anal., Liber Amicorum for Moshe Zakai. Academic Press (1991) 371–382.
[27] Naturality of quasi-invariance of some measure. In: Proc. of the Lisbon conf. (A.B. Cruzeiro ed.). Birkhäuser, Prog. Probab. 26 (1991) 144–154.
[28] Stochastic analysis in infinite dimension, Saint-Chéron Round Tables 1992 (M. Yor ed.). Bull. Sci. Math. **117** (1993) 63–90.
[29] Universal Wiener spaces. In: Proc. Barcelona Sém. on Stoch. Anal. (D. Nualart ed.). Birkhäuser, Prog. Probab. 32 (1993) 77–103.

[30] A Bootstrap proof of the limit theorem for linear SDE. In: Proc. Symp. Pure and Appl. Math. **57** (M. Cranston and M. Pinsky, eds.) Cornell 1993, Amer. Math. Soc. (1995) 361–371.
[31] Abstract Wiener space versus filtered Wiener space. In: MIT Wiener centennial Conference. Amer. Math. Soc. 1997.

Malliavin, P., and Nualart, D.
[1] Quasi-sure analysis and Stratonovitch anticipative SDE. Probab. Theory Relat. Fields **96** (1993) 45–55.
[2] Quasi-sure analysis of stochastic flow and Banach-space-valued smooth functionals on the Wiener space. J. Funct. Anal. **112** (1993) 287–317.

Malliavin, P., and Stroock, D.
[1] Short time behaviour of heat kernel and its derivative. J. Differ. Geom. (1997).

Malliavin, P., and Taniguchi, S.
[1] Analytic functions, Cauchy formula and stationary phase on a real abstract Wiener space. J. Funct. Anal. **143** (1997) 470–528

Malliavin, P., and Vauthier, J.
[1] Plurisubharmonic measures on loop spaces. J. Anal. Math. **60** (1993) 213–226.

Manabe, S.
[1] On the intersection number of the path of a diffusion and chains. Proc. Japan Acad., Ser. A **55** (1979) 23–26.
[2] Large deviation for a class of current-valued processes. Osaka J. Math. **29** (1992) 89–103.
[3] Stochastic intersection number and homological behaviour of a diffusion process on a Riemannian manifold. Osaka J. Math. **19** (1982) 429–457.

Manabe, S., and Ochi, Y.
[1] The central limit theorem for current-valued processes induced by the geodesic flow. Osaka J. Math. **26** (1989) 191–204.

Manabe, S., and Shigekawa, I.
[1] A comparison theorem for eigenvalues of the covariant Laplacian. J. Funct. Anal. **94** (1990) 349–357.

Martias, C.
[1] Une formule de Itô-Ventsel pour les processus anticipatifs. C.R. Acad. Sci., Paris, Sér. I **307** (1988) 675–678.

Masuda, T.
[1] Absolute continuity of distributions of solution of anticipative stochastic differential equations. J. Funct. Anal. **95** (1992) 414–432.

Matsumoto, H., and Shigekawa, I.
[1] Limit theorem for stochastic flows of diffeomorphisms of jump type. Z. Wahrsch. verw. Geb. **69** (1985) 507–540.

Matsumoto, H., and Ueki, N.
[1] Spectral analysis of Schrödinger operators with magnetic fields. J. Funct. Anal. **140** (1996) 218–255.

Mayer-Wolf, E.
[1] Preservation of measure continuity under conditioning. J. Funct. Anal. **114** (1993).

[2] The variational distance between random fields and application to conditional densities. In: Stoch. Anal. and Rel. Probl. (B. Oksendale and A. Üstünel eds.). Gordon & Breach, Stochastic Monogr. (1993) 203–215.
[3] A central limit theorem in nonlinear filtering. Stoch. Stoch. Rep. **35** (1991) 191–211.
[4] Covariance inequalities induced by biorthogonal operators. J. Funct. Anal. **140** (1996) 170–193.

Mayer-Wolf, E., and Zeitouni, O.
[1] Onsager-Machlup functionals for non-trace-class SPDE. Probab. Theory Relat. Fields **95** (1993) 199–216.

Mazya, V.G., and Khavin, V.P.
[1] Non-linear potential theory. Usp. Mat. Nauk **27** (1973) 67–138.

Mazziotto, G., and Millet, A.
[1] Absolute continuity of the law of an infinite-dimensional Wiener functional with respect to the Wiener probability. Probab. Theory Relat. Fields **85** (1990) 403–411.

McKean Jr., H.P.
[1] Stochastic integrals. Acad. Press (1969).
[2] An upper bound to the spectrum of the laplacian on a manifold of negative curvature. J. Differ. Geom. **4** (1970) 359–380.

Mellouk, M.
[1] Support des diffusions dans les espaces de Besov-Orlicz. C.R. Acad. Sci., Paris, Sér. I **319** (1994) 261–266.

Méritet, A.
[1] Théorème d'annulation pour la cohomologie absolue d'une variété Riemannienne à bord. Bull. Sci. Math. **103** (1979) 379–400.

Meyer, P.A.
[1] Note sur les processus d'Ornstein-Uhlenbeck, Sém. Probab., XVI. Springer, Lecture Notes in Math. 920 (1982) 95–132.
[2] Transformations de Riesz pour les lois gaussiennes, Sém. Probab. XVII. Springer, Lecture Notes in Math. 1059 (1984) 179–193.
[3] Quelques résultats analytiques sur le semi-groupe d'Ornstein-Uhlenbeck en dimension infinie, Bangalore 1982, (G. Kallianpur ed.). Springer, Lecture Notes in Control and Inf. Sci. 49 (1983).
[4] Flow of stochastic differential equations (following Malliavin, Bismut, Kunita), Sém. Probab. XV. Springer, Lecture Notes in Math. 850 (1981) 103–117.

Michel, D.
[1] Formule de Stokes stochastique. Bull. Sci. Math. **103** (1979) 193–240.
[2] Régularité des lois conditionnelles en théorie du filtrage non-linéaire et calcul des variations stochastique. J. Funct. Anal. **41** (1981) 1–36.
[3] Conditional laws and Hörmander's condition. In: Proc. Symp. on Stoch. Anal. (K. Itô ed.) Katata/Kyoto 1982. North-Holland, Math. Library 32 (1984) 387–408.

Michel, D., and Pardoux, E.
[1] An introduction to Malliavin calculus and some of its applications. In: Recent Adv. in Stoch. Calculus (J.S. Baras ed.). Progr. in Autom. and Inform. Syst., Springer (1989) 65–104.

Millet, A., and Nualart, D.
[1] Support theorem for a class of anticipating stochastic differential equations. Stoch. **39** (1992) 1–24.

Millet, A., Nualart, D., and Sanz, M.
[1] Integration by parts and time reversal of diffusion processes. Ann. Probab. **17** (1989) 208–238.
[2] Time reversal for infinite-dimensional diffusion. Probab. Theory Relat. Fields **82** (1989) 315–347.
[3] Composition of large deviations, principles and applications. In: Liber Amicorum for Moshe Zakai. Academic Press (1991) 383–396.
[4] Large deviation for a class of anticipating stochastic differential equations. Ann. Prob. **20** (1992) 1902–1931.

Millet, A., and Sanz-Sole, M.
[1] Un théorème de support pour une équation aux dérivées partielles stochastique hyperbolique. C.R. Acad. Sci., Paris, Sér. I **315** (1992) 615–619.
[2] On the support of a Skorokhod anticipating stochastic differential equation. In: Proc. San Felipe Conference (D.Nualart ed.). Birkhäuser, Prog. Probab. 32 (1993) 103–131.
[3] A simple proof of the support theorem for diffusion processes, Sém. Probab. XXVIII. Springer, Lecture Notes in Math. 1583 (1994) 36–48.

Millet, A., and Smolenski, W.
[1] On the continuity of Ornstein-Uhlenbeck processes in infinite dimension, Probab. Theory Relat. Fields **92** (1992) 529–547.

Mizumachi, H., and Sato, H.
[1] Continuity of quasi-invariant measures and zero-one laws on groups. J. Funct. Anal. **112** (1994) 188–200.

Morimoto, Y.
[1] Criteria for hyperellipticity of differential operators. Publ. Res. Inst. Math. Sci. **22** (1986) 1129–1154.

Moulinier, J.M.
[1] Absolue continuité de probabilités de transition par rapport à une mesure gaussienne dans un espace de Hilbert. J. Funct. Anal. **64** (1985) 275–295.
[2] Fonctionnelles oscillantes stochastiques et hypoellipticité. Bull. Sci. Math. **109** (1985) 37–60.

Mountford, T.S.
[1] Double points and the Ornstein-Uhlenbeck process on Wiener space, Ill. J. Math. **34** (1990) 38–48.

Nguyen, M.D., and Nualart, D.
[1] Stochastic processes possessing a stochastic integral representation. Stoch. **30** (1990) 47–60.

Nguyen, M.D., Nualart, D., and Sanz, M.
[1] Application of the Malliavin calculus to a class of stochastic differential equations. Probab. Theory Relat. Fields **84** (1990) 549–571.

Norin, N.V.
[1] Itô formula for extended stochastic integral with anticipating kernel. Theory Prob. Appl. **39** (1995) 573–593.

Norris, J.
[1] Simplified Malliavin calculus, Sém. Probab. XX. Springer, Lecture Notes in Math. 1204 (1986) 101–130.
[2] Integration by parts for jump processes, Sém. Probab. XXII. Springer, Lecture Notes in Math. 1321 (1988) 271–315.

[3] A complete differential formalism for stochastic calculus on manifolds, Sém. Probab. XXVI. Springer, Lecture Notes in Math. 1526 (1992) 189–209.
[4] Path integral formulae for heat kernels and derivatives. Probab. Theory Relat. Fields **94** (1993) 525–541.
[5] Twisted sheets. J. Funct. Anal. **132** (1995) 273–334.

Norris, J.R., and Stroock, D.W.
[1] Estimate of the fundamental solution to heat flow with uniformly elliptic coefficient. Proc. London Math. Soc. **62** (1991) 373–402.

Nualart, D.
[1] Application du calcul de Malliavin aux équations différentielles stochastiques sur le plan, Sém. Probab. XX. Springer, Lecture Notes in Math. 1204 (1986) 379–395.
[2] Malliavin calculus and stochastic integrals. Springer, Lecture Notes in Math. 1221 (1986) 182–192.
[3] Generalized multiple stochastic integrals and representation of Wiener functional. Stoch. **23** (1988) 311–330.
[4] Non-causal stochastic integrals and calculus. Springer, Lecture Notes in Math. 1316 (1988) 80–126.
[5] Non-linear transformation of the Wiener measure and application, Stoch. Anal., Liber Amicorum for Moshe Zakai. Academic Press (1991) 397–431.
[6] Une remarque sur le développement en chaos d'une diffusion, Sém. Probab. XXIII. Springer, Lecture Notes in Math. 1372 (1989) 165–169.
[7] Anticipating stochastic differential equations, Saint-Chéron Round Tables 1992 (M. Yor ed.). Bull. Sci. Math. **117** (1993) 49–62.
[8] Markov fields and transformations of the Wiener measure, Coll. Stoch. Analysis Rel. Topics (Oslo 1992). Gordon & Breach, Stochastic Monogr. 8 45–88.
[9] Malliavin Calculus and Related Topics. Springer 1995, 266 pp.

Nualart, D., and Pardoux, E.
[1] Stochastic calculus with anticipating integrand. Probab. Theory Relat. Fields **78** (1988) 535–581.
[2] Boundary value problems for stochastic differential equations. Ann. Probab. **19** (1991) 1118–1144.

Nualart, D., and Sanz, M.
[1] Malliavin calculus for two-parameter Wiener functionals. Z. Wahrsch. verw. Gebiete **70** (1985) 573–590.
[2] Stochastic differential equations on the plane: Smoothness of the solution. J. Multivariate Anal. **31** (1989) 1–29.

Nualart, D., and Üstünel, A.S.
[1] Geometric analysis of conditional independence on Wiener space. Probab. Theory Relat. Fields **89** (1991) 407–422.

Nualart, D., Üstünel. A.S., and Zakai, M.
[1] On the moments of a multiple Wiener Itô integral and the space induced by the polynomials of the integrals. Stoch. **25** (1988) 233–240.
[2] Some relation among classes of σ-fields on Wiener spaces. Probab. Theory Relat. Fields **85** (1990) 119–129.
[3] Some remarks on independence and conditioning on Wiener spaces. Springer, Lecture Notes in Math. 1444 (1990) 122–127.

Nualart, D., and Vives, J.
[1] Continuité absolue de la loi du maximum d'un processus gaussien. C.R. Acad. Sci., Paris, Sér. I **307** (1988) 349–354.

[2] Anticipative calculus for the Poisson process based on the Fock space, Sém. Probab. XXIV. Springer, Lecture Notes in Math. 1426 (1990) 154–165.
[3] Chaos expansion and local time. Publ. Math. **36** (1992) 827–836.
[4] Smoothness of Brownian local time and related functionals. Potential Anal. **1** (1992) 257–263.

Nualart, D., and Wschebor, M.
[1] Intégration par parties dans l'espace de Wiener et approximation du temps local. Probab. Theory Relat. Fields **90** (1991) 83–109.

Nualart, D., and Zakai, M.
[1] Generalized stochastic integrals and the Malliavin calculus. Probab. Theory Relat. Fields **73** (1986) 255–280.
[2] Generalized multiple stochastic integrals and the representation of Wiener functionals. Stoch. **23** (1988) 311–330.
[3] Generalized Brownian functionals and the solution to a stochastic partial differential equation. J. Funct. Anal. **84** (1989) 279–296.
[4] The partial Malliavin calculus, Sém. Probab. XXIII. Springer, Lecture Notes in Math. 1372 (1989) 362–381.
[5] On the relation between Stratanovitch and Ogawa integrals. Ann. Probab. **17** (1989) 1536–1540.
[6] A summary of some identities of the Malliavin calculus. Springer, Lecture Notes in Math. 1390 (1989) 192–196.
[7] Multiple Wiener-Itô integrals possessing a continuous extension. Probab. Theory Relat. Fields **85** (1990) 131–145.
[8] Positive and strongly positive Wiener functionals, Barcelona Sem. Stoch. Anal. 1991 (D. Nualart and M. Sanchez, eds.). Birkhäuser, Prog. Probab. 32 (1993) 132–146.
[9] Multiple Wiener-Itô integrals possessing a continuous extension. Probab. Theory Relat. Fields **85** (1990) 131–145.

Obata, N.
[1] Rotation-invariant operators on white noise functionals. Math. Z. **210** (1992) 69–89.

Ochi, Y.
[1] Limit theorem for a class of diffusion processes. Stoch. **15** (1985) 251–269.

Ocone, D.
[1] Malliavin's calculus and stochastic integral representations of functionals of diffusion processes. Stoch. **12** (1984) 161–185.
[2] Stochastic calculus of variations for stochastic partial differential equation. J. Funct. Anal. **79** (1988) 231–288.
[3] A guide to stochastic calculus of variations. In: Proc. Stoch. Anal. and Rel. Topics, Silivri 1988 (A.S. Üstünel ed.). Springer, Lecture Notes in Math. 1444 (1990) 1–79.
[4] Existence of densities for the statistics of the cubic sense problem. In: Stochastic differential equations, IMA Vol. Math. Appl. **10**. Springer 1988, 409–428.

Ocone, D., and Karatzas, I.
[1] A generalized Clark representation formula with application to optimal portfolio. Stoch. **34** (1990) 187–220.

Ocone, D., Karatzas, I., and Jinlu Li
[1] An extension of Clark's formula. Stoch. **37** (1991) 127–131.

Ocone, D., and Pardoux, E.
[1] A Lie algebraic criterium for non-existence of finite-dimensionally computable filters. In: Stoch. Part. Diff. Equations II (Da Prato ed.). Springer, Lecture Notes in Math. 1390 (1989) 197–204.

[2] A generalized Itô-Ventzell formula. Application to a class of anticipating stochastic differential equations. Ann. Inst. Henri Poincaré, Probab. Stat. **25** (1989) 39–71.
[3] Linear stochastic differential equations with boundary conditions. Probab. Theory Relat. Fields **82** (1989) 489–526.

Ogawa, S.
[1] Quelques propriétés de l'intégrale stochastique de type non-causal. Japan J. Appl. Math. **1** (1984) 405–416.
[2] Une remarque sur l'approximation de l'intégrale stochastique du type non-causal par une suite d'intégrale de Stieltjes. Tohoku Math. J. **36** (1984) 41–48.
[3] Sur la question d'existence de solutions d'une équation différentielle stochastique de type non-causal. J. Math. Kyoto Univ. **24** (1984) 699–704.
[4] Topics in the theory of non-causal stochastic integrals. In: Diff. Proc. on Rel. Probl. in Anal. (M. Pinsky ed.). Birkhäuser (1990) 411–420.

Øksendal, B.
[1] Stochastic differential equations. Springer 1985; 4th. ed. 1995, 271pp.

Pardoux, E.
[1] Application of anticipating stochastic flow to stochastic differential equation. Springer, Lecture Notes in Math. 1444 (1990) 63–105.
[2] Filtrage non-linéaire et équations aux dérivées partielles stochastiques associées. In: Ecole d'été de Probab., Saint-Flour. Springer, Lecture Notes in Math. 1464 (1988) 67–163.
[3] Stochastic partial differential equations, Saint-Chéron Round Tables 1992 (M. Yor ed.). Bull. Sci. Math. **117** (1993) 29–48.

Pardoux, E., and Protter, P.
[1] A two-sided stochastic integral and its calculus. Probab. Theory Relat. Fields **76** (1987) 15–49.

Pardoux, E., and Zhang, T.
[1] Absolute continuity of the law of the solution of a parabolic PDE. J. Funct. Anal. **112** (1993) 373–441.

Peters, G.
[1] Symplectic Geometry on an abstract Wiener space and the Sine-Gordon equation. C.R. Acad. Sci., Paris, Sér. I **320** (1995) 1123–1128.
[2] Flows on the Wiener space generated by vector fields with low regularity. C.R. Acad. Sci., Paris, Sér. I **320** (1995) 1003–1008.

Picard, J.
[1] Approximation of stochastic differential equations and application of the stochastic calculus of variations to the rate of convergence. In: Proc. Stoch. Anal. and Rel. Topics, Silivri 1986. Springer, Lecture Notes in Math. 1316 (1988) 267–287.
[2] On the existence of smooth densities for jump processes. Preprint.

Piech, M.A.
[1] A fundamental solution of the parabolic equation on Hilbert spaces. J. Funct. Anal. **3** (1969) 85–114.
[2] The Ornstein-Uhlenbeck semi-group in an infinite-dimensional L^2-setting. J. Funct. Anal. **18** (1975) 271–285.
[3] The exterior algebra for Riemann manifolds. J. Funct. Anal. **28** (1978) 279–308.

Pinsky, M.
[1] Multiplicative operator functionals and their asymptotic properties. In: Adv. Probab. Relat. Top. **3** (P. Ney and S. Port ed.). M. Dekker (1974) 1–100.

[2] Homogenization in stochastic differential geometry. Publ. Res. Inst. Math. Sci. **17** (1981) 235–244.
[3] Homogenization and stochastic parallel displacement. In: Stoch. Integrals, Durham 1980, (D. William ed.). Springer, Lecture Notes in Math. 851 (1981) 271–284.
[4] Lectures on random evolution. World Scientific (1991) 191 pp.

Pisier, G.
[1] Riesz transform: a simpler analytic proof of P.A. Meyer's inequality, Sém. Probab. XXII. Springer, Lecture Notes in Math. 1321 (1988) 485–501.

Pitman, J., and Yor, M.
[1] Asymptotic law of planar Brownian motion. Ann. Probab. **14** (1986) 733–779.

Pontier, M., and Szpirglas, J.
[1] Uniqueness of the solutions of the filtering equations with observation on a riemannian symmetric space. In: Stoch. Anal. and Rel. Topics (A.S. Üstünel ed.). Springer, Lecture Notes in Math. 1316 (1988) 328–340, and SIAM J. Control Optimization **26** (1988) 609–627.

Pontier, M., and Üstünel, A.S.
[1] Analyse stochastique sur l'espace de Lie-Wiener. C.R. Acad. Sci., Paris, Sér. I **311** (1991) 313–316.

Potthoff, J.
[1] White noise approach to Malliavin's calculus. J. Funct. Anal. **71** (1987) 207–217.
[2] Littlewood-Paley theory on Gaussian spaces. Nagoya Math. J. **109** (1988) 47–61.

Potthoff, J., and Streit, L.
[1] A characterization of Hida distribution. J. Funct. Anal. **101** (1991) 212–229.

Prat, J.J.
[1] Equation de Schrödinger matricielle et stabilité des surfaces minima. J. Funct. Anal. **87** (1989) 370–391.
[2] Equation de Schrödinger: analyticité transverse de la densité de la loi d'une fonctionnelle additive. Bull. Sci. Math. **115** (1991) 131–176.
[3] Gevrey regularity of $C3$ functionals and stochastic calculus of variations. Stoch. **44** (1993) 167–211.
[4] Résolution de l'équation de Schrödinger par intégration stochastique du champ magnétique. Probab. Theory Relat. Fields **95** (1993) 217–237.

Privault, N.
[1] Calcul chaotique et calcul variationnel pour le processus de Poisson. C.R. Acad. Sci., Paris, Sér. I **316** (1993) 597–600.
[2] Décompositions chaotiques sur l'espace de Poisson et applications. C.R. Acad. Sci., Paris, Sér. I **317** (1993) 385–388.
[3] A transfer principle from Wiener to Poisson space. J. Funct. Anal. **132** (1995) 335–360.
[4] Girsanov theorem for anticipative shifts on Poisson space. Probab. Theory Relat. Fields **104** (1996) 61–76.
[5] Calcul des variations stochastiques pour les martingales. C.R. Acad. Sci., Paris, Sér. I **321** (1995) 923–928.

Ramer, R.
[1] On non-linear transformation of gaussian measure. J. Funct. Anal. **15** (1974) 166–187.

Ren, J.
[1] Formule de Stokes et estimation explicite d'une intégrale stochastique oscillante. Bull. Sci. Math. **112** (1988) 453–471.

[2] Analyse quasi-sûre des équations différentielles stochastiques. Bull. Sci. Math. **114** (1990) 187–214.
[3] Topologie p-fine sur l'espace de Wiener et Théorie des fonctions implicites. Bull Sci. Math. **114** (1990) 99–114.
[4] Analyse quasi-sûre de martingales régulières. C.R. Acad. Sci., Paris, Sér. I **317** (1993) 299–303.
[5] On smooth martingales. J. Funct. Anal. **120** (1994) 72–81.
[6] Analytic functionals on the Wiener spaces. Acta Math. Sin. **8** (1992) 394–405.
[7] Quasi-sure quadratic variations of smooth martingales. J. Math. Kyoto Univ. **34** (1994) 191–206.

Ren, J., and Watanabe, S.
[1] A convergence theorem for probability densities and conditional expectation of Wiener functionals. In: Dirichlet processes, Beijing 1993 (Ma Zhi-Ming, M. Röckner and J.A. Yan eds.). de Gruyter (1995) 335–344.

Revuz, D., and Yor, M.
[1] Continuous martingales and Brownian motion. Grundlehren der math. Wiss., Springer 1991.

Röckner, M., and Schmuland, B.
[1] Tightness of general $C(1, p)$-capacities on Banach spaces. J. Funct. Anal. **106** (1992) 1–13.

Roëlly, S., and Zessin, H.
[1] Une caractérisation des diffusions par le calcul des variations stochastiques. C.R. Acad. Sci., Paris, Sér. I **313** (1991) 309–312.
[2] Une caractérisation des mesures de Gibbs sur $(c([0, 1]))^{\mathbb{Z}d}$ par le calcul des variations stochastiques. Ann. Inst. Henri Poincaré, Probab. Stat. **29** (1993) 327–378.
[3] Une caractérisation des champs gibbsiens sur un espace de trajectoires. C.R. Acad. Sci., Paris, Sér. I **321** (1995) 1377–1382.

Rossignol, S.
[1] Développements asymptotiques d'intégrales de Laplace sur l'espace de Wiener dans le cas dégénéré. C.R. Acad. Sci., Paris, Sér. I **317** (1993) 971–974.

Rovira, C., and Sanz-Sole, M.
[1] Anticipating stochastic differential equations: regularity of the law. J. Funct. Anal. **142** (1996).

Russek, A.
[1] On series expansions of stochastic integrals, Probab. on vector spaces, IV (Lancut 1987). Springer, Lecture Notes in Math. 1391 (1989) 337–352.

Russo, F., and Vallois, P.
[1] Forward, backward symmetric stochastic integration. Probab. Theory Relat. Fields **97** (1993) 403–421.

Sato, H.
[1] Characteristic functional of a probability measure absolutely continuous with the respect of a Gaussian measure. J. Funct. Anal. **61** (1985) 222–245.

Segal, I.E.
[1] Abstract probability spaces and a theorem of Kolmogorov. Am. J. Math. **76** (1954) 721–732.
[2] Tensor algebras over Hilbert spaces I. Trans. Am. Math. Soc. **81** (1956) 106–134.

Sekiguchi, T., and Shiota, Y.
[1] Theory of non-causal stochastic integrals. Math. Rep. Toyama Univ. **8** (1985) 119–195.

Sevljakov, A.
[1] The Itô formula for the extended stochastic integral. Theory Probab. Math. Stat. **22** (1981) 163–174.

Shigekawa, I.
[1] Absolute continuity of probability laws of Wiener functionals. Proc. Japan Acad., Ser. A **54** (1978) 230–233.
[2] Derivatives of Wiener functionals and absolute continuity of induced measures. J. Math. Kyoto Univ. **20** (1980) 263–289.
[3] Transformation of the Brownian motion on a Riemannian symmetric space. Z. Wahrsch. verw. Gebiete **65** (1984) 493–522, and in: Taniguchi Symp. on Stoch. Anal. (K. Itô ed.) Katata/Kyoto 1982. North-Holland, Math. Library 32 (1984) 409–422.
[4] The de Rham-Hodge-Kodaira decomposition on an abstract Wiener space. J. Math. Kyoto Univ. **26** (1986) 191–202.
[5] On quasi-everywhere existence of local time of 1-dimensional Brownian motion. Osaka J. Math. **21** (1987) 126–139.
[6] Existence of invariant measures of diffusions on an abstract Wiener space. Osaka J. Math. **24** (1987) 37–59.
[7] Itô-Wiener expansion of holomorphic functions on the complex Wiener space. In: Liber Amicorum for Moshe Zakai. Academic Press (1991) 459–473.
[8] On stochastic horizontal lifts. Z. Warsch. verw. Gebiete **59** (1982) 211–221.
[9] Transformation of the Brownian motion on a Lie group. In: Taniguchi Symp. on SDtoch. Anal. (K. Itô ed.) Katata/Kyoto 1982. North Holland (1984) 409–422, and Z. Wahrsch. Grenzgeb. **65** (1984) 493–522.
[10] Eigenvalue problem for the Schrödinger operator with magnetic field on a compact Riemannian manifold. J. Funct. Anal. **75** (1987) 92–127.
[11] Spectral properties of Schrödinger operators with magnetic fields for a spin 1/2 partial. J. Funct. Anal. **101** (1991) 256–264.
[12] Sobolev spaces over the Wiener space based on an Ornstein-Uhlenbeck process. J. Math. Kyoto Univ. **32** (1992).
[13] A quasi-homeomorphism of the Wiener space. In: Proc. Symp. Pure and Appl. Math. **57** (M. Cranston and M. Pinsky eds.) Cornell, 1993, Am. Math. Soc. (1995) 473–487.
[14] Sobolev spaces of Banach valued functions associated to a Markov process. Probab. Theory Relat. Fields **99** (1994) 425–443.

Shigekawa, I., and Taniguchi, S.
[1] A Kähler metric on a based loop group and a covariant differentiation. In: Itô's Stochastic Calculus and Probability Theory (N. Ikeda, S. Watanabe, M. Fukushima, H. Kunita eds.). Springer Tokyo 1996, 327–347.

Shigekawa, I., and Ueki, N.
[1] A stochastic approach to Riemann-Roch theorem. Osaka J. Math. **25** (1988) 759–784, and **32** (1992) 731–748.

Shigekawa, I., Ueki, N., and Watanabe, S.
[1] A probabilistic proof of the Gauss-Bonnet-Chern theorem for manifolds with boundary. Osaka Math. J. **26** (1989) 497–530.

Shigekawa, I., and Yoshida, N.
[1] Littlewood-Paley-Stein inequality for a symmetric diffusion. J. Math. Soc. Japan **44** (1992) 251–280.

Shiota, Y.
[1] A linear stochastic integral equation containing the extended Itô integral. Math. Rep. Toyama Univ. **9** (1986) 43–65.

Sintes-Blanc, A.
[1] On the density of some Wiener functionals; an application of Malliavin calculus. Publ. Math. **36** (1993) 981–987.

Skorohod, A.V.
[1] Integration in Hilbert space. Ergeb. Math. Grenzgeb. 79, Springer (1974).
[2] On a generalisation of stochastic integral. Theory Probab. Appl. **20** (1975) 219–233.

Smolyanov, O.G., and Weizsäcker, H.V.
[1] Change of measures and their logarithmic derivatives under smooth transformations. C.R. Acad. Sci., Paris, Sér. I **321** (1995) 103–108.

Smorodina, N.V.
[1] Differential calculus in a configuration space, and stable measures. Teor. Veroyatn. Primen. **33** (1988) 522–534.

Solé, J.LI, and Utzet, F.
[1] Stratonovich integral and trace. Stoch. Stoch. Rep. **29** (1990) 203–220.

Song, S.
[1] Inégalités relatives aux processus d'Ornstein-Uhlenbeck à n-paramètres et capacités gaussienne $C_{n,2}$, Sém. Probab. XXVII. Springer, Lecture Notes in Math. 1557 (1993) 276–301.

Stafford, S.
[1] A stochastic criterion for Yang-Mills connection. In: Diff. Proc. and Rel. Probl. in Anal., vol.1 (M. Pinsky ed.). Birkhäuser (1990) 313–322.

Stein, E.M.
[1] Abstract Littlewood-Paley theory. Princeton Univ. Press (1969) 150 pp.

Stoltz, W.
[1] Une méthode élémentaire pour l'évaluation de petites boules Browniennes. C.R. Acad. Sci., Paris, Sér. I **31** (1993) 1217–1223.

Stroock, D.W.
[1] The Malliavin calculus, a functional analytic approach. J. Funct. Anal. **44** (1981) 212–257.
[2] The Malliavin calculus and its application to second order parabolic differential equations I. Math. Syst. Theory **14** (1981) 25–65.
[3] The Malliavin calculus and its application to second order parabolic differential equations II. Math. Syst. Theory **14** (1981) 141–171.
[4] The Malliavin calculus and its applications. In: Stochastic Integrals, Proceedings, Durham 1980 (D. Williams, ed.). Springer, Lecture Notes in Math. 851 (1981) 394–432, and Proc. Intern. Congress Math., Warsaw 1983 (P. Malliavin, ed.). PWN, Warszawa (1984).
[5] Some applications of stochastic calculus to partial differential equations. In: Ecole d'Eté de Saint-Flour XI, 1981. Springer, Lecture Notes in Math. 976 (1983) 267–382.
[6] Stochastic analysis and regularity properties of certain partial differential operators. In: Proc. Intern. Congress Math., Warsaw 1983 (P. Malliavin ed.). PWN, Warszawa (1984) 1107–1116.
[7] Malliavin calculus and applications to PDE. IMA, Minneapolis (1986).
[8] Homogeneous chaos revisited, Sém. Probab. XXI. Springer, Lecture Notes in Math. 1247 (1987) 1–8.

[9] Probability theory: An analytic view. Cambridge Univ. Press (1993) 510 pp.
[10] An estimate of the Hessian of the heat kernel. In: Itô's Stochastic Calculus and Probability Theory (N. Ikeda, S. Watanabe, M. Fukushima, H. Kunita, eds.). Springer Tokyo 1996, 355–371.

Stroock, D.W., and Taniguchi, S.
[1] Diffusions as integral curves of vector fields, Stratonovich diffusion without Itô integration, Vol. in the honour of E. Dynkin (1994) (Freidlin ed.). Birkhäuser, Prog. Probab. 34, 333–369.

Stroock, D.W., and Varadhan, S.R.S.
[1] On the support of diffusion processes with application to the strong maximum principle, Proc. Sixth Berkeley Symp. Math. Stat. Probab. III. Univ. of California Press (1972) 333–368.
[2] Multidimensional diffusion processes. Grundlehren der math. Wiss. 233, Springer 1979.

Stroock, D.W., and Zegarlinski, B.
[1] The logarithm Sobolev inequality for continuous spin systems on a lattice. J. Funct. Anal. **104** (1992) 299–326.

Stroock, D.W., and Zeitouni, O.
[1] Variations on a theme by Bismut. Preprint

Stroock, D.W., and Zheng, W.A.
[1] Markov chains approximation to symmetric diffusion. Preprint.

Sugita, H.
[1] Sobolev spaces of Wiener functionals and Malliavin's calculus. J. Math. Kyoto Univ. **25** (1985) 31–48.
[2] On a characterization of the Sobolev spaces over an abstract Wiener space. J. Math. Kyoto Univ. **25** (1985) 717–725.
[3] Positive generalized functions and potential theory over an abstract Wiener space. Osaka J. Math. **25** (1988) 665–696.
[4] Hu-Meyer multiple Stratanovitch integrals and essential continuity of multiple Wiener integrals. Bull. Sci. Math. **113** (1989) 463–474.
[5] Various topologies in the Wiener space and Lévy's stochastic area. Probab. Theory Relat. Fields **91** (1992) 283–296.
[6] Properties of holomorphic Wiener functions, skeleton contraction and local Taylor expansion. Probab. Theory Relat. Fields **100** (1994) 117–126.
[7] Regular version of holomorphic Wiener functions. J. Math. Kyoto Univ. **34** (1994) 849–857.

Sunada, T.
[1] Homology, Wiener integrals and integrated densities of states. J. Funct. Anal. **106** (1992) 50–58.

Suzuki, M.
[1] Applications of the Malliavin calculus to McKean equations. J. Fac. Sci. Univ. Tokyo, Sect. I A **37** (1990) 129–149.

Takanobu, S.
[1] Diagonal short time asymptotics of heat kernels for certain degenerate second order differential operators of Hörmander type. Publ. Res. Inst. Math. Sci. **24** (1988) 169–203.
[2] Diagonal estimate of transition probability densities of certain degenerate diffusions. J. Funct. Anal. **91** (1990) 221–236.

Takanobu, S., and Watanabe, S.
[1] Asymptotic expansion formulas of the Schilder type for a class of functional integration. Proc. Taniguchi Symp. on Asymp. Probl. in Probab. Th. Sanda and Kyoto (K.D. Elworthy and N. Ikeda eds.). Pitman Res. Notes Math. Ser. 254 (1993).

Takeda, M.
[1] (r, p)-capacity on the Wiener space and properties of Brownian motion. Z. Wahrsch. verw. Gebiete **68** (1984) 149–162.
[2] On the uniqueness of Markovian self-adjoint extension of diffusion operators on infinite-dimensional spaces. Osaka J. Math. **22** (1985) 733–742.
[3] On a martingale method for symmetric diffusion processes and its applications. Osaka J. Math. **26** (1989) 605–623.
[4] On the conservativeness of the Brownian motion on a Riemannian manifold. Bull. London Math. Soc. **23** (1991) 86–88.

Taniguchi, S.
[1] Malliavin's stochastic calculus of variations for manifold valued Wiener functionals and its applications. Z. Wahrsch. verw. Gebiete **65** (1983) 269–290.
[2] Applications of Malliavin's calculus to time-dependent systems of heat equations. Osaka J. Math. **22** (1985) 307–320.
[3] Holomorphic functions on balls in an almost complex Wiener space. J. Math. Soc. Japan (1994).
[4] On Ricci curvature of hypersurfaces in abstract Wiener spaces. J. Funct. Anal. **136** (1996) 226–244.

Taylor, J.C.
[1] Brownian motion on a symmetric space of non-compact type asymptotic behaviour in polar coordinates. Canadian J. Math. **43** (1991) 1065–1085.

Thieullen, M.
[1] Calcul stochastique non adapté pour des processus à deux paramètres: formule de changement de variable. Probab. Theory Relat. Fields **89** (1991) 457–486.
[2] Second order stochastic differential equations and non-gaussian reciprocal diffusions. Preprint

Thieullen, M., and Zambrini, J.C.
[1] Symmetries in stochastic calculus of variations. Preprint

Ueki, N.
[1] A stochastic approach to the Poincaré-Hopf theorem. Probab. Theory Relat. Fields **82** (1989) 271–293.
[2] Hypoellipticity for the $\bar{\partial}$-Neumann problem on certain domains with infinite degeneracy. Preprint (1994).
[3] Lower bounds for the spectra of Schrödinger operators with magnetic fields. J. Funct. Anal. **120** (1994) 344–379.
[4] On the spectra of random Schrödinger operators with magnetic fields. Osaka J. Math. **31** (1994) 177–187.

Uemura, H.
[1] On short time expansion of the fundamental solution of the heat equation by the method of Wiener functionals. J. Math. Kyoto Univ. **27** (1987) 417–431.
[2] Off-diagonal short time estimate of the heat kernel on a certain nilpotent Lie group. J. Math. Kyoto Univ. **30** (1990) 403–450.

Üstünel, A.S.
[1] Une extension du calcul de Itô via le calcul des variations stochastiques. C.R. Acad. Sci., Paris, Sér. I **300** (1985) 277–279, and Stoch. **23** (1988) 353–376.

[2] Representation of the distributions on Wiener space and stochastic calculus of variations. J. Funct. Anal. **70** (1987) 126–139.
[3] The Itô formula for anticipative processes with non-autonomous time scale via the Malliavin calculus. Probab. Theory Relat. Fields **79** (1988) 249–269.
[4] Intégrabilité exponentielle de fonctionnelles de Wiener. C.R. Acad. Sci., Paris, Sér. I **315** (1992) 997–1000.
[5] An explicit representation of local time of an anticipative process. Stoch. **36** (1991) 65–69.
[6] La formule de changement de variables pour l'intégrale anticipante de Skorokhod. C.R. Acad. Sci., Paris, Sér. I **303** (1986) 329–331.
[7] Some comments on the filtering of diffusion. In: Stoch. Anal. and Rel. Topics (A.S. Üstünel ed.). Springer, Lecture Notes in Math. 1316 (1988) 247–266.
[8] An explicit representation of local time of anticipative processes. Stoch. Stoch. Rep. **36** (1991) 65–69.
[9] Exponential tightness of Wiener functionals. In: Stoch. Anal. and Rel. Topics (B. Øksendal and A. Üstünel eds.). Gordon & Breach, Stochastic Monogr. **8** (1993) 265–274.
[10] Some exponential moment inequalities for the Wiener space. J. Funct. Anal. **136** (1996) 154–171.
[11] An introduction to stochastic analysis on the Wiener space. Springer, Lecture Notes in Math. 1610 (1996) 93 pp.

Üstünel, A.S. and Zakai, M.
[1] Caractérisation géométrique de l'indépendance forte sur l'espace de Wiener. C.R. Acad. Sci., Paris, Sér. I **306** (1988) 199–201, 487–489.
[2] On independence and conditioning on the Wiener space. Ann. Probab. **17** (1989) 1441–1453.
[3] On the structure of independence on the Wiener space. J. Funct. Anal. **90** (1990) 113–137.
[4] Une approche variationnelle pour une formule de Cameron-Martin non-linéaire. C.R. Acad. Sci., Paris, Sér. I **312** (1991) 619–623.
[5] Transformation of Wiener measure under anticipative flow. Probab. Theory Relat. Fields **93** (1992) 91–136.
[6] The Wiener chaos expansion of certain Radon-Nikodym derivatives. Birkhäuser, Prog. Probab. 31 (1992) 355–400.
[7] Applications of the degree theorem to absolute continuity on Wiener space. Probab. Theory Relat. Fields **95** (1993) 509–520.
[8] The composition of Wiener functionals with non-absolute continuous shifts. Probab. Theory Relat. Fields **98** (1994) 163–184.
[9] Random rotation of the Wiener path. Probab. Theory Relat. Fields **103** (1995) 409–430.
[10 Absolute continuity on the Wiener space and some applications. In: Proc. Symp. Pure and Appl.Math. **57** (M. Cranston and M. Pinsky eds.), Cornell 1993, Amer. Math. Soc. (1995) 487–495.
[11] The construction of filtrations on abstract Wiener space. J. Funct. Anal. **142** (1996).

Van Biesen, J.
[1] The divergence on submanifolds of the Wiener space. J. Funct. Anal. **113** (1993) 426–461.

Veretennikov, A.Yu
[1] Probabilistic approach to hypoellipticity. Usp. Mat. Nauk **38** (1983) 113–125.
[2] Probabilistic problems in the theory of hypoellipticity. Izv. Akad. Nauk n°6 **48** (1984) 1151–1170.

Vintschger, R.
[1] The existence of invariant measures for $C([0, 1])$-valued diffusion. Probab. Theory Relat. Fields **82** (1989) 307–313.

Watanabe, S.
[1] Malliavin's calculus in terms of generalized Wiener functionals. Springer, Lecture Notes in Control and Inf. Sci. 49, Kyoto (1983) 284–290.
[2] Lectures on stochastic differential equations and Malliavin calculus, Lect. Math. Phys., Tata Inst. Fundam. Res. 73 (1984) 111 pp.
[3] Analysis of Wiener functionals (Malliavin calculus) and its applications to heat kernels. Ann. Probab. **15** (1987) 1–39.
[4] Short time asymptotic problems in Wiener functional integration, Application to heat kernel and index theorem. In: Proc. Stoch. Anal. and Rel. Topics, Silivri 1988 (A.S. Üstünel ed.). Springer, Lecture Notes in Math. 1444 (1990) 1–62.
[5] Donsker's δ-function in Malliavin's calculus, Stoch. calculus. In: Liber Amicorum for Moshe Zakai. Academic Press (1991) 495–502.
[6] Some refinement on conditional expectation on the Wiener space by means of Malliavin calculus. In: Proc. of the 6th URSS/Japan Symp. on Probab. (A.N. Shiryaev ed.). World Scientific (1992) 414–421.
[7] Fractional order Sobolev spaces on Wiener space. Probab. Theory Relat. Fields **95** (1993) 175–198.
[8] Desintegration problems in Wiener functional integration. Preprint
[9] Lévy's stochastic area formula and Brownian motion on compact Lie groups. In: Itô's Stochastic Calculus and Probability Theory (N. Ikeda, S. Watanabe, M. Fukushima, H. Kunita, eds.). Springer, Tokyo 1996, 401–411.

Wiener, N.
[1] Differential spaces. J. Math. Phys. **2** (1923) 131–174.
[2] Homogeneous chaos. Amer. J. Math. **60** (1938) 879–936.

Wong, E., and Zakai, M.
[1] On the convergence of ordinary integrals to stochastics integrals. Ann. Math. Stat. **36** (1965) 1560–1564.

Wschebor, M.
[1] Surfaces aléatoires. Springer, Lecture Notes in Math. 1147 (1985).

Wu, L.
[1] Construction de l'opérateur de Malliavin sur l'espace de Poisson, Sém. Probab. XXI. Springer, Lecture Notes in Math. 1247 (1987) 100–113.
[2] Un traitement unifié de la représentation des fonctionnelles de Wiener, Sém. Probab. XXIV, 1988/89. Springer, Lecture Notes in Math. 1426 (1990) 166–187.

Yamada, T.
[1] A comparison theorem for solutions of SDE and its applications. J. Math. Kyoto Univ. **13** (1973) 497–512.

Yamada, T., and Ogura, Y.
[1] On the strong comparison theorem for solutions of stochastic differential equations. Z. Wahrsch. verw. Gebiete **56** (1981) 3–19.

Yan, J.A.
[1] On the existence of density of the law of a Wiener functional. Acta Math. Sin. **5** (1989) 97–100.

Yasue, K.
[1] Stochastic calculus of variations. J. Funct. Anal. **41** (1981) 327–346.

[2] A variational principle for the Navier-Stokes equations. J. Funct. Anal. **51** (1983) 133–140.

Yor, M.
[1] Représentation des martingales de carré intégrable relative aux processus de Wiener et de Poisson à n-paramètres. Z. Wahrsch. verw. Gebiete **35** (1976) 121–129.

Yoshida, Nakahiro
[1] Asymptotic expansions of maximum likehood estimations via the theory of Malliavin-Watanabe. Probab. Theory Relat. Fields **92** (1992) 275–313.
[2] Malliavin calculus and higher order statistical inference, Coll. Statistical Science and Data Analysis (Tokyo 1991). VSP Utrecht (1993) 175–180.
[3] Asymptotic expansions for statistics related to small diffusions. J. Jap. Stat. Soc. **22** (1992) 139–159.
[4] Asymptotics of Bayes estimation for small diffusions. Probab. Theory Relat. Fields **95** (1993) 429–450.
[5] Asymptotic expansion for local density of martingales. Preprint
[6] Malliavin calculus and asymptotic expansions for martingales. Preprint

Yoshida, Nobuo
[1] A large deviation principle for capacities on the Wiener space. Probab. Theory Relat. Fields **94** (1993) 473–488.
[2] The Littlewood-Paley-Stein inequality on an infinite-dimensional manifold. J. Funct. Anal. **122** (1994) 402–427.

Zakai, M.
[1] On the optimal filtering of diffusion processes. Z. Wahrsch. verw. Gebiete **11** (1969) 230–243.
[2] Malliavin derivatives and derivatives of functionals of the Wiener process with respect to a scale parameter. Ann. Probab. **13** (1985) 609–615.
[3] The Malliavin calculus, Acta Appl. Math. **3** (1985) 175–207.
[4] Stochastic integration, trace and the skeleton of Wiener functionals in stochastics. Stoch. Stoch. Rep. **32** (1990) 93–108.

Zakai, M., and Zeitouni, O.
[1] When does the Ramer formula look like the Girsanov formula. Ann. Probab. **20** (1992) 1436–1440.

Zegarlinski, B.
[1] Dobrushin uniqueness and logarithmic Sobolev inequalities. J. Funct. Anal. **105** (1992) 77–111.
[2] Ergodicity of Markov semi-groups on infinite-dimensional spaces. In: Dirichlet forms, Beijing 1993 (Ma Zhi Ming, M. Röckner, and J.A. Yan eds.). de Gruyter (1995) 404–415.

Zeitouni, O.
[1] Infinite-dimensionality result for MAP estimation. In: Liber Amicorum for Moshe Zakai. Acad. Press (1991) 513–532.

Zhiyan, H.
[1] A comparison theorem for solutions of stochastic differential equations and its application. Proc. Amer. Math. Soc. **91** (1984).

Zhou-Xian-Yin
[1] Infinite-dimensional Malliavin calculus and its applications. Acta Math. Appl. Sin. **8** (1992) 97–114.

Index

Index of Notations

Grundlehren der mathematischen Wissenschaften

A Series of Comprehensive Studies in Mathematics

A Selection

210. Gihman/Skorohod: The Theory of Stochastic Processes I
211. Comfort/Negrepontis: The Theory of Ultrafilters
212. Switzer: Algebraic Topology – Homotopy and Homology
215. Schaefer: Banach Lattices and Positive Operators
217. Stenström: Rings of Quotients
218. Gihman/Skorohod: The Theory of Stochastic Procrsses II
219. Duvaut/Lions: Inequalities in Mechanics and Physics
220. Kirillov: Elements of the Theory of Representations
221. Mumford: Algebraic Geometry I: Complex Projective Varieties
222. Lang: Introduction to Modular Forms
223. Bergh/Löfström: Interpolation Spaces. An Introduction
224. Gilbarg/Trudinger: Elliptic Partial Differential Equations of Second Order
225. Schütte: Proof Theory
226. Karoubi: K-Theory. An Introduction
227. Grauert/Remmert: Theorie der Steinschen Räume
228. Segal/Kunze: Integrals and Operators
229. Hasse: Number Theory
230. Klingenberg: Lectures on Closed Geodesics
231. Lang: Elliptic Curves. Diophantine Analysis
232. Gihman/Skorohod: The Theory of Stochastic Processes III
233. Stroock/Varadhan: Multidimensional Diffusion Processes
234. Aigner: Combinatorial Theory
235. Dynkin/Yushkevich: Controlled Markov Processes
236. Grauert/Remmert: Theory of Stein Spaces
237. Köthe: Topological Vector Spaces II
238. Graham/McGehee: Essays in Commutative Harmonic Analysis
239. Elliott: Probabilistic Number Theory I
240. Elliott: Probabilistic Number Theory II
241. Rudin: Function Theory in the Unit Ball of C^n
242. Huppert/Blackburn: Finite Groups II
243. Huppert/Blackburn: Finite Groups III
244. Kubert/Lang: Modular Units
245. Cornfeld/Fomin/Sinai: Ergodic Theory
246. Naimark/Stern: Theory of Group Representations
247. Suzuki: Group Theory I
248. Suzuki: Group Theory II
249. Chung: Lectures from Markov Processes to Brownian Motion
250. Arnold: Geometrical Methods in the Theory of Ordinary Differential Equations
251. Chow/Hale: Methods of Bifurcation Theory
252. Aubin: Nonlinear Analysis on Manifolds. Monge-Ampère Equations
253. Dwork: Lectures on p-adic Differential Equations
254. Freitag: Siegelsche Modulfunktionen
255. Lang: Complex Multiplication
256. Hörmander: The Analysis of Linear Partial Differential Operators I
257. Hörmander: The Analysis of Linear Partial Differential Operators II
258. Smoller: Shock Waves and Reaction-Diffusion Equations
259. Duren: Univalent Functions
260. Freidlin/Wentzell: Random Perturbations of Dynamical Systems
261. Bosch/Güntzer/Remmert: Non Archimedian Analysis – A System Approach to Rigid Analytic Geometry

262. Doob: Classical Potential Theory and Its Probabilistic Counterpart
263. Krasnosel'skiĭ/Zabreĭko: Geometrical Methods of Nonlinear Analysis
264. Aubin/Cellina: Differential Inclusions
265. Grauert/Remmert: Coherent Analytic Sheaves
266. de Rham: Differentiable Manifolds
267. Arbarello/Cornalba/Griffiths/Harris: Geometry of Algebraic Curves, Vol. I
268. Arbarello/Cornalba/Griffiths/Harris: Geometry of Algebraic Curves, Vol. II
269. Schapira: Microdifferential Systems in the Complex Domain
270. Scharlau: Quadratic and Hermitian Forms
271. Ellis: Entropy, Large Deviations, and Statistical Mechanics
272. Elliott: Arithmetic Functions and Integer Products
273. Nikol'skiĭ: Treatise on the Shift Operator
274. Hörmander: The Analysis of Linear Partial Differential Operators III
275. Hörmander: The Analysis of Linear Partial Differential Operators IV
276. Liggett: Interacting Particle Systems
277. Fulton/Lang: Riemann-Roch Algebra
278. Barr/Wells: Toposes, Triples and Theories
279. Bishop/Bridges: Constructive Analysis
280. Neukirch: Class Field Theory
281. Chandrasekharan: Elliptic Functions
282. Lelong/Gruman: Entire Functions of Several Complex Variables
283. Kodaira: Complex Manifolds and Deformation of Complex Structures
284. Finn: Equilibrium Capillary Surfaces
285. Burago/Zalgaller: Geometric Inequalities
286. Andrianov: Quadratic Forms and Hecke Operators
287. Maskit: Kleinian Groups
288. Jacod/Shiryaev: Limit Theorems for Stochastic Processes
289. Manin: Gauge Field Theory and Complex Geometry
290. Conway/Sloane: Sphere Packings, Lattices and Groups
291. Hahn/O'Meara: The Classical Groups and K-Theory
292. Kashiwara/Schapira: Sheaves on Manifolds
293. Revuz/Yor: Continuous Martingales and Brownian Motion
294. Knus: Quadratic and Hermitian Forms over Rings
295. Dierkes/Hildebrandt/Küster/Wohlrab: Minimal Surfaces I
296. Dierkes/Hildebrandt/Küster/Wohlrab: Minimal Surfaces II
297. Pastur/Figotin: Spectra of Random and Almost-Periodic Operators
298. Berline/Getzler/Vergne: Heat Kernels and Dirac Operators
299. Pommerenke: Boundary Behaviour of Conformal Maps
300. Orlik/Terao: Arrangements of Hyperplanes
301. Loday: Cyclic Homology
302. Lange/Birkenhake: Complex Abelian Varieties
303. DeVore/Lorentz: Constructive Approximation
304. Lorentz/v. Golitschek/Makovoz: Constructive Approximation. Advanced Problems
305. Hiriart-Urruty/Lemaréchal: Convex Analysis and Minimization Algorithms I. Fundamentals
306. Hiriart-Urruty/Lemaréchal: Convex Analysis and Minimization Algorithms II. Advanced Theory and Bundle Methods
307. Schwarz: Quantum Field Theory and Topology
308. Schwarz: Topology for Physicists
309. Adem/Milgram: Cohomology of Finite Groups
310. Giaquinta/Hildebrandt: Calculus of Variations I: The Lagrangian Formalism
311. Giaquinta/Hildebrandt: Calculus of Variations II: The Hamiltonian Formalism
312. Chung/Zhao: From Brownian Motion to Schrödinger's Equation
313. Malliavin: Stochastic Analysis
314. Adams/Hedberg: Function Spaces and Potential Theory
315. Bürgisser/Clausen/Shokrollahi: Algebraic Complexity Theory
316. Saff/Totik: Logarithmic Potentials with External Fields

Printing: Druckhaus Beltz, Hemsbach
Binding: Buchbinderei Schäffer, Grünstadt